WONDERS IN THE SKY

UNEXPLAINED AERIAL OBJECTS FROM ANTIQUITY TO MODERN TIMES

and Their Impact on Human Culture, History, and Beliefs

WAYNE PUBLIC LIBRARY JUN - 8 2011

WONDERS IN THE SKY

Unexplained Aerial Objects from Antiquity to Modern Times

and Their Impact on Human Culture, History, and Beliefs

JACQUES VALLEE

and CHRIS AUBECK

JEREMY P. TARCHER/PENGUIN

a member of Penguin Group (USA) Inc.

New York

JEREMY P. TARCHER/PENGUIN
Published by the Penguin Group
Penguin Group (USA) Inc., 375 Hudson Street, New York, New York 10014, USA • Penguin Group (Canada), 90 Eglinton Avenue East, Suite 700, Toronto, Ontario M4P 2Y3, Canada (a division of Pearson Penguin Canada Inc.) • Penguin Books Ltd, 80 Strand, London WC2R 0RL, England • Penguin Ireland, 25 St Stephen's Green, Dublin 2, Ireland (a division of Penguin Books Ltd) • Penguin Group (Australia), 250 Camberwell Road, Camberwell, Victoria 3124, Australia (a division of Pearson Australia Group Pty Ltd) • Penguin Books India Pvt Ltd, 11 Community Centre, Panchsheel Park, New Delhi–110 017, India • Penguin Group (NZ), 67 Apollo Drive, Rosedale, North Shore 0632, New Zealand (a division of Pearson New Zealand Ltd) • Penguin Books (South Africa) (Pty) Ltd, 24 Sturdee Avenue, Rosebank, Johannesburg 2196, South Africa

Penguin Books Ltd, Registered Offices: 80 Strand, London WC2R 0RL, England

Copyright © 2009 by Chris Aubeck and Documatica Research, LLC
First Jeremy P. Tarcher/Penguin edition 2010
All rights reserved. No part of this book may be reproduced, scanned, or distributed in any printed or electronic form without permission. Please do not participate in or encourage piracy of copyrighted materials in violation of the authors' rights. Purchase only authorized editions.
Published simultaneously in Canada

Most Tarcher/Penguin books are available at special quantity discounts for bulk purchase for sales promotions, premiums, fund-raising, and educational needs. Special books or book excerpts also can be created to fit specific needs. For details, write Penguin Group (USA) Inc. Special Markets, 375 Hudson Street, New York, NY 10014.

Library of Congress Cataloging-in-Publication Data

Vallee, Jacques.
Wonders in the sky : unexplained aerial objects from antiquity to modern times and their impact on human culture, history, and beliefs / Jacques Vallee and Chris Aubeck.—
1st Jeremy P. Tarcher/Penguin ed.
p. cm.
Includes bibliographical references and index.
ISBN 978-1-58542-820-5
1. Unidentified flying objects—Sightings and encounters—History.
2. Unidentified flying objects—Psychological aspects.
3. Unidentified flying objects—Religious aspects. I. Title.
TL789.3.V354 2010 201024720
001.942—dc22

Printed in the United States of America
1 3 5 7 9 10 8 6 4 2

While the authors have made every effort to provide accurate telephone numbers and Internet addresses at the time of publication, neither the publisher nor the authors assume any responsibility for errors, or for changes that occur after publication. Further, the publisher does not have any control over and does not assume any responsibility for author or third-party websites or their content.

In memoriam:

Janine Vallee

I will show wonders in the sky above, and signs on the earth beneath; blood, and fire, and billows of smoke.

—Acts 2:19

There shall be Signs in the Sun, and in the Moon, and in the Stars.

—Luke 21:25

The most beautiful thing we can experience is the mysterious. It is the source of all true art and science.

—Albert Einstein, “What I Believe,” *Forum*, October 1930

ABLE OF CONTENTS

Wonders in the Sky

Unexplained Aerial Objects from Antiquity to Modern Times

and Their Impact on Human Culture, History, and Beliefs

FOREWORD

by David J. Hufford, Ph.D.
Professor Emeritus of Humanities and Psychiatry
Penn State College of Medicine
Adjunct Professor of Religious Studies
University of Pennsylvania
Author, *The Terror That Comes in the Night*

In 1969 I was a graduate student at the University of Pennsylvania, pursuing a Ph.D. in the field of Folklore. My primary interest was in what was called "folk belief." This term was, and still is, generally reserved for beliefs that are at odds in some way with the official modern worldview. I was taught that such beliefs were both non-empirical and non-rational, that they were cultural fictions that reflected local concerns and functioned to support community values and psychological needs. The experiences on which they claimed to be based were, to use the term popularized by Thomas Kuhn's landmark work, *The Structure of Scientific Revolutions* (1962), "anomalies."

From seeing a ghost to the alleged cures of folk medicine, the events described in folklore seemed to contradict the paradigm of science, the gold standard of modern rationality. For this reason they were, as Charles Fort had said, "damned" (1919), forbidden entry to the corpus of valid knowledge. However, I was pursuing the heretical idea that folk belief traditions might actually incorporate accurate observations, and that if they did they might point to important new knowledge.

I was already frustrated by the way that widely held folk beliefs, beliefs common to many distinct cultures, were dismissed without investigation or argument. I had, in fact, already seen that investigation of the possible validity of folk belief claims was subject to an intimidating array of sanctions. I was thrilled, therefore, to find Jacques Vallee's book, *Passport to Magonia: From Folklore to Flying Saucers* (1969).

I considered UFOs to be a part of contemporary folk belief and, given my questions about valid anomalous observations, I had been reading the UFO literature. I had read Vallee's *Anatomy of a Phenomenon* (1965) and knew him to be both scientific and open-minded. More than most of the popular UFO literature, Vallee's *Anatomy...* provided a convincing case for the objective reality of anomalous aerial phenomena. In *Passport to Magonia* he continued to strengthen the case for there being real phenomena behind UFO reports, but linked these reports to older reports of fairies, ghosts, angels, demons, and so forth in a compelling and fascinating way. He recognized the difference between the core phenomenology of reports and the local language and interpretations that clothed that core in traditional accounts.

This is a sophisticated distinction that I had rarely found among scholars of folk belief, and in *Magonia* Vallee laid out the conceptual basis for using this distinction in the cross-cultural analysis of reports of strange aerial phenomena and the events often associated with them. Criticizing conventional UFO investigators for "confusing appearance and reality" he said that "The phenomenon has stable, invariant features, some of which we have tried to identify and label clearly. But we have also had to note carefully the chameleonlike character of the secondary attributes of the sightings: the shapes of the objects, the appearances of their occupants, their reported statements, vary as a function of the cultural environment..." (1969: 149).

In 1971 I traveled to Newfoundland, Canada, where I spent four years teaching and doing fieldwork for my doctoral dissertation on folk belief. Vallee's ideas went with me and were repeatedly confirmed by the folklore that I studied there. Ghost ships, Jackie-the-Lanterns, and weather lights comprised a very old set of folk traditions and were constantly reported around the island, often in very UFO-like terms. In one small village a series of strange aerial sightings was described and interpreted in old fashioned terms by older residents, while the young people in the community simply called the lights UFOs. In Newfoundland I also found the tradition that they call "the Old Hag," a terrifying nocturnal paralysis accompanied by a frightening entity that Newfoundlanders associated with witches or ghosts.

Using Vallee's approach I was able to immediately recognize in the Old Hag the "bedroom invader" experience that I had encountered in popular UFO literature (Keel 1970). This phenomenon, known to sleep researchers as "sleep paralysis," has "stable, invariant features" that in reports are surrounded by culturally shaped language and interpretations. Among the stable core features of sleep paralysis is the anomalous presence of a frightening entity. This experience, like the experience of strange lights and aerial objects, has wandered through a great variety of traditions around the world: witchcraft, ghosts, vampires, and UFOs. In the 1992 booklet *Unusual Personal Experiences* (Hopkins et al.) UFO abduction investigators Hopkins, Mack and Jacobs report a large national survey intended to determine how many humans have been abducted by aliens—their number one index question asks whether the respondent recalls "Waking up paralyzed with a sense of a strange ... presence ... in the room" (p. 26): sleep paralysis.

Anomalies are a threat to the intellectual status quo. They are powerfully resisted, and that resistance often seems to co-opt the efforts of those bravely investigating the anomalous just as much as it recruits the efforts of intransigent skeptics. As Thomas Kuhn's ground-breaking work showed, this cultural dynamic is inseparable from more obvious data in the effort to make—and to understand—scientific progress. The initial response of a paradigm to anomalies is to ignore or, when reports become too numerous, to assimilate. Both of these strategies are facilitated by the distribution of anomalous reports across a large number of apparently disparate conceptual categories. This process is facilitated by investigators who rush to theories, such as the extraterrestrial spaceship explanation of UFOs, that divide large sets of anomalous reports into smaller and more numerous subdivisions.

UFOs do not seem like Newfoundland weather lights or Will-o'-the-Wisp or the burning ship of Ocracoke Island—until you strip away the culturally elaborated language and secondary interpretations, leaving "anomalous aerial phenomena." Just as "sleep paralysis," "the Old Hag" and UFO abductions don't appear similar—until you strip away the cultural layers and find "Waking up paralyzed with a sense of a strange ... presence ... in the room." This is the beauty of the approach pioneered by Vallee in *Magonia*. *Wonders in the Sky* extends this with the huge corpus of additional early reports assembled by Chris Aubeck and his colleagues through The Magoniax Project. The willingness of these authors to cast a very wide net, and not to allow the particular cultural interpretations of events to limit their view, offers us a remarkable opportunity to seek patterns that may lead to new understandings.

Those with a view of these matters narrowly focused on a particular interpretation, especially the extraterrestrial idea, may be annoyed by the mixing of the aerial and the religious, the political and the mystical and more. Enthusiastic advocates of various anomalous phenomena tend to oppose, even to be offended by, the kind of rigorous methodology found in *Wonders in the Sky*. Not only does this method refuse to accept particular theories as a starting point; it also has much in common with the method of debunkers. When Dr. Hynek invented the "marsh gas" explanation for UFOs (which he later recanted) he implied that he was stripping away layers of cultural elaboration to find the "stable core" of the phenomenon, just as skeptics have used "*just* sleep paralysis" to debunk UFO abduction reports (and a variety of other anomalous events). The work of Jacques Vallee and Chris Aubeck is especially steadfast and courageous in two respects. While seeking a core phenomenology that requires the stripping away of layers of cultural elaboration, they nonetheless systematically attend to the data. After they have removed "spaceship" as a core feature of an observation, they do not proceed to remove all anomalous features. The problem with "spaceship" is not that it is anomalous; it is that it is an interpretation rather than an observation. This is true open-mindedness, and it suggests that we are seeking to understand aspects of the world that are deeply strange.

Their rigorously scientific insistence allows Vallee and Aubeck to retain the most challenging and interesting aspects of these events without the distraction of premature commitment to any particular interpretation. That, I believe, is true science: to follow the data wherever they lead, and to move away from established theory when it fails to deal adequately with the data. As philosopher of science Paul Feyerabend pointed out (1975), what he called the "consistency principle"—judging a theory or hypothesis on the basis of its fit with well established prior theory—ensures the survival of the oldest theory, not the best theory.

The other beautiful innovation in Vallee and Aubeck's work is the combination of science and scholarship. A willingness to combine documentary research, the heart of humanities scholarship, with physical and astrophysical knowledge is rare. To do this in an open-ended search for elusive truth without needing to offer a theory of their own is rarest of all. To do this in a way that harnesses the possibilities of international scholarly collaboration through the Internet offers a view of truly 21st century inquiry.

When I met Jacques Vallee for the first time at Esalen, almost 40 years after I read *Passport to Magonia*, it was truly a peak experience. To have learned that with Chris Aubeck he was preparing the successor to *Magonia* just added to my delight. When Jacques asked me to write a foreword to the new book I felt the sense of completion when an aspect of life comes full circle.

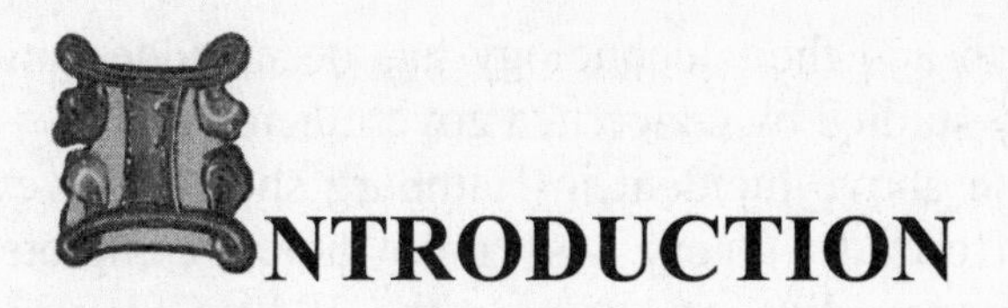

INTRODUCTION

Imagine that we have been transported back in time to Hamburg, Germany, on the 15th day of December in the year of the Lord 1547. Historian Simon Goulart, in his *Trésors Admirables et Mémorables de notre Temps* (1600) writes that on that day the sailors who were aboard ships in the harbor of Hamburg saw in the air, at midnight, a glistening globe as fiery as the Sun. It rolled towards the north, emitting so much heat that people could not remain inside the ships, but were forced to take cover, thinking the vessels were about to burn up.

A meteor? The behavior of this aerial phenomenon is not typical of meteors, which are too high in the atmosphere for their heat to be felt on the ground. In any case a meteor would have passed overhead in seconds, never giving people aboard the ships time to run away from the heat. Globular lightning? Unlikely in the absence of thunder or stormy conditions. Lacking more information, we have to classify the incident as an unidentified flying object.

Thousands of such incidents have been recorded in the last 60 years or so, giving rise to much speculation about flying saucers, visitors from other planets, and alien abductions. Influenced by books and movies, most people have jumped to hasty conclusions: they believe that unidentified flying objects are spaceships from another planetary civilization that became aware of us when we exploded the first atom bombs at the end of World War Two. Understandably concerned about the irresponsible antics of our young species, the theory goes, these aliens decided to come over and take a closer look. According to this interpretation, some of the spaceships even crashed on the earth and their technology has been hidden away and secretly studied by concerned governments.

As the above incident in Hamburg shows, however, the extraterrestrial theory is not quite complete: The phenomenon did not begin in the 1940s, or even in the nineteenth century. It is much older than that. Further, it has some definite physical features – such as the heat felt and reported by witnesses – that have not changed much over the centuries.

The evening of September 3, 1965, two law enforcement officers, Sheriff McCoy and Robert Goode, were patrolling the highways around Angleton, Texas, when they observed a huge object, estimated at 70 meters long and 15 meters high with a bright violet light at one end and a pale blue light at the other. It flew within 30 meters of them, and cast a large shadow when it intercepted the moonlight. They felt a heat wave that scared them, prompting them to hastily drive away. Just like the sailors of Hamburg in 1547.

A robust phenomenon

Such similarities between ancient sightings and modern reports are the rule rather than the exception. In this book we will examine 500 selected reports of sightings from antiquity to the year 1879, when the industrial revolution deeply changed the nature of human society.

We selected the cutoff date of 1880 for our study because it marked a turning point in the technical and social history of the advanced nations. We wanted to analyze aerial phenomena during a period that was entirely free of those modern complications represented by airplanes, dirigibles, rockets and the often-mentioned opportunities for misinterpretation represented by military prototypes. There may have been a few balloons in the sky towards the end of our period, but the first dirigible able to return to its starting point was not demonstrated until the celebrated flight of French Captains Renard and Krebs on August 9, 1884, and the first airplane (equipped with a steam engine) would not fly until Clément Ader's feat at Satory on October 14, 1897.

Even more important than technical achievement were the social changes that marked the end point of our study. It is in 1879 that the world's first telephone exchange is established in London and the first electric tram exhibited by Siemens in Berlin. The following year, both Edison and Swan devise the first practical electric lights, Carnegie develops the first large steel furnace, and New York streets are first lit by electricity. Any study of unidentified flying objects after that date has to adopt the standards of a world where communications, social interaction, travel patterns, and the attitudes of people in everyday life have been deeply altered by the impact of technical progress.

We will show that unidentified flying objects have had a major impact not only on popular culture but on our history, on our religion, and on the models the world humanity has formed since it has evolved a culture that includes writing, science, and the preservation of historical records in stone, clay, parchment, paper, or electronic media.

So why hasn't science taken notice? Given the robust nature of the phenomenon, and the enormous interest it elicits among the public, you would think that inter-disciplinary teams of historians, anthropologists, sociologists, and physical scientists would rush to study it.

The answer lies in the arrogance of academic knowledge and in the fact that our best and brightest scientists have never bothered to inform themselves about the extent and reliability of the sightings. In a recent interview (for www.ted.com, April 2008) the celebrated astrophysicist Stephen Hawking flatly stated he didn't believe in flying saucer stories: "I am discounting reports of UFOs. Why would they appear only to cranks and weirdos?" were his exact words.

He later asserted that we were the only form of technologically evolved life in a 200 light-year radius, thus out of reach of interplanetary travelers.

Unfortunate and ill-informed as they are, these statements by one of the brightest scientists of our time reflect the general view of academic researchers. Back in 1969 the U.S. Academy of Sciences put its stamp of approval on a report by a commission headed up by physicist Edward Condon, stating that science had nothing to gain by a study of unidentified flying objects, even though fully one third of all the cases studied by the commission had remained unexplained after investigation! Clearly, we are dealing with a belief system here, not with rational science.

There are two obvious problems with Stephen Hawking's statement: first, as we will show, most of our 500 cases come from known witnesses who represent a cross-section of human society, including numerous astronomers, physical scientists, military officers and even emperors–hardly the motley crew of cranks and weirdos rashly hypothesized by Hawking. Second, even if the witnesses were of unknown background, the fact would remain that an unexplained phenomenon has played and continues to play a fantastically important role in shaping our belief systems, the way we view our history and the role of science.

Consider the following incident, which transports us to the year 438. An earthquake has destroyed Constantinople; famine and pestilence are spreading. The cataclysm has leveled the walls and the fifty-seven towers. Now comes a new tremor, even stronger than all the previous ones.

Nicephorus, the historian, reports that in their fright the inhabitants of Byzantium, abandoning their city, gathered in the countryside, "They kept praying to beg that the city be spared total destruction: they were in no lesser danger themselves, because of the movements of the earth that nearly engulfed them, when a miracle quite unexpected and going beyond all credence, filled them with admiration."

In the midst of the entire crowd, a child was suddenly taken up by a strong force, so high into the air that they lost sight of him. After this he came down as he had gone up, and told Patriarch Proclus, the Emperor himself, and the assembled multitude that he had just attended a great concert of the Angels hailing the Lord in their sacred canticles.

Angels or Aliens? Many contemporary reports of abductions involve ordinary humans caught up by a strange force that alters their reality in drastic ways and causes them to report contact with other forms of consciousness, or even with a totally alien world.

Acacius, bishop of Constantinople, states, "The population of the whole city saw it with their eyes." And Baronius, commenting upon this report, adds the following words:

"Such a great event deserved to be transmitted to the most remote posterity and to be forever recorded in human memory through its mention every year in the ecclesiastical annals. For this reason the Greeks, after inscribing it with the greatest respect into their ancient Menologe, read it publicly every year in their churches."

Over the centuries many extraordinary events have taken place and chroniclers have transmitted them to "the most remote posterity."

We are that posterity.

It is our responsibility to assess the data they have transmitted to us. Upon their authority and their accuracy rest our concept of history and our vision of the world.

Four major conclusions

The authors of the present book have performed such a study. While we make no claim that any of the events we have uncovered "proves" anything about flying objects from alien worlds, or influence by non-human intelligences, we have emerged with four major observations:

1. **Throughout history, unknown phenomena variously described as prodigies or celestial wonders, have made a major impact on the senses and the imagination of the individuals who witnessed them.**

2. **Every epoch has interpreted the phenomena in its own terms, often in a specific religious or political context. People have projected their worldview, fears, fantasies, and hopes into what they saw in the sky. They still do so today.**

3. **Although many details of these events have been forgotten or pushed under the colorful rug of history, their impact has shaped human civilization in important ways.**

4. **The lessons drawn from these ancient cases can be usefully applied to the full range of aerial phenomena that are still reported and remain unexplained by contemporary science.**

Whether we like it or not, history and culture are often determined by exceptional *incidents*. Stories about strange beings and extraordinary events have always influenced us in an unpredictable fashion. Our vision of the world is a function of the old myths with which we have grown familiar, and of new myths we pick up along the way.

The importance and antiquity of myths was noted by anthropologist of religion Mircea Eliade in *Myths, Dreams and Mysteries: The Encounter between Contemporary Faiths and Archaic Realities*:

> "What strikes us first about the mythology and folk-lore of the "magical flight" are their primitivity and their universal diffusion. The theme is one of the most ancient motifs in folk-lore: it is found everywhere, and in the most archaic of cultural

> strata. . . . Even where religious belief is not dominated by the "ouranian" gods [those of the sky], the symbolism of the ascent to heaven still exists, and always expresses the transcendent."

Yet the lessons from the past are often forgotten. An examination of contemporary cults centered on the belief in extraterrestrial visitations shows that the modern public is still willing to jump to conclusions every time a UFO incident is reported, anxious as people are to follow instructions that appear to come from above. Even in these early years of the 21st century, we observe a continuing process through which the myths of humankind become implemented as social and political realities. We are the witnesses and the victims of that process.

Alien contact: mankind's oldest story

Most "experts" in the study of UFOs in the context of popular culture, state that visitations by "flying saucers" started after World War II. It is traditional for UFO books and television documentaries to begin with the statement that the *Flying Saucer Era* began on June 24, 1947, when an American businessman and pilot named Kenneth Arnold reported a series of unidentified flying objects over Mount Rainier, in the State of Washington. Even some well informed researchers have posed as an axiom (without citing any evidence) that the UFO phenomenon is a recent historical occurrence—"apparently no more than two centuries old" in the words of one American writer. This late date is consistent with the idea that UFOs are extraterrestrial spacecraft bent on studying or inspecting the Earth, perhaps as a result of the atomic explosions of Hiroshima and Nagasaki.

In contrast, if the phenomenon has existed in fairly constant form for a very long time, it becomes harder to hold to a simplistic "ET visitation" scenario to explain it.

Indeed, many documents point to the very ancient nature of the observations. In a recent book on abductions a Canadian researcher, Dr. Persinger, has observed that "for thousands of years and within every known human culture, normal individuals have reported brief and often repeated 'visitations' by humanoid phenomena whose presence produced permanent changes within the psychological organization of the experient. When these phenomena were labeled as deities the "messages" were employed to initiate religious movements that changed the social fabric of society."

Historical scholarship reinforces the latter view. In a book entitled *Out of this World: Otherworldly Journeys from Gilgamesh to Albert Einstein* (Boston: Shambhala, 1991) Professor Couliano, editor in chief of the journal *Incognita* and professor of the history of religions at the University of Chicago, has made it clear that the observation of UFOs and abductions by beings from beyond the Earth is mankind's oldest story. Couliano asserts on the basis of ethnosemiotics that "human beings had beliefs concerning other worlds long before they could write" and that "the most ancient documents of humankind and the study of its most 'primitive' cultures...both show that visits to other worlds were top priorities." He defines the basic question in terms similar to those used by modern abductionists: "Where did those people who pretended to travel to another world actually go?"

It is impossible to catalogue the information accumulated by Couliano, who cautions us that he barely scratched the surface: "To collect all historical documents referring to otherworldly journeys is a gigantic task, a task that has never been undertaken before." Clear examples of this material cover every culture, from eastern Melanesia (where living people had access to a netherworld called Panoi, either in body or in spirit) to Mesopotamia, the source of abundant material about otherworldly journeys. In a typical example Etana, king of Kish, makes an ascent to the sky in order to bring down a plant that cures childlessness—that reference to the theme of reproduction again. "Along with Etana we move from heaven to heaven and see the land underneath becoming smaller and smaller, and the wide sea like a tub," a classic abductee statement.

Otherworldly beings, celestial vehicles

While some individuals in antiquity have allegedly left the Earth by non-physical means, many were said to be taken away by beings that actually used flying vehicles, variously described in the language of their time and culture. Taoists often describe such vehicles involved with "dragons." Thus K'u Yuan, about 300 BC, wrote a poem about the experience of flying over the Kun-lun Mountains of China in a chariot drawn by dragons and preceded by Wang-Shu, the charioteer of the moon. Modern ufologists might characterize this description as a "screen memory," where the mind of the percipient is assumed to replace the awesome vision of a space being with a more familiar human or animal. Under their interpretation, such a story resembles a classic abduction, in which a human is captured by space beings who take their victim away in an interplanetary craft. But the Taoist literature goes further, describing a ritual in which otherworldly entities are actually invited and come down to Earth in order to meet the celebrant.

At the end of the ritual "they mount the cloud chariot, and the team of cranes takes off." The cloud chariots are reminiscent of the "cloudships" seen over southern France in the ninth century, to which Archbishop Agobard of Lyon devoted part of one of his books. Saint Agobard had to preach to the crowd to dissuade the citizens of Lyon from killing four individuals, "three men and one woman" who had alighted from one of these cloudships, alleged to have come from Magonia, a magical land in the sky.

The Middle East is one of the most fertile sources for such early stories. Ezekiel was transported by the "wheels within wheels" of his vision to a far away mountain in a state of stupor. The testament of Abraham tells us he was given a heavenly tour by Archangel Michael in his chariot. In Jewish mysticism such descriptions sound like actual physical observations, witness the experience of Rabbi Nehuma ben Hakana: "When I caught sight of the vision of the Chariot I saw a proud majesty, chambers of chambers, majesties of awe, transparencies of fear, burning and flaming, their fires fire and their shaking shakes."

In the words of Couliano, "All Jewish apocalypses (a word that means revelation, uncovering) share a framework in which the individual is accompanied by an angelic guide, the revelation is obtained in dialogue form, multiple levels of heaven are visited..."

Enoch ascends through the sky in a chariot of fire. The Slavonic Book of Enoch gives additional details about his abduction: Enoch was asleep on his couch when two angels looking like oversized men came and took him on a heavenly trip. Similarly, Elijah goes to heaven without dying. Couliano adds that "a third one might have been abducted to heaven as well, for 'no one knows the place of his burial to this day', that one is Moses." Also in the Mediterranean region, Muslim stories of the Mi'Raj recount the ascent of Prophet Muhammad to heaven, while the Greeks have preserved the records of the travels in space of Phormion of Croton and Leonymus of Athens. Heraclides himself (circa 350 BC) was fascinated by air travel, otherworldly journeys and knowledge of previous incarnations.

Similar imagery can be found (under the guise of a "journey of the soul") in the Mithraic Paris codex, where we are told that the great God Helios Mithra "ordered that it be revealed by his archangel, that I alone may ascend into heaven as an inquirer and behold the universe... It is impossible for me, born mortal, to rise with the golden brightnesses of the immortal brilliance. Draw in breath from the rays, drawing up three times as much as you can, and you will see yourself being lifted and ascending to the height, so that you seem to be in mid-air."

The text goes on: "The visible gods will appear through the disk of gold… and in similar fashion the so-called 'pipe,' the origin of the ministering wind. For you will see it hanging from the sun-disk like a pipe… and when the disk is open you will see the fireless circle, and the fiery doors shut tight. Then open your eyes and you will see the doors open and the world of gods which is within the doors."

An invocation follows: "Hail, o Guardians of the pivot, o sacred and brave youths, who turn at one command the revolving axis of the vault of heaven, who send out thunder and lightning, and jolts of earthquakes and thunderbolts …"

Similar beliefs appear throughout American Indian cultures. Thus Lowell John Bean reports (in the book *California Indian Shamanism*, Menlo Park: Ballena Press 1992) that "souls and ghosts transcended the space between worlds," while "some humans, through ecstatic experience, were able to transport themselves to the other worlds or to bring from them supernatural power."

Physical interpretations

Couliano spends more time speculating about possible physical interpretations of the material he studies than ufologists preoccupied with modern abduction claims. In a chapter entitled "A Historian's Kit for the Fourth Dimension," he cites Charles Howard Hinton, Robert Monroe, Charles Tart, Ouspensky, and Einstein, and observes that "Physics and mathematics are to be held responsible to a large extent for the return of interest in mystical ways of knowledge."

If the soul is a "space shuttle," as religious tradition and folklore seem to suggest, does it follow special laws of physics yet to be discovered? And what conclusion can we draw from the multiplicity of current representations of other worlds? Simply that we live in a state of advanced other-world pluralism, where the "coarse hypothesis of a separable soul" is becoming obsolete. New models of mind, "inspired by cybernetics and artificial intelligence, are replacing the old ones."

Later in his analysis Couliano remarks that "science itself has opened amazing perspectives in the exploration of other worlds, and sometimes in other dimensions in space. Accordingly, our otherworldly journeys may lead to parallel universes or to all sorts of possible or even impossible worlds."

It is to such a journey that we invite the reader.

Return to Magonia

Forty years ago a book entitled *Passport to Magonia* (subtitled "From Folklore to Flying Saucers") documented the parallels between contemporary sightings of "Aliens" and the behavior of beings mentioned in ancient times, often interpreted as gods, angels, or devils. They were the "Daimons" of Greek antiquity, the "Little People" of the Celtic fairy-faith, the elves and gnomes of Paracelsian tradition, the familiars of the witchcraft era. They flew through the air in various devices such as spheres of light. They abducted humans, had sexual intercourse with them, showed them visions of parallel worlds, and gave them messages that changed history.

Passport to Magonia shocked many UFO believers, because it questioned the simplistic "extraterrestrial" origin of the phenomenon, calling for a more complex interpretation where symbolic and cultural factors added another layer to the mythical dimension of the observations. Yet the book was based on preliminary data and scanty documentary evidence. Its claims were subject to interpretation and criticism from many angles.

In the last 40 years much has happened to strengthen this research. Several teams of historians, anthropologists, folklore specialists and philologists have entered the field. Their work has deepened and broadened the investigation of these ancient themes. The advent of powerful Internet search engines, followed by a worldwide movement to make historical archives available online, has amplified the ability of interested amateurs and professionals alike to make important contributions to the work. The result of this massive cooperative effort is astounding.

Anyone who doubts that descriptions of unusual aerial phenomena and the entities associated with them have made a major impact on human history and culture only has to browse through this book – purposely restricted to 500 prominent cases between Antiquity and the Age of Flight – to realize what wonderful events they've been missing.

Historical references suggest that in the absence of claims of unknown aerial phenomena that amazed and inspired their people, Pharaoh Amenophis IV would not have taken the name Akhenaton and introduced the cult of the Sun Disk into Egypt and Emperor Constantine might not have established Christianity in Rome in 312 AD. Ancient chronicles assure us that beings from celestial realms (referred to as Magonia, Nirvana, Heaven, or Walhalla) were responsible for telling Mary she would bear the son of God, for instructing Japanese emperor Amekuni to honor the Supreme God, for inspiring Mohammed to found Islam in Medina in 612, for saving the life of a priest named Nichiren shortly before his execution in 1271, for helping Henry V of England win a decisive battle over French knights at Agincourt in 1415 and for convincing Charles Quint to abandon the siege of Magdeburg in 1551.

Other episodes – whether or not we believe in their actual physical reality – have acquired a colorful place in human history: Emperor Charlemagne was thrown from his horse when an unknown object flew over him in 810 AD; Joan of Arc was inspired to take the leadership of French armies and throw the English out of France after getting her instructions from beings of light in 1425; Christopher Columbus saw a strange light as he approached America; and the claim of an apparition in Guadalupe was responsible for converting millions of Mexican Indians to Catholicism in 1531.

Among great scientists and scholars who carefully recorded sightings of aerial phenomena they could not identify and did not hesitate to publish their observations were mathematician Facius Cardan, Sir H. Sloane (president of the Royal Society), Charles Messier, Cromwell Mortimer (secretary of the Royal Society), and such illustrious literary figures as Casanova and Goethe. So much for Stephen Hawking's "cranks and weirdos."

Structure of this work

Part I, ***A Chronology of Wonders***, contains 500 selected events that give, in varied detail, descriptions of aerial phenomena that have remained unidentified after we exhausted analysis with the means at our disposal. For convenience of the reader, it is divided in six distinct periods, with commentaries about the social and historical characteristics of each period, as it affects the context and reporting of unusual events in the sky.

We stopped the compilation before 1880, at the beginning of a new era when man, thanks to newly-invented balloons and lighter-than-air devices, had begun to fly at last.

Before that date, human observers were often confused by atmospheric effects, optical illusions, meteors, and comets, and the visionary experiences common to prophets and excited crowds, but there were no man-made craft in the sky until 1783, when Louis XVI of France granted permission for the first human balloon flight, and of course no heavier-than-air machine at all over the period we cover.

We have tried to recognize common errors, only keeping in our catalogue truly intriguing descriptions suggestive of actual physical anomalies. During the period we study there were no airplanes, no searchlights playing on cloud banks, no rockets fired into space, and none of the shenanigans of secret prototypes or clandestine operations of psychological warfare often recruited by skeptics to "explain" modern UFOs. After 1879, while the sky is still somewhat pristine, research into unidentified aerial phenomena becomes more complex with the frequent reporting of balloons, "airships," and the hoaxes typical of the new Western media, including competing journalists with blurred standards of accuracy.

Part II, ***Myths, Legends, and Chariots of the Gods***, draws the lessons from the larger body of physical data that has come to form man's view of the universe. By restricting ourselves to a period stretching from Antiquity to the Age of Flight, we were able to apply systematic standards to reports of unknown things in the sky. In the process, we had to make our way through much material that did not fit our criteria for valid entries as aerial phenomena, yet provided considerable insight into cultural, religious, or social attitudes of the time.

Some of the rejected material is assembled in this section of the book but our assessment of it is not final. We recognize that much is still to be uncovered about the literature of this phenomenon. Further study of this material by other researchers may, in time, yield revised data that should be included in future catalogs of aerial phenomena.

Part III, ***Sources and Methods,*** discloses our selection criteria and the process through which we assembled the chronology. It also delves into the difficult issues of scholarship, when the problem is to decide which version of a particular historical event is worthy of being retained, and which is inaccurate, deceptive or frivolous.

In this section we also explain how the emergence of the Internet has changed the methodology of research into ancient material by making previously inaccessible documents searchable, and, equally importantly, by allowing the building of networks of communication among interested researchers and scholars in many countries.

In our ***Conclusion*** we will review the major patterns we were able to extract from the historical unfolding of the observations, and we will ask how they relate to the phenomenon as it continues to be observed today.

Symbols

In order to facilitate future reference, we have used the following symbols to indicate the nature of each account:

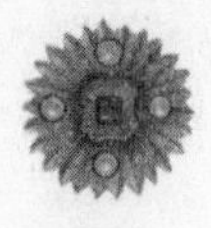

Unidentified Aerial Light

Unidentified Aerial Object

Abduction

Phenomenon with physical evidence

Entity (alone)

Entity associated with an aerial phenomenon

Communication

These symbols are extracted from the Dover Publications collection of Medieval Ornaments (copyright 2002).

PART I

A Chronology of Wonders

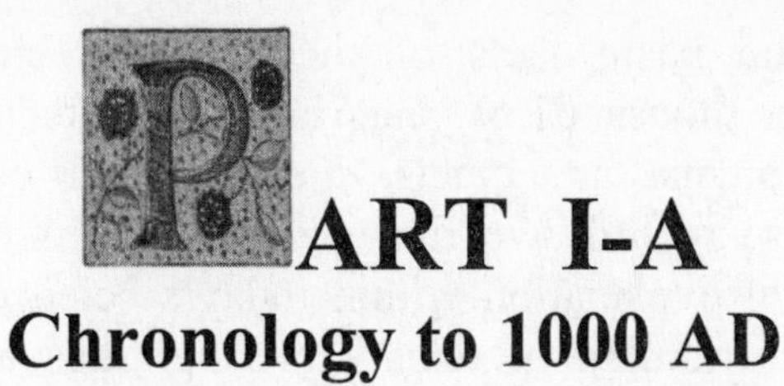

PART I-A

Chronology to 1000 AD

Ancient records of unknown phenomena in the sky pose special challenges. Unlike modern accounts, they are often kept in remote libraries, neglected by scholars, written in little-understood languages and seldom translated with accuracy. Indications of their existence are found in secondary sources, typically slanted to particular belief systems in religious or political terms, and are therefore doubtful. When they are quoted on the Internet or in popular literature they are often so garbled as to become unrecognizable.

The study of such cases has to begin with the search for a primary source, often a chronicler, a historian, or an astronomer, together with an assessment of the context in which the sighting was made. Not surprisingly, ancient civilizations with the most advanced astrology and meteorology have produced good records of this type. China and the Roman Empire, in particular, have given us valuable astronomical reports, often with precise dates. Japan and the Middle East are also prominent.

Given the lack of knowledge at the time about the nature of celestial objects such as meteors or comets, observations of such phenomena were often reported as "portents" or "omens." Chroniclers generally pointed to specific historical events that followed the observation, attributing a cause-and-effect relationship to the sighting. This was a natural tendency, with two unintended consequences: on the one hand, it has contributed to slanting the narrative to special political or religious viewpoints; on the other hand, the association with historical records has served to preserve the basic facts of the sighting, enabling us, hundreds or thousands of years later, to better understand such phenomena as comets, meteors, and novae. And among these records we find accounts that still have no conventional explanation within today's science. In some cases, the reframing of remarkable sightings as mystical events has probably resulted in the loss of accounts that would interest us today as physical anomalies.

In extreme cases, this process has led to the popular belief that "the Gods" were intervening in human affairs through celestial manifestations. Indeed, it was convenient for secular or clerical rulers to claim that divine powers were supporting their views or guided them in battle.

In selecting cases for inclusion in this Chronology we have paid special attention to such biases in order to steer clear of the suggestion that aerial phenomena intervened directly in terrestrial history. Of course, as the reader will see, the societal and psychological impact was a real and lasting one, but only because of the interpretations witnesses and their contemporaries gave to the events.

This process continues today in the many heated controversies about unidentified flying objects, their origin, their nature, and their possible technological implications. For this reason, the study of the oldest records is crucial to an understanding of unidentified aerial phenomena that are still commonly reported.

As we go further back in time, our unidentified cases owe more to mythology than to history. Yet we wish to show the reader the rich variety of experiences that were reported throughout the ages. Accordingly, in this initial section we have relaxed our selection standards in terms of date and contents, while providing critical comments when appropriate.

The symbol ⓘ denotes cases whose nature or source, in our opinion, needs new information because it is vague, unreliable, or insufficiently documented. We included them for illustration purposes, and to stimulate further research.

1.

Ca. 1460 BC, Upper Retjenu, Lebanon
A "star" defeats the Nubians

The stela of Gebel Barkal, erected in honor of Thutmosis III, describes a fantastic celestial event during a war: "A star fell to their South position. It struck those opposed to him (the Nubians). None could stand..." (Lines 33-36).

"[The star] positioned itself above them as if they didn't exist, and then they fell upon their own blood. Now [the star] was behind them (illuminating) their faces with fire; no man amongst them could defend himself, none of them looked back. They had not their horses as [these] had fled into the mountain, frightened... Such is the miracle that Amon did for me, his beloved son in order to make the inhabitants of the foreign lands see the power of my majesty."

Source: this document, of undisputable authenticity, was first published in 1933, in a German Egyptological journal, *Zeitschrift fur Agyptischen Sprache und Altertumskunde* 69: 24-39.

The text, now on display in the Museum of Jardum, Sudan, was found by archaeologists excavating in the Temple of Amon, located at the bottom of the Gebel Barkal Mountain in the great Bayunda desert. The stela, which is made of granite and measures 173 cm by 97 cm, was erected on 23 August 1457 BC in honor of Thutmosis III's important victories in Asia.

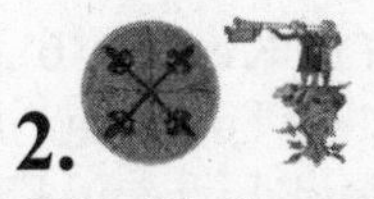

2.

1347 BC, El-Amarna, Nile Valley, Egypt ⓘ
Akhenaton's flying disk

Pharaoh Akhenaton (Amenophis IV) had a unique experience that was to shape Egyptian history. According to inscriptions on the 'Frontier Stelae' found on the circumference of El-Amarna, Akhenaton was strolling along the river admiring the splendors of nature one summer morning when he looked up and saw “a shining disc” descend from the sky.

He heard the voice of the Solar Disc itself tell him that he was to build a new capital for Egypt, and give it the name Akhetaton, “The Horizon of the Solar Disc.”

During the time of Amenophis IV, Egypt’s capital became the City of Akhetaton. The ideographic symbol for the word “horizon” was a disc floating over a mountain range.

Akhenaton also founded a new religion based on the worship of the Solar Disc, thus assuring his immortality in our history books as the most powerful heretic of ancient Egypt. Although it refers here to the shape of the sun itself, it is interesting to find that the basic disc shape often mentioned in art and ancient manuscripts has been quoted (or misquoted) as evidence of “flying saucers” by contemporary writers.

Source: David P. Silverman, Josef William Wegner, and Jennifer Houser Wegner. *Akhenaten and Tutankhamun: Revolution and Restoration* (Philadelphia: University of Pennsylvania Museum of Archaeology and Anthropology, 2006), 44-47.

3.

Circa 850 BC, shores of the Jordan River, Israel ⓘ
Abduction of Elijah

The prophet Elijah (1 Kings 16:29 to 2 Kings 2:18) practiced his ministry in Israel during the reigns of King Ahab (874-853 BC) and his son King Abaziah (853-852 BC). The trouble began when Ahab married the pagan princess Jezebel and erected an altar to Baal in Samaria.

Baal was the Sun-God of the Phoenicians, and a Sacred Pole was used to chart Baal’s journey through the twelve signs of the zodiac. According to the Old Testament, by embracing the religion of a "false God," Ahab did more to anger "the Lord" than any of the kings of Israel before him.

The prophet Elijah (Elias) the Tishbite then delivered a divine message to King Ahab that God would bring a drought to his kingdom. The significance of this message

Fig. 1: The abduction of Elijah (Gustave Doré)

was that Baal was worshipped for his supposed power over the sky and the weather, so God's message was a direct challenge to Baal and a sign of displeasure.

The second chapter of 2 Kings mentions an episode when Elijah revealed he was about to be taken away:

"And it came to pass, when they were gone over, that Elijah said unto Elisha, Ask what I shall do for thee, before I be taken away from thee. And Elisha said, I pray thee, let a double portion of thy spirit be upon me. And he said, Thou hast asked a hard thing: nevertheless, if thou see me when I am taken from thee, it shall be so unto thee; but if not, it shall not be so.

"And it came to pass, as they still went on, and talked, that, *behold, there appeared a chariot of fire, and horses of fire, and parted them both asunder; and Elijah went up by a whirlwind into heaven.*" – 2 Kings 2:11.

Elijah was the only Old Testament prophet who did not die, but was said to be *taken up* to heaven. Even today, Jews are waiting for Elijah to return. An empty chair and a goblet of wine are set at the Passover feast table as a reminder of this belief. The Mormons, on the other hand, believe that Elijah came back on April 3rd 1836, appearing before Joseph Smith.

Source: Unless otherwise indicated, we are using the King James Version of the Bible.

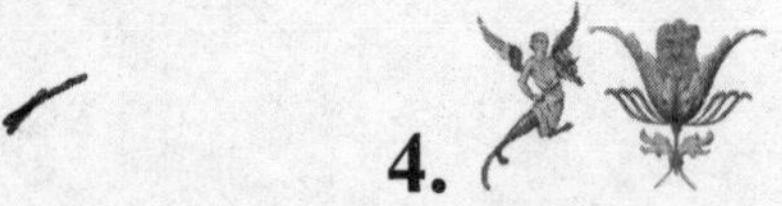

4.

Ca. 593 BC, Chaldea, near the river Chebar, Iraq ⓘ
Ezekiel's abduction

The Bible states that Prophet Ezekiel saw a strange craft appear in the sky above him. It consisted of "wheels within wheels," a brilliant dome, and four beings. He was transported to a mountaintop, without knowing how he got there, and remained stunned, an experience reminiscent of numerous contemporary reports by people claiming abductions.

We are not so naïve as to believe that the Ezekiel account in the Bible, which was written down centuries after the life of the prophet by that name, represents a first-hand report of an observation, any more than the abduction of Elijah in the previous account. Wikipedia notes that "the academic community has been split into a number of different camps over the authorship of the book. W. Zimmerli proposes that Ezekiel's original message was influenced by a later school that added a deeper understanding to the prophecies. Other groups, like the one led by M. Greenberg, still tend to see the majority of the work of the book done by Ezekiel himself. Some scholars have suggested that the person described by the Book of Ezekiel may have suffered from temporal lobe epilepsy, which has several characteristic symptoms that are apparent from his writing. These symptoms include hypergraphia, hyperreligiosity, fainting spells, mutism, and pedantism, often collectively ascribed to a condition known as Geschwind syndrome."

Even with these qualifications, the account is remarkable for Ezekiel's description of a phenomenon that would resonate with the people of his time, and still strikes us by its awesome imagery:

> *"Then I looked, and behold, a whirlwind was coming out of the north, a great cloud with raging fire engulfing itself; and brightness was all around it and radiating out of its midst like the colour of amber, out of the midst of the fire. Also from within it came the likeness of four living creatures. And this was their appearance: they had the likeness of a man. Each one had four faces, and each one had four wings."*

The text goes on: "Now as I looked at the living creatures, behold, a wheel was on the earth beside each living creature with its four faces. The appearance of the wheels and their workings was, as it were, a wheel in the middle of a wheel. When they moved, they went toward any one of four directions; they did not turn aside when they went.

"When the living creatures went, the wheels went beside them; and when the living creatures were lifted up from the earth, the wheels were lifted up. Wherever the spirit wanted to go, they went, because there the spirit went; and the wheels were lifted together with them, for the spirit of the living creatures was in the wheels. When those went, these went; when those stood, these stood; and when those were lifted up from the earth, the wheels were lifted up together with them, for the spirit of the living creatures was in the wheels" (Ezekiel 1: 4-21).

Fig. 2: The abduction of Ezekiel

Later Ezekiel describes what today would be characterized as "abduction":

> **2:9** And when I looked, behold, a hand was stretched out to me, and a written scroll was in it
> **3:12** Then the spirit lifted me up, and as the glory of the Lord arose from its place, I heard behind me the sound of a great earthquake.
> **3:13** It was the sound of the wings of the living creatures as they touched one another, and the sound of the wheels beside them that sounded like a great earthquake.

> **3:14** *The spirit lifted me up and took me away, and I went in bitterness in the heat of my spirit, the hand of the Lord being strong upon me.*
> **3:15** *And I came to the exiles at Tel-Abib, who dwelt by the river Chebar. And I sat there overwhelmed among them seven days.*

It is noteworthy that the description includes some words that appear only once in Ezekiel's writing and some that only appear once in the entire Old Testament, an indication that the prophet was indeed looking for ways to express a vision that surpassed his understanding – and the ability of translators to adequately convey his experience.

5.

464 BC, Rome, Italy: Prodigious shapes and figures ⓘ

According to fourth-century Roman writer Julius Obsequens' *Liber de Prodigiis* (Book of Prodigies),

"In the consulate of Aulo Postumio Albino Regillense and Spurio Furio Medullino Fusco, once again and with great splendor a burning in the sky and many other prodigies appeared with shapes and strange figures, frightening the spectators."

Such accounts are frequently found in old texts, yet they are of only marginal interest to us, in spite of their tantalizing context, because they give no hint of a description of an actual event. A "burning in the sky" could be a common meteor or an auroral display, and there is no evidence that the "shapes and strange figures" were seen in the air. These considerations, well understood by most scholars of ancient texts, have led us to exclude many such references from our Chronology.

It is important to note that the version of Obsequens' chronicle containing the reference cited here was not the original. In 1552, humanist Conrad Wolffhart (1518-1561), who took the Greek name of Lycosthenes, edited the chronicle and added illustrations from wood-cuts. Obsequens' *Liber de prodigiis* (Book of Prodigies) was an account of the portents observed in Rome between 190 BC and 12 BC. As some of the original text had not survived, Lycosthenes reconstructed the missing parts himself, starting at 749 BC, from other historical sources. Therefore, the records attributed to Obsequens from prior to 190 BC were possibly not in the Latin original.

Source: Lycosthenes, *Julii Obsequentis Prodigiorum Liber...per Conradum Lycosthenem Rubeaquensem integrati suae restitutus* (Basel, 1552).

6.

404 BC, Attica, Greece
Guided by a glowing pillar in the sky

"When Thrasybulus was bringing back the exiles from Phyla, and wished to elude observation, a pillar became his guide as he marched over a trackless region... The sky being moonless and stormy, a fire appeared leading the way, which, having conducted them safely, left them near Munychia, where is now the altar of the light-bringer."

Note: We have found no comet recorded for that period, and the observation remains unexplained.

Source: Clement of Alexandria, *Stromata*, Book I, Chapter 24. Cited in *The Ante-Nicene Fathers, translations of the writings of the Fathers down to AD 325*, by Rev. Alexander Roberts and James Donaldson (eds.) revised and arranged by A. Cleveland Coxe, Vol. II: Fathers of the Second Century (Edinburgh reprint, 2001).

7.

Circa 343 BC, Near Sicily, Italy: a blazing light

In Diodorus Siculus' first century text *Historical Library*, (book 16, 24-5) we read that the voyage of Timoleon from

Corinth to Sicily was guided by one or more blazing lights referred to as *lampas*: "Heaven came to the support of his venture and foretold his coming fame and the glory of his achievements, for *all through the night he was preceded by a torch blazing in the sky* up to the moment when the squadron made harbor in Italy."

Note: This might have been a comet, but it has never been matched with any known cometary object, according to Gary Kronk's *Cometography*. P. J. Bicknell, writing in *The Classical Quarterly* ("The Date of Timoleon's Crossing to Italy and the Comet of 361 BC" in *New Series*, Vol. 34, No. 1, 1984, 130-134) argues that "a cometary hypothesis is barely compatible with the implication of Diodorus' account that the *lampas* were visible in the east at nightfall and therefore in opposition to the sun... All in all it is difficult to resist the conclusion that Diodorus (or his source) elaborated on the *lampas* for dramatic effect…"

Bicknell leans towards the interpretation of the objects as a spectacular meteor shower, possibly the Lyrids, which would put the date of his voyage at 21 March 344 BC However this does not account for a phenomenon seen "all through the night" in a fixed direction.

Source: Gary Kronk. *Cometography–A Catalog of Comets, Volume 1 Ancient-1799* (Cambridge: Cambridge University Press, 1999), 511.

8.

218 BC, Amiterno, Italy: phantom ships

"During this winter many portents occurred in Rome and the surrounding area, or at all events, many were reported and easily gained credence, for when once men's minds have been excited by superstitious fears they easily believe these things... A phantom navy was seen shining in the sky; in the territory of Amiternum beings in human shape and clothed in white were seen at a distance, but no one came close to them."

There is no evidence that the aerial sightings had any connection with the other reports, so the mystery only seems compounded by the juxtaposition of strange events. In their chronological chapters, both Pliny and Livy appended a list of all prodigies reported for a given year, which were compiled in the *Annales Maximi* for the Consuls. These Annals, which were lost even before the time of Livy and Pliny, are now lost. This explains why the Roman prodigies that have reached us are only dated by their year, with an odd juxtaposition of unrelated events.

Source: *The History of Rome Vol III by Livy*, trans. Reverend Canon Roberts (Montana: Kessinger Publishing 2004), 51.

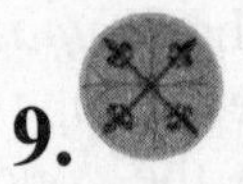

9.

216 BC, Arpi, Apulia, Italy: Shields ⓘ

"At Arpi shields had been seen in the sky and the sun had appeared to be fighting with the moon; at Capena two moons were visible in the daytime."

This description from Livy suggests disk-shaped flying objects but could also refer to meteors, as we do not know the duration of the observation.

Source: *The History of Rome Vol III by Livy*, trans. Reverend Canon Roberts (Montana: Kessinger Publishing 2004), 54.

10.

2 August 216 BC, Cannae, Apulia, Italy ⓘ
Round objects, white figures

During the famous battle won by Hannibal in Cannae (2 August, 216 BC), in the Apulian plain near Barletta, which saw the largest defeat in the history of Rome, a mysterious phenomenon was observed: "On the day of the battle, in the sky of the Apulia, round objects in the shape of ships were seen. The prodigies carried on all night long. On the edge of such objects were seen men dressed in white, like clergymen around a plow."

Source: Italian magazine *Cielo e Terra* (August 1967): 2. We were unsuccessful in tracking down an original source. We include this quote from a popular magazine with reservations, given the abundance of fictional historical material in that period, and acknowledge a possible confusion with case 8 above.

11.

June 213 BC, Hadria, Gulf of Venice, Italy ⓘ
Men seen in the sky

"At Hadria an altar was seen in the sky and about it the forms of men in white clothes."

Fig. 3: An interpretation of the Hadria sighting

This illustration attempts to capture the scene, which suggests an event remarkable enough for historians to have noted it, and for a record to have been preserved. We suspect, however, that a confusion of locations may exist with the case of 218 BC in Amiterno. White clothes are indicative of sacerdotal garments.

Source: Lycosthenes, *Julii Obsequentis Prodigiorum Liber...per Conradum Lycosthenem Rubeaquensem integrati suae restitutus* (Basel, 1552).

12.

173 BC, Lanuvium, Albano Laziale (Lanuvio), Italy
Aerial fleet

"As it was fully expected that there would be war with Macedonia, it was decided that portents should be expiated and prayers offered to win 'the peace of the Gods,' of those deities, namely, those mentioned in the Books of Fate. *At Lanuvium the sight of a great fleet had been witnessed in the heavens....*"

Source: *The History of Rome Vol III by Livy*, trans. Reverend Canon Roberts (Montana: Kessinger Publishing 2004), 72.

13. ⓘ

163 BC, Cassino, Lazio Province, Italy
Nocturnal lights, sounds

A "sun" shone at night for several hours. The original text reads: "Consulship of Tiberius Gracchus and Manius Juventus: at Capua the sun was seen during the night. At Formice two suns were seen by day. The sky was afire... In Cephallenia a trumpet seemed to sound from the sky... By night *something like the sun* shone at Pisaurum."

These phenomena are grouped together by a chronicler, but they were not observed at the same time or in the same region. It is frustrating for us not to have more detail.

Note that this is the last of Lycosthenes' restored cases; all further references from Obsequens' book were also in the original.

Source: Lycosthenes, *Julii Obsequentis Prodigiorum Liber...per Conradum Lycosthenem Rubeaquensem integrati suae restitutus* (Basel, 1552).

14.

122 BC, Ariminium, Italy: Three "moons" at once

A huge luminous body lit up the sky, and three moons rose together.

Pliny writes in his *Natural History*, Book II, Chapter XXXII: "Three moons have appeared at once, for instance in the consulship of Gnaeus Domitius and Gaius Fannius."

Another citation from Dio Cassius (Roman History, Book I) states: "At Ariminium a bright light like the day blazed out at night; in many portions of Italy three moons became visible in the night time."

The observation of triple moons in the night sky is a rare but explainable atmospheric phenomenon. We include the case because of the ambiguity about the coincidence of several phenomena making a strong enough impression to be recorded by serious authors.

Source: Pliny the Elder, *Natural History*, trans. Harris Rackham (Harvard University Press, 1963), vol. 10, 243.

15.

103 BC, Amelia and Todi, Italy
Shields clashing in the sky

During the War with the Cimbri, "from Amelia and Todi, cities of Italy, it was reported that at night there had been seen in the heavens flaming spears, *and shields which at first moved in different directions, and then clashed together*, assuming the formations and movements of men in battle, and finally some of them would give way, while others pressed on in pursuit, and all streamed away to the westward."

The description of the objects' behavior is puzzling, radically different from what would be expected in the case of a meteor shower. Nor does it fit well with an aurora borealis. Note that Obsequens locates the sighting at Rimini in the Emilia-Romagna region of Italy.

Source: Plutarch, *Plutarch's Lives,* trans. Bernadotte Perrin (Harvard University, 1950) v.9, 509. Also see: Lycosthenes, *Julii Obsequentis Prodigiorum Liber...per Conradum Lycosthenem Rubeaquensem integrati suae restitutus* (Basel, 1552).

16.

99 BC, Tarquinia, Viterbo Province, Italy ⓘ
Another flying shield

In Tarquinia, over a wide area, a fiery meteor was seen, which flew away quickly. "*At sunset a round shield (orbis clypeus) flew west to east.*"

As noted before, the reference to "flying shields" appears several times in the old chronicles. In the absence of additional detail, it is impossible to determine whether the object was a meteor that seemed disk-shaped. We mention such cases with reservation. The document does specify that the object was "round", suggesting a defined structure.

Tarquinia was 52 Roman miles Northwest of Rome.

Source: Lycosthenes, *Julii Obsequentis Prodigiorum Liber...per Conradum Lycosthenem Rubeaquensem integrati suae restitutus* (Basel, 1552).

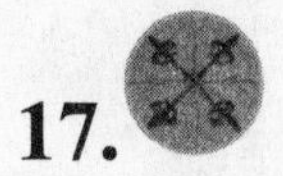

17.

91 BC, Spoletium in Umbria, N. Rome, Italy
Globe, flying up!

"Near Spoletium a gold-colored fireball rolled down to the ground, increased in size; seemed to *move off the ground toward the east and was big enough to blot out the sun.*"

Such an object does not match the pattern of a meteor. We considered the possibility that it might have been a rare form of ball lightning, but this idea is contradicted by the observation that it could "blot out the sun."

Source: Obsequens, *Prodigiorum*, op. cit., ch. 114; Paulus Orosius, *Historiarum Adversum Paganos*, Book V.

18.

Circa June 76 BC, China, exact location unknown
Mysterious candle star

"The fifth year of the Yüan-feng reign period, in the fourth month (12th May to 9th June, 76 BC), *a candle star appeared between K'uei and Lou.*" Astronomers have no idea what it could have been. Some suggest it was a nova, others a comet or meteor.

Chapter 26: 1292 of the same History defines the term thus: "A candle star resembles Venus. It remains stationary from sight right after its appearance. Riot is expected in cities and districts over which it shone." A candle star was one of the 18 irregular "stars" defined in Chinese records.

Source: *History of the Han Dynasty*, ch. 26: 1307; quoted by Y. L. Huang, "The Chinese Candle Star of 76 BC," *The Observatory* 107 (1987): 213. *The History of the Han Dynasty* was part of "Astrological Treatise," compiled by Ma Hsü around 140 AD.

19.

76 BC, Rome, Italy: Maneuvering "torch" in the sky

A group of witnesses with Proconsul Silenus: A spark fell from a star, became as big as the moon, and went up again, which contradicts natural explanations.

The original text reads: "In the consulship of Gnaeus Octavius and Gaius Scribonius a spark was seen to fall from a star and increase in size as it approached the earth, and after becoming as large as the moon it diffused a sort of cloudy daylight, and then returning to the sky changed into a torch; this is the only record of this occurring. It was seen by the proconsul Silanus and his entourage."

Source: Pliny the Elder, *Natural History*, trans. Harris Rackham (Harvard University Press, 1963).

20.

48 BC, Thessaly and Syria: Fiery bombardment ⓘ

Another example of a sighting where the object appears to favor one camp over another in battle: "Thunderbolts had fallen upon Pompey's camp. *A fire had appeared in the air over Caesar's camp and then fell upon his own.*"

In other cases of ancient battles, such fiery objects turned out to be primitive incendiary missiles, so we include this case with reservations.

Source: Cassius Dio Cocceianus, *Dio's Rome: An Historical Narrative Originally Composed in Greek During the Reigns of Septimus Severus, Geta and Caracalla, Macrinus, Elagabalus and Alexander Severus*, trans. Herbert Baldwin Foster (Troy, New York, 1905), vol. 2, 227.

21.

24 May 12 BC, China, exact location unknown
A large hovering object, fire rain

"In the first year of the Yuen-yen period, at the 4th Moon, between 3 P.M. and 5 P.M., by clear sky and serene weather, a sound similar to thunder was heard repeatedly. A meteor (sic) appeared, *the front part the size of a vase, over 100 feet long. Its light was red-whitish. It stood far to the SE of the sun. It threw off fiery sparks on four sides, some as large as a pail, others the size of an egg. They fell like rain. This phenomenon lasted until the evening."*

This is an unexplained episode. Meteors do not linger for two hours, and do not shower the landscape with fiery rain.

Source: Edouard Biot, *Catalogue des étoiles filantes et des autres météores observés en Chine pendant 24 siècles* (Paris: Imprimerie Royale, 1846), 9-10. This book provides an extremely valuable record of astronomical observations in China during much of its history.

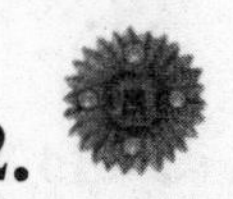

22.

10 February 9 BC, Kyushu, Japan: Nine evil suns ⓘ

The Kumaso people were prospering, until nine "Suns" were seen in the sky, followed by great chaos.

We considered the hypothesis that the phenomenon was a sun-dog, but we found no record of a refraction effect producing nine images of the sun. This is one of numerous items for which it is difficult to locate Asian sources in translation. We mention such cases, fragmentary as they are, in the hope of encouraging future researchers to seek complete sources. This story may originate in the ancient Chinese legend of the nine suns shot down from the sky by Yao dynasty hero Yi when Earth's original ten suns were making life insufferable, in which case it should be regarded as legend rather than fact.

Source: *Brothers Magazine* (Japan) No. III, 1964. This magazine was one of the earliest publications about UFOs in Japan. Unfortunately, it did not provide a quote from an actual source.

23.

April 34 AD, China, exact location unknown
Squadron of flying intruders

A white, round object accompanied by 10 small stars flies overhead. This could refer to a train of meteors, but the pattern is unusual if "accompanied" means that the ten small stars were flying in some sort of formation with the main object.

Source: Edouard Biot, *Catalogue des étoiles filantes et des autres météores observés en Chine pendant 24 siècles* (Paris: Imprimerie Royale, 1846).

24.

61 AD, China ⓘ
A golden apparition is said to have spread Buddhism

Han Emperor Mingti, who had heard of Buddhism, had a vision of a golden figure floating in a halo of light – interpreted as a flying Apsara (Buddhist angel). Some sources present this vision as a dream, others as an "apparition." Arthur Lillie mentions it as a "golden man, a spirit named Foe," while Gray calls it "a foreign god entering his palace."

Whatever it was, the visionary being was interpreted by the Emperor's wise men, including Minister Fu Yi, to be the Buddha himself. Consequently, an envoy was sent to India to learn about the new religion, returning with sacred Buddhist texts and paintings as well as Indian priests to explain the teachings of the Buddha to the Emperor.

The narrative we have does not mention any physical object in the sky at the time, which strictly takes the event out of the realm of aerial phenomena. As will be seen in Part III of this book ("Sources and Methods") the two co-authors have had extensive debate over the wisdom of including such cases in this catalogue, especially from sources steeped in myth and symbolism.

Fig. 4: Flying apsara: painting from the Mogao caves, China

On the one hand, there are thousands of instances where witnesses describe "entities" similar to those typically associated with UFOs, in connection with a "halo of light" that can hardly be considered as a material object. These are often reported in the contemporary literature as "bedroom visitations" or apparitions. We did agree that a difficult line must be drawn between such events and reports of ghosts, ape-men or monsters, which belong in a related but separate study. On the other hand, we find continuity between the interpretation of "signs in the sky," aerial objects with entities aboard, and flying or luminous entities seen by themselves. Accordingly, we have included a limited number of prominent cases of this kind.

Source: John Henry Gray, *China, a History of the Laws, Manners and Customs of the People* (Courier Dover: 2003), 106; Arthur Lillie,

Buddhism in Christendom or Jesus the Essene (London: K. Paul, Trench, 1887), 188.

25.

21 May 70 AD, Jerusalem
Flying chariots surround the city

Flavius Josephus writes: "On the one and twentieth day of the month Artemisius, a certain prodigious and incredible phenomenon appeared: I suppose the account of it would seem to be a fable, were it not related by those that saw it, and were not the events that followed it of so considerable a nature as to deserve such signals; for, *before sun-setting, chariots and troops of soldiers in their armor were seen running about among the clouds, and surrounding cities.*"

Source: Flavius Josephus, *History of the Destruction of Jerusalem, Jewish Wars*, Book CXI, quoted in "*The Genuine Works of Flavius Josephus, the Jewish Historian,*" translated by William Whiston, (London, 1737). See also: Sara Schechner, *Comets, Popular Culture and the Birth of Modern Cosmology* (Princeton University Press: 1999), 32.

26.

Winter 80 AD, Caledon Wood, Scotland
Fast-moving airship

"When the Roman Emperor, Agricola was in Scotland (Caledonia), wondrous flames were seen in the skies over Caledon wood, all one winter night. Everywhere the air burned, and on many nights, *when the weather was serene, a ship was seen in the air, moving fast.*"

The passage goes on to describe another staple of Fortean literature: "In Athol, shower of stones fell from the sky into one place, and a shower of paddocks (frogs) fell on one day from the sky. And high in the air, at night, there raged a burning fire, as if knights in armor and on foot or horse fought with great force."

Here again, these phenomena were not simultaneous or even in the same region but they provide us with a treasure-trove of anomalies, from the fall of frogs to the mention of an aerial ship. Caledon Wood appears in Geoffrey of Monmouth's rendition of the King Arthur legend (in *The History of the Kings of Britain*). Arthur defeated the Saxons at Caledon Wood, among other places.

Source: Hector Boece, *Historia Gentis Scotorum* (1527).

27.

187, Rome, Italy: Hovering stars in daylight

"We read in Herodian that in the time of Commodus stars were seen all the day long, and that some stretched in length, hanging as it were in the midst of the air, which was a token of a cloud not kindled but driven together: for it seemed kindled in the night, but in the day when it was far off it vanished away."

Source: Lycosthenes, *Julii Obsequentis Prodigiorum Liber... per Conradum Lycosthenem Rubeaquensem integrati suae restitutus* (Basel, 1552).

28.

January 195, Rome, Italy: Bright stars around the sun

"I shall now speak of what happened outside, and of the various rebellions. For three men at this time, each commanding three legions of citizens and many foreigners besides, attempted to secure the control of affairs – Severus, Niger, and Albinus. The last-named was governor of Britain, Severus of Pannonia, and Niger of Syria. These, then, were the three men portended by the three stars that suddenly came to view surrounding the sun when Julianus in our presence was offering the Sacrifices of Entrance in front of the senate-house.

"*These stars were so very distinct that the soldiers kept continually looking at them and pointing them out to one another, while declaring that some dreadful fate would befall the emperor.* As for us, however much we hoped and prayed that it might so prove, yet the fear of the moment would not permit us to gaze up at them except by furtive glances."

Were the "bright stars" a case of parhelia or false suns? The description here does not suggest simple refraction.

Source: Cassius Dio Cocceianus, *Dio's Rome: An Historical Narrative Originally Composed in Greek During the Reigns of Septimus Severus, Geta and Caracalla, Macrinus, Elagabalus and Alexander Severus*, trans. Herbert Baldwin Foster (Troy, New York, 1905), vol. 9, 151.

29.

235, Weinan, China ⓘ
A red object flies above the Emperor's army

The army of Emperor Hou Chu saw a red object with pointed rays that flew over them three times.

This case is reported in a compilation of "shooting stars and meteors," but the notion of an ordinary meteor returning three times to fly over an army stretches credulity.

Source: Edouard Biot, *Catalogue des étoiles filantes et des autres météores observés en Chine pendant 24 siècles* (Paris: Imprimerie Royale, 1846).

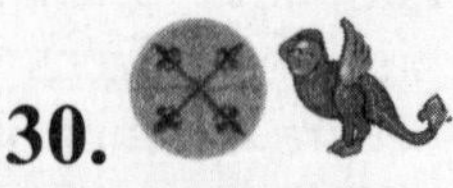

30.

240, Che-chiang Province, China
The dragon and the little blue boy

"Under the Emperor Ta Ti of the Wu dynasty (AD 228-251), in the seventh month of the third year of the Ch'ih-wu era, there was a certain Wang Shuh who gathered medicinal herbs on T'ien Tai Mountain. At the hottest time of the day he took a rest under a bridge, when suddenly he saw a little blue boy, over a foot long, in the brook.

"*The boy held a blue rush in his hand and rode on a red carp. The fish entered a cloud and disappeared little by little.*

"After a good while Shuh climbed upon a high mountain top and looked to all four sides. He saw wind and clouds arising above the sea, and in a moment a thunderstorm broke forth. Suddenly it was about to reach Shuh, who terrified hid himself in a hollow tree. When the sky cleared up, he again saw the red carp on which the boy rode and the little boy returning and entering the brook. It was a black *kiao*!"

We include this case, clearly unexplained in terms of ordinary phenomena, because it illustrates characteristics ascribed to "dragons" in the Chinese literature.

Source: Dr. M. W. De Visser, *The Dragon in China and Japan* (Amsterdam: Johannes Müller, 1913), 80-81. Visser quotes from "the *Wu ki*." A Kiao is a "scaled dragon."

31.

Circa March 260, China, exact location unknown
A child from Mars flies away

At a time when the government of Wu faced critical dangers, during the reign of Sun Hsiu (258 to 263) the generals of border garrisons used to leave their wives and children (known as "hostage children") as pledges of loyalty. It was not unusual for a dozen of these children to play together. The record goes on:

"A strange child suddenly joined the hostage children in their play. He was less than four feet tall, dressed in dark clothes, and appeared to be between six and seven years old. None of the other children recognized the newcomer, so they asked him, "To what family do you belong, that you should suddenly appear among us?"

"I came only because you seemed to be enjoying yourselves so much," was the reply. On closer examination, it was noticed that light rays from the stranger's eyes flashed brilliantly, and the other children began to be afraid. They asked him about his past. "Do you fear me, then?" he asked. "Don't. Though I am not human, I am the star-god Yung-huo (Mars) and have come to deliver a message to you: 'The Three Lords will return to Ssu-ma.'

"The children were startled, and some ran off to tell their parents. The adults arrived in haste to witness all this, but the visitor said, 'I must leave you.' So saying, he propelled his body upward and transformed himself.

"The children looked up and watched him rise to the heavens leaving what appeared to be a great train of flowing silk behind him. Some of the adults arrived in time to watch him drifting gradually higher. A moment later, he vanished."

Given the political crisis, nobody reported this at the time. Four years later Hsiu was overthrown; in 21 years Wu was put down, and the power fell to Ssu-ma.

Source: *In the Wu Kingdom during the Three Kingdoms Period* (222-280), cited in *In Search of the Supernatural: The Written Record*, trans. Kenneth J. DeWoskin and J. I. Crump (Stanford University Press: 1996), 110.

32.

January 314, China, exact location unknown ⓘ
Three suns, flying east

The Sun came down to the ground and three other suns rose together over the western horizon and "flew together towards the East." This is yet another frustrating example of partial information which, taken literally, indicates a most unusual phenomenon. Only reference to the original text could permit a fuller interpretation.

Source: Shi Bo, *La Chine et les Extraterrestres* (Paris: Mercure de France: 1983), 47. We have not been able to find an original source for this case.

33.

Circa 334, Antioch, Turkey
An object emitting smoke for hours

"In Antioch a star appeared in the eastern part of the sky during the day, *emitting much smoke as though from a furnace, from the third to the fifth hour.*" The duration of the phenomenon precludes a comet, but it was seen too long for a meteor.

Source: Theophanes, *Chronographia*, trans. C. Mango & R. Scott, with G. Greatrex, *The Chronicle of Theophanes Confessor: Byzantine and Near Eastern History AD 284-813* (Oxford: Clarendon Press, 1997), 49.

34.

Circa 350, Emesa, Syria: Dialogue with a globe of fire

In Ancient Greece, where meteorology played an important role in religion and scientific philosophy, claims involving strange aerolites abound. Damaskios, in his book, *The Life of Isidorus*, relates that one sacred baitylos (meteorite) was kept by a man named Eusebios, who acquired it in strange circumstances. A Byzantine scholar called Photios, who lived in the 9th century A.D., described the story in his own writings. The following is from Arthur Bernard Cook's *Zeus, A Study in Ancient Religion,* Vol. III, 888:

"This man stated that there had once come upon him a sudden and much unexpected desire to roam at midnight away from the town of Emesa as far as he could get towards the hill on which stands the ancient and magnificent temple of Athena. So he went as quickly as possible to the foot of the hill, and there sat down to rest after his journey. Suddenly he saw a globe of fire leap down from above, and a great lion standing beside the globe. The lion vanished immediately, but *he himself ran up to the globe as the fire died down and found it to be the baitylos. He took it up and asked it to which of the gods it might belong. It replied that it belonged to Gennaios, the 'Noble One.' He took it home the self-same night, traveling, so he said, a distance of over 210 furlongs.…It was, he says, an exact globe, whitish in color, three hand-breadths across. But at times it grew bigger, or smaller; and at others it took on a purple hue. He showed us, too, letters that were written into the stone, painted in the pigment called cinnabar."*

Cinnabar was employed widely in antiquity as a pigment for calligraphy and for decorating precious objects, such as in jewelry. The bright red pigment, whose name has been traced to the Persian zinjifrah ("dragon's blood"), was held in extraordinary esteem in ancient times.

Meteorites are very unlikely to contain enough cinnabar to mislead even the most superstitious priests. Meteoric stone is dark, not white, and any trace of mercury sulphide is unlikely to be visible to the naked eye.

In nature, certain stones, such as opal and limestone, can display narrow veins of cinnabar that could possibly be interpreted as esoteric writing, but this would not explain the anomalies in Eusebios' baitylos.

Source: Arthur Bernard Cook, *Zeus, a study in ancient religion* (Cambridge University Press, 1914), vol. 3, 888.

35.

7 May 351, Jerusalem
A luminous cross terrifies witnesses

Hermias Sozomen, in his *Ecclesiastical History*, notes that "At the time Cyril succeeded Maximus in the government of the church of Jerusalem, the sign of the cross appeared in the heavens; its radiance was not feeble and divergent like that of comets, but splendid and concentrated. Its length was about fifteen stadia from Calvary to the Mount of Olives, and its breadth was in proportion to its length.

"So extraordinary a phenomenon excited universal terror."

He also stated it was visible for several days and was brighter than the sun.

Source: *The Ecclesiastical History of Philostorgius, compiled by Photius, Patriarch of Constantinople*, trans. Edward Walford (London: Henry G. Bohn, 1855), 49. The Byzantine text *Chronicon Paschale* provides the date of May 7th.

36.

Ca. 393, Rome, Italy: A sign in the sky

A "new and strange star was seen in the sky, announcing the arrival of major disasters on Earth."

This oft-quoted sighting listed in the UFO literature seems to have been a comet. The original text describes the "star" being seen for the first time at midnight towards the east: "It was big and bright and the light was not much less than the morning star [Venus]. After that, a cloud of stars gathered around it on the same side, like a swarm of bees, clustering together around their queen."

Later it took the form of "a double-blade sword, great and terrible." Its movement was very different from the rest of the stars: it began to rise and came next to the Morning Star.

Later it moved to the North. Finally, after completing this trip in forty days, it came inside the Big Dipper and was last seen at the center of it, where it became extinct.

We only include this case in the Chronology because we have not found confirmation of a cometary observation about this date in Gary Kronk's extensive *Cometography*, but we suspect the object was indeed a comet.

Source: *The Ecclesiastical History of Philostorgius, compiled by Photius, Patriarch of Constantinople*, trans. Edward Walford (London: Henry G. Bohn, 1855).

37.

396, Constantinople (Istanbul), Turkey
Sulphurous fire from Heaven

St. Augustine wrote that "At the beginning of the night as the world was being darkened, a *fiery cloud was seen from the East, small at first then, as it approached the city, gradually enlarging, until it hung terribly over the whole city. All fled to the Church; the place did not hold the people.* But after that great tribulation, when God had accredited His word, the cloud began to diminish and at last disappeared.

"The people, freed from fear for a while, again heard that they must migrate, because the whole city would be destroyed on the next Sabbath. The whole people left the city with the Emperor; no one remained in his house."

The city was saved. "What shall we say?" adds Augustine. "Was this the anger of God, or rather His mercy?"

Source: Albert Barnes, *Minor Prophets I* (Michigan: Baker Books, 1985), 414. Augustine doesn't give a date, but 16th century ecclesiastical historian Cesare Baronius said it was 396. It isn't known how he reached this conclusion.

38.

438, Constantinople (Istanbul), Turkey
A child abducted to Heaven

An earthquake has destroyed Constantinople; famine and pestilence are spreading. The cataclysm has leveled the walls and the fifty-seven towers. Now comes a new tremor, even stronger than all the previous ones. Nicephorus, the historian, reports that in their fright the inhabitants of Byzantium, abandoning their city, gathered in the countryside: "They kept praying to beg that the city be spared total destruction: they were in no lesser danger themselves, because of the movements of the earth that nearly engulfed them, when a miracle quite unexpected and going beyond all credence filled them with admiration."

"*In the midst of the entire crowd, a child was suddenly taken up by a strong force, so high into the air that they lost sight of him. After this, he came down as he had gone up, and told Patriarch Proclus, the Emperor himself, and the assembled multitude that he had just attended a great concert of the Angels hailing the Lord in their sacred canticles.*

"Acacius, the bishop of Constantinople, states, 'The population of the whole city saw it with their eyes.' And Baronius, commenting upon this report, adds the following words: 'Such a great event deserved to be transmitted to the most remote posterity and to be forever recorded in human memory through its mention every year in the ecclesiastical annals. For this reason the Greeks, after inscribing it with the greatest respect into their ancient Menologe, read it publicly every year in their churches.'"

Source: This story has been collected and published by writers for many centuries. The version quoted here is by 14th century chronicler Nicephorus Callistus, but versions can be found in a letter by Acacius, Patriarch of Constantinople (d.489) to Peter Fullo, Patriarch of Antioch, and also in a letter by Pope Felix III (483-492) to the same Peter Fullo. The story in itself serves as the founding story for the origin of the Trisagion hymn of the Greek Church. The different versions agree on most details except the precise year and the fate of the raised child.

39.

497, British Isles: Globe in the sky and two light beams

An immense globe appeared in the sky. A second ball of fire came from its rays, projecting two beams: "During these transactions at Winchester, there appeared a star of wonderful magnitude and brightness, darting forth a ray, at the end of which was a globe of fire in the form of a dragon, out of whose mouth issued forth two rays; one of which seemed to stretch out itself beyond the extent of Gaul, the other towards the Irish sea, and ended in seven lesser rays."

There is some doubt about the date here because Geoffrey of Monmouth wrote it coincident with Ambrosius' death. Scholars disagree about the date of this event, suggesting either 473 or, according to Roger of Wendover, 497.

Source: Geoffrey of Monmouth, *Historia Regum Britanniae*, VIII, ch. 14; Lycosthenes, *Julii Obsequentis Prodigiorum Liber...per Conradum Lycosthenem Rubeaquensem integrati suae restitutus* (Basel, 1552).

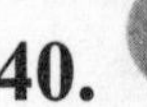

40.

507, Poitiers, France
King Clovis guided by a light in the sky

A peculiar phenomenon took place when French king Clovis defeated the Visigoths, killing their king Alaric II, and taking over their French lands, including Aquitaine.

"The decisive moments were wasted in idle deliberation. The Goths too hastily abandoned, perhaps, an advantageous post; and the opportunity of a secure retreat was lost by their slow and disorderly motions.

Fig. 5: An interpretation of the sighting by Clovis

"After Clovis had passed the ford, as it is still named, of the Hart, he advanced with bold and hasty steps to prevent the escape of the enemy. His nocturnal march was directed by a flaming meteor, suspended in the air above the cathedral of Poitiers; and this signal, which might be previously concerted with the orthodox successor of St. Hilary, was compared to the column of fire that guided the Israelites in the desert."

Shortly thereafter Clovis moved the Frankish capital to Paris.

Source: Gibbon, Edward. *The History of the Decline and Fall of the Roman Empire*. London: T. Cadell (1837), 594.

41.

Circa 540, Rome, Italy: A spark grows into a disk

"Often a little spark has seemed to come down from the sky to the Earth; then, *having grown into a kind of orb like the*

Moon, it has been seen as disc-like. This very thing recently happened and foretold a danger of seditions and misfortunes beyond measure."

Source: John Lydus, *On Portents*, 6. Quoted in "The Roman Fireball of 76 BC" by Richard B. Stothers, NASA Goddard Space Flight Center, New York. *The Observatory* 107 (1987): 211.

42.

540, Monte Cassino, Italy: A Fiery Globe

At dawn, Saint Benedict of Nursia observed a glittering light that became a fiery globe. He had time to call a second witness:

"The man of God, Benedict, being diligent in watching, rose early up before the time of matins (his monks being yet at rest) and came to the window of his chamber, where he offered up his prayers to almighty God. Standing there, all on a sudden in the dead of the night, as he looked forth, he saw a light, which banished away the darkness of the night, and glittered with such brightness, that the light which did shine in the midst of darkness was far more clear than the light of the day.

"Upon this sight a marvelous strange thing followed, for, as himself did afterward report, the whole world gathered as it were together under one beam of the sun was presented before his eyes, and while the venerable father stood attentively beholding the brightness of that glittering light, he saw the soul of Germanus, Bishop of Capua, in a fiery globe to be carried up by Angels into heaven.

"Then desirous to have some witness of this so notable a miracle, he called with a very loud voice Servandus the Deacon twice or thrice by his name, who, troubled at such an unusual crying out of the man of God, went up in all haste, and looking forth saw not anything else, but a little remnant of the light, but wondering at so great a miracle, the man of God told him all in order what he had seen, and sending by and by to the town of Cassino, he commanded the religious man Theoprobus to dispatch one that night to the city of Capua, to learn what was become of Germanus their Bishop: which being done, the messenger found that reverent Prelate had departed this life, and enquiring curiously the time, he understood that he died at that very instant, in which the man of God beheld him ascending up to heaven."

Researcher Yannis Deliyannis, who located this case, adds: "The account of the vision of Saint Benedict of Nursia is interesting enough in its description. While needing to be cautious, we cannot rule out completely the idea that it was eventually based on a (more or less) factual event which was, later on or by extrapolation, given a mystical signification."

Source: Dialogues of Gregory the Great, book II, chap. xxxv. *Sancti Gregorii Papae Dialogorum Libri IV*, as published in Migne's *Patrologia Latina,* Vol. 77.

43.

November 596, Nara prefecture, West Japan ⓘ
Descending canopy

An object like a canopy or lotus flower descends and appears suspended above the Houryuji temple. It changes color and shape.

Source: *Fusouriyatuki* Vol. 3 (Buddhist history), quoted in: Takao Ikeda, *Nihon nu ufo* (Tokyo: Tairiki shobo, 1974). We have not seen the book and give references from it with reservations.

44.

9 June 597, Ireland: An immense pillar of fire

"Another vision also given at the same hour under a different form was related to me, Adomnan, who was a young man at the time, by one of those who had seen it, and who solemnly assured me of its truth...He said:

"'On that night when St. Columba, by a happy and blessed death, passed from earth to heaven, while I and others with me were engaged in fishing in the valley of the river Fend, which abounds in fish, *we saw the whole vault of heaven become suddenly illuminated. Struck by the suddenness of the miracle, we raised our eyes and looked towards the east, when, lo! there appeared something like an immense pillar of fire, which seemed to us, as it ascended upwards at that midnight, to illuminate the whole earth like the summer sun at noon*; and after that column penetrated the heavens darkness followed, as if the sun had just set.

"'And not only did we, together in the same place, observe with intense surprise the brightness of this remarkable luminous pillar, but many other fishermen also, who were engaged in fishing here and there in different deep pools along the same river, were greatly terrified, as they afterwards related to us, by an appearance of the same kind.'"

Source: William Reeves, ed. *Life of Saint Columba, founder of Hy. Written by Adamnan, Ninth Abbott of that Monastery* (Edinburgh: Edmonston & Douglas, 1874).

45.

May 655, Katuragi Mountain near Nara, W. Japan ⓘ
Dragon rider

A man who rides a Dragon is seen in the sky. The figure is said to look "like a Chinese man." We have no other data about this curious description, so our first inclination is to reject it. We mention it here in the hope to stimulate more research into ancient sources in Asia.

Source: Takao Ikeda, *Nihon nu ufo* (Tokyo: Tairiki shobo, 1974). The author quotes *Fusouriyatuki,* vol. 3 (Buddhist history).

46. 

664, Kent, England
Pillar of light, splendid globe

"In the dead of night there appeared from God a glittering pillar of light shining over the hall of the king's [Ecgbert I, king of Kent] palace, which by its unwonted illumination aroused many of the king's household; and they in their great astonishment uttering loud cries, the king was awakened, and, ignorant of what had occurred, arose from his bed, and set out to go to the hymns of matins while it was yet night. On leaving the house, he saw a globe of extraordinary splendor burning with a white flame, the origin of which proceeded from the aforesaid wonderful seat of light. [...]"

Source: Symeon of Durham, *Opera et Collectanea,* Vol. 1 (Durham: Andrews and Co., 1868).

47. 

6 September 670, Nara Prefecture, Western Japan ⓘ
Flying umbrella

A cloud like an umbrella appeared, accompanied with a strange sound, over the Nara prefecture.

This does not provide enough information to understand the full circumstances of the phenomenon. Although meteors have been known to emit sounds, they are not described in terms of "clouds with the shape of an umbrella." Therefore the phenomenon has to remain unidentified, at least until a detailed reference is provided in the course of future research.

Source: Takao Ikeda, *Nihon nu ufo* (Tokyo: Tairiku shobo, 1974). The author quotes from the Teiohennenki.

48.

675, Berecingum Convent, near London, England
Circling light

A large light came down over praying nuns At Berecingum (Barking) convent, circled their location, and flew up. The description suggests that the light came from a well-defined object:

"For one night, after matins had been sung, and those handmaids of Christ had gone out of their chapel to the tombs of the brothers who had departed this life before them, and were singing the customary songs of praise to the Lord, *on a sudden a light from heaven, like a great sheet, came down upon them all,* and struck them with such amazement, that, in consternation, they even left off singing their hymn.

"But that resplendent light, in comparison wherewith the sun at noon-day might seem dark, soon after, rising from that place, removed to the south side of the monastery, that is, to the westward of the chapel, and having continued there some time, and rested upon those parts, in the sight of them all withdrew itself again to heaven, leaving no doubt in the minds of all, but that the same light, which was to lead or to receive the souls of those handmaids of Christ into Heaven, also showed the place in which their bodies were to rest and await the day of the resurrection."

We note that, although the "great sheet" of light could have been caused by a meteor, the later behavior of the phenomenon (rising and circling) seems to exclude this explanation.

Source: J. A. Giles, D.C.L., ed. *The Venerable Bede's Ecclesiastical History of England,* Book IV, ch. VII (London: Henry G. Bohn, 1867).

49.

21 November 684, Japan, location unknown ⓘ
Seven drifting stars

At dusk, seven stars are said to have "drifted together" to the north-east, after which they sank below the horizon. The information is too sketchy to reach any conclusion regarding the nature of the unusual "drifting stars."

Source: W. Raymond Drake, *Gods and Spacemen in the Ancient East* (New York: Signet, 1968), 106. The original source has not emerged.

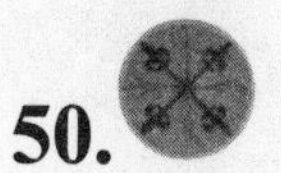

50.

Circa May 698, Ireland, location unknown
Three flying shields

A passage extracted from a 17th century transcription of an older but undated manuscript offers another example of the use of the term 'shield' in connection with a phenomenon in the sky. As noted by researcher Yannis Deliyannis, "it is interesting and unusual in medieval records. It is reminiscent of course of the 'clipei' of the authors of the Roman period."

The text reads: "Three shields were seen in the heavens, as it were warring from the east to the west, after the manner of undulating waves on a very calm night, being that of the Ascension of the Lord. The first was snowy, the second fiery, the third bloody; which prefigured, as is thought, three succeeding evils: for in the same year the herds of cows throughout Ireland were nearly destroyed, and not only in Ireland, but also throughout the whole of Europe."

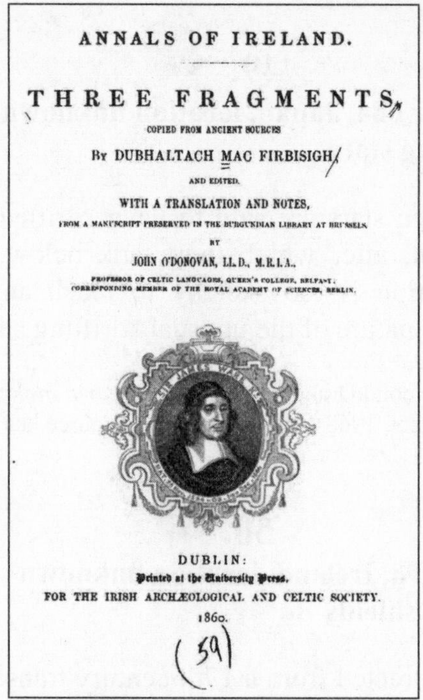

ANNALS OF IRELAND.

THREE FRAGMENTS,

COPIED FROM ANCIENT SOURCES

BY DUBHALTACH MAC FIRBISIGH;

AND EDITED,

WITH A TRANSLATION AND NOTES,

FROM A MANUSCRIPT PRESERVED IN THE BURGUNDIAN LIBRARY AT BRUSSELS,

BY

JOHN O'DONOVAN, LL.D., M.R.I.A.,

PROFESSOR OF CELTIC LANGUAGES, QUEEN'S COLLEGE, BELFAST;
CORRESPONDING MEMBER OF THE ROYAL ACADEMY OF SCIENCES, BERLIN.

DUBLIN:
Printed at the University Press,
FOR THE IRISH ARCHÆOLOGICAL AND CELTIC SOCIETY.
1860.

(59)

Fig. 6: Annals of Ireland

Source: *Annals of Ireland, three fragments copied from ancient sources by Dubhaltach Mac Firbisigh,* trans. John O'Donovan (Dublin: Irish Archaeological and Celtic Society, 1860). The date would have been 40 days following that year's celebration of Easter.

51. 

June 741, Constantinople (Istanbul), Turkey
Hovering crescents and fire

In the reign of Constantine, Copronymus, son of Leo, Emperor of Byzantium, *three columns of fire and flame appeared in the sky during the month of June*. The same phenomenon was also seen in the month of September: "There appeared a thing, also in 735 AD, like a half-moon, in the northern quadrant of the sky, and little by little, over a rather long time, it passed to the southern quarter, and then returned to the north, and finally descended under the Earth." (i.e., presumably dropped down below the horizon).

Source: Jean-Baptiste Chabot, *Chronique de Michel le Syrien, patriarche jacobite d'Antioche 1166-1199* (Paris, 1899-1910).

52.

749, Ulster, Ireland
Aerial ships, seen along with their crews

The 15th century Annals of Ulster, which cover the period AD 431 to AD 1540, state that "Ships, with their crews, were seen in the air above Cluain Moccu."

Source: *The Annals of Ulster* (Corpus of Electronic Texts Edition www.ucc.ie/celt/published/T100001A/index.html)

53.

Circa 760, France: Abductions and aerial ships ⓘ

During the reign of Pépin le Bref (715-768) many extraordinary phenomena are said to have appeared in the French skies. The air was filled with human figures, ships with sails and battling armies. Several individuals stated they had been abducted by aerial beings.

A contemporary source has never been found and there is a strong suspicion that it originated with the *Comte De Gabalis* (1670), by Abbé N. de Montfaucon de Villars.

Source: Jules Garinet, *Histoire de la Magie en France* (Paris, 1818).

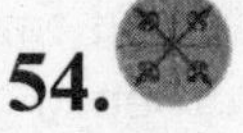

54.

776, Syburg Castle, Germany
Two flying objects stop a war

In 776 the Saxons rebelled against Charlemagne and attacked the castle of Syburg with continued lack of success, finally deciding to storm the castle. They reportedly "saw the likeness of *two shields red with flame wheeling over the church. When the heathens outside saw this miracle, they were at once thrown into confusion and started fleeing to their camp in terror.* Since all of them were panic-stricken, one man stampeded the next and was killed in return, because those who looked back out of fear impaled themselves on the lances carried on the shoulders of those who fled before them. Some dealt each other aimless blows and thus suffered divine retribution."

Source: *Carolingian Chronicles: Royal Frankish Annals and Nithard's Histories*, trans. Bernhard Walter Scholz (Ann Arbor: University of Michigan Press, 1970), 53, 55.

55.

811, Near Aachen on Via Aquisgrana, Germany
Great flaming globe

Emperor Charlemagne sees a great flaming globe descending from east to west and is thrown from his horse. Although the horse may have been frightened by an especially bright meteor, the situation suggests either that the object was close to the emperor's party, or that the meteor was very spectacular indeed: "One day in his last campaign into Saxony against Godfred, King of the Danes, Charles himself saw a ball of fire fall suddenly from the heavens with a great light, just as he was leaving camp before sunrise to set out on the march. It rushed across the clear sky from right to left, and everybody was wondering what was the meaning of the sign, when the horse which he was riding gave a sudden plunge, head foremost, and fell, and threw him to the ground so heavily that his cloak buckle was broken and his sword belt shattered; and after his servants had hastened to him and relieved him of his arms, he could not rise without their assistance. He happened to have a javelin in his hand when he was thrown, and this was struck from his grasp with such force that it was found lying at a distance of twenty feet or more from the spot."

Source: *Einhard: The Life of Charlemagne,* trans. Samuel Epes Turner (New York: Harper & Brothers, 1880).

56.

813, Santiago de Compostela, Spain ⓘ
Mysterious star

One night, a hermit named Pelayo heard music in a wood and saw a peculiar shining star above Mount Libredon, a former Celtic sacred site. Because of this sighting the place was called, in Latin, "Campus Stellae," *field of the star*, a name that was later turned into *Compostela*.

A modern brochure adds: "Bishop *Teodomiro*, who received notice of that event, instituted an investigation, and so the tomb of the Apostle was discovered. King Alphonse II declared Saint James the patron of his empire and had built a chapel at that place (…) More and more pilgrims followed the way of Santiago, the *'Path of Saint James,'* and the original chapel soon became the cathedral of the new settlement, Santiago de Compostela."

Source: *Santiago, History and Legends* (http://www.red2000.com/spain/santiago/history.html).
To the best of our knowledge, the story first appeared in the *Concordia de Antealtares*, a text dated from 1077.

57.

814, China, exact location unknown
Stars emerge from an object

A luminous object rises, lights up the ground. Many small "stars" emerge from it.

Source: Biot, *Catalogue des étoiles filantes en Chine* (1846), op. cit.

58.

Circa 815, Lyons, France
Saint Agobard and the abductees from Magonia

Saint Agobard was born about 769 in Languedoc, came to Lyons at age 20, was ordained in 804 and succeeded Archbishop Leidrade when the latter retired in 814.

Archbishop Agobard was an enlightened, intelligent man who took an active role in the political debates of his time: he became involved on Lothaire's side in his fight against his father and even wrote a book supporting him. This cost him his position when Louis the Pious came to power, but he was reinstated two years later, in 837. He died in 840.

A serious philosopher and early-day "rationalist," Agobard left no less than 22 books, including several treatises against superstitions and heretical beliefs, along with political pamphlets and volumes of poetry. The anonymous French translator of his work (actually Antoine Péricaud, Sr.) entitled *De Grandine et Tonitruis* or "About Hail and Thunder" notes in his introduction:

"All of his writings, whose style is consistently correct and often elegant, deserve the honor of being translated, for they make known to us the mores and customs of the first half of the ninth century, better than those of any other writer of the time. In particular one must acknowledge that he fought the prejudices and superstitions of his time more strongly and with a higher sense of reason than anyone else. It is against one of these prejudices that he compiled "About Hail and Thunder".

The book was first partially translated from the Latin as a piece published in *L'Annuaire de Lyon* for 1837. The translation was then revised and reprinted as an essay, with very limited distribution, in 1841 (Lyon: Imprimerie de Dumoulin, Ronet et Sibuet, Quai St. Antoine). It is this volume we have studied in the Lyons municipal library.

The main purpose of *De Grandine et Tonitruis* is to debunk popular misconceptions about the weather. In particular, the Archbishop of Lyons fought against the idea that winds and storms were due to the influence of sorcerers (appropriately named "tempestari" by the vulgar people): his main argument is that "Whoever takes away from God His admirable and terrible works, and attributes them to Man, is a false witness against God Himself." It is in this context that he raises his voice against those who are insane enough to believe that there could be ships ("naves") flying through the clouds: "*Plerosque autem vidimus et audivimus tanta dementia obrutos, tanta stultitia alienates, ut credant et dicant: quandam esse regionem, quae dicatur MAGONIA, ex qua naves veniant in nubibus...*" which our translator renders as follows:

"We have seen and heard many people crazy enough and insane enough to believe and to state that there exists a certain region called MAGONIA, out of which ships come out and sail upon the clouds; these ships (are said to) transport to that same region the products of the earth that have fallen because of the hail and have been destroyed by the storm, after the value of the wheat and other products of the earth has been paid to the 'Tempestaires' by the aerial navigators who have received them."

Saint Agobard continues: "We have even seen several of these crazy individuals who, believing in the reality of such absurd things, *exhibited before an assembled crowd four people in chains, three men and one woman, said to have [fallen] down from one of these ships. They had been holding them bound for a few days when they brought them before me, followed by the multitude, in order to lapidate them.* After a long argument, truth having prevailed at last, those who had shown them to the people found themselves, as a prophet says, in the same state of confusion as a robber who has been caught." (Jerem. 2:26)

What distinguishes this episode from many folklore tales of ships sailing in the sky is the availability of a precise reference, the authority of a known and respected historical figure who has written extensively on many other subjects, and the fact that the Archbishop, while he testified to the authenticity of a first-hand report, remained a skeptic about the reality of the objects themselves.

Since we do not have access to the statements made on the other side of the argument, we will never know what the "cloudships" looked like, or why the witnesses thought that the three men and one woman had in fact come from these ships and should be stoned to death. Naturally the mere fact of alighting from a "cloudship" may have been proof of sorcery. In one of his books French physicist Arago states that until the time of Charlemagne it was a common custom to erect long poles in the fields to protect them from the hail and the thunderstorms. These poles were not lightning rods, as one might suppose, but magical devices which were only effective when they held aloft certain parchments. In his *Capitularies*, published in 789, Emperor Charlemagne forbade this "superstitious" practice. His statement teaches us that interaction between us and the ships that sail through the clouds is not a new phenomenon. It also indicates that the vision of these "ships" was linked, in the popular mind, to atmospheric disturbances and to the stealing of fruits, plants and animals by beings from the sky.

The reference to animals comes from a passage in a book by J. J. Ampère (in *Histoire Littéraire de la France*): "it was believed that certain men, called 'Tempestaires,' raised storms in order to sell the fruits hit by hail and the animals who had died as a result of storms and floods to mysterious buyers who came by way of the air."

Most importantly, Agobard's book shows that as early as the ninth century there was a belief in a separate region from whence these vessels sailed, and about the possibility for men and women to travel with them. We must be thankful to him for saving the lives of these four poor people, an episode that shows that the skeptics, in this field, can do some good after all.

59.

817, China, exact location unknown: Slow-flying globe

A globe appeared at the zenith, followed by a tail. It flew slowly West under the moon, while witnesses heard something like the sound of birds. We include the case because of the slow motion noted in the report. It is rare, yet not impossible, to hear a sound in connection with a meteor, but they do not fly "slowly."

Source: Abel Remusat, "Bolides en Chine," *Journal de Physique* (1819): 358.

60.

827, Barcelona, Spain: Terrible lights ⓘ

Eginard writes that "terrible things in the sky" were observed during the night while Pepin I was at war in Spain. The objects emitted lights, pale or red in color. Here again, the interpretation is difficult: meteors are not "pale and red."

Source: Michel Bougard, *Inforespace* 22 (August 1975):36, citing the *Vita Hludowici Pii* by Astronomus (835 AD).

61.

840, China, exact location unknown
Two "sacred lamps" astound the crowd

"Early that night, we saw a sacred lamp on top of the ridge, on the other side of a valley East of the terrace. Our whole group saw it and admired it. The light was about the size of a begging bowl at first, but it expanded to the size of a small house. Deeply moved, the crowd sang with full voice the name of His Holiness. Then another lamp appeared, near the valley. That one, too, only was the size of a rain straw hat at first, and then it grew gradually. The two lights, when seen from afar, seemed about 100 steps apart. They were shining ardently. At midnight they died, becoming invisible."

Source: Ennin, *Journal d'un voyageur en Chine au IXième siècle*, trans. Roger Lévy (Paris: Albin Michel, 1961), 206.

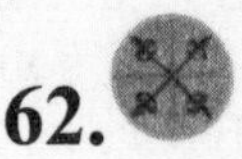

62.

November 879, China, exact location unknown
Two suns fighting

Two "suns" fought energetically in the sky. On another day of the same month, two Suns fought, and then merged together. Note that a similar phenomenon is described in Japan in case 69 below.

Source: Shi Bo, *La Chine et les Extraterrestres*, op.cit., 47.

63.

Circa 25 April 880, Montserrat–Santa Cova, Spain
Magical light

Towards the end of April in the year 880 seven young children from Monistrol in Barcelona saw a strange light descend from the sky and head towards a small grotto on the mountain of Montserrat, accompanied by a soft melody. A week later a group of priests headed by the Bishop of Manresa returned to the spot, and saw it again.

On four Saturdays in a row *the light reappeared in the sky and dropped towards the mountain grotto*. In the end seven men were sent to the place the light seemed to indicate, which was in an area called Santa Cova. When they entered the cave they discovered an image of a black virgin, surrounded by a magical light and giving off a pleasant aroma. The locals tried to carry the sculpture to Manresa but, according to their story, the further they moved it, the heavier it became. It grew so heavy that they had to leave it in the middle of the fields, where they decided to erect a hermitage in the name of St. Mary. The hermitage is still there today.

Source: Josep Guijarro, *Guía de la Cataluña Mágica* (Barcelona: Ediciones Martínez Roca, 1999), 42-43.

64.

3 September 881, Japan, exact location unknown ⓘ
Stellar maneuvers

Two stars appeared in the sky, and went through strange movements as if merging and separating. This is the same pattern as in case 67, reported in China.

Source: *Nihon-Tenmonshiriyou*, by Morihiro Saito. We have looked in vain for an exact reference. Perhaps one of our Japanese readers can research this case further?

65.

March 900, China, exact location unknown ⓘ
Two huge flying objects with complex shapes

The *New Book of the Tang* records that during the year of Guang Hua, "a fat star, as large as 500 meters square, yellow in color, flew towards the southwest. It had a pointed head and the rear was cylindrical…"

The same book records another "star-like object" that was five times bigger than the above one and flew in a north-westerly direction. When it descended to a point some thirty meters from the ground the witnesses could see its upper part emit red-orange flames. "It moved like a snake, accompanied by numerous small stars that disappeared suddenly."

Source: Shi Bo, *La Chine et les Extraterrestres*, op.cit., 31.

66.

905, China, location unknown
A globe and stars hover in the sky

A large fiery globe appeared at the zenith and flew towards the northwest. It stopped 100 feet away as many tiny stars moved above it. It left a greenish vapor.

Source: Abel Remusat, "Bolides en Chine" *Journal de Physique* (1819), 358.

67.

919, Hungary: Bright spheres in the sky ⓘ

People saw bright spherical objects shining like stars, along with a bright torch, moving to and fro in the sky.

Source: Antonio Ribera, *El gran enigma de los platillos volantes* (Barcelona: Editorial Plaza & Janés, 1974).

68.

12 May 922, Bulgaria: The Jinni in the red clouds

In July 921 the ambassador Susan ar-Rassi headed a mission that left Baghdad for Vulga Bulgaria with the aim of seeking support from King Almush to form a military alliance against Khazar Kaganate. On 12 May 922, the first night they spent in Almush's country, a strange phenomenon was seen in the sky. The secretary of the mission, Ahmad ibn-Fadlan described the sighting thus:

"I saw that before the final disappearance of sunlight, at the usual hour of prayer, the sky horizon reddened considerably. *And I heard in the air loud sounds and a strong hubbub. Then I lifted up my head and lo! A cloud [was seen] not far from me, red like fire, and this hubbub and these sounds came from it. And lo! [there were] seen in it something like men and horses, and in the hands of some figures inside it, similar to men, [there were] bows, arrows, spears, and naked swords.* And they seemed to me sometimes absolutely clear, sometimes just apparent. And lo! [there appeared] near them another similar armed group, a black one, in which I also saw men, horses and weaponry. And this detachment began to attack the other one, as a cavalry troop attacks another cavalry troop."

The phenomenon lasted for some time. Ibn-Fadlan writes that the men asked the king what it may have meant. The king replied that his forefathers, who were accustomed to seeing such things, believed the riders were "Jinni" (Jinns). Ambassador ar-Rassi also gave his version of the event. His account coincided with that of the secretary, except in that he noticed two red clouds rather than one. He also noted that the sighting began an hour before sunset and ended at around 1:00 A.M.

Source: A. Kovalevskiy, *The book by Ahmad ibn-Fadlan about his voyage to the Volga in the years 921-922. Papers, Translations, and Comments* (Kharkov, 1956).

69.

March 927, Reims, France: Armies of fire

"An army of fire was seen in the sky in Reims on a Sunday morning in the month of March. After this sign a pestilence followed." This citation from Flodoard is an example of a frequent description for which we have no precise correlation in terms of optical or atmospheric phenomena.

Source: *Flodoardi Annales*, in *Monumenta Germaniae Historiae*, trans. G. H. Pertz, Tome III (Hanover, 1839).

70.

944, Trans-Rhenan Germany: Iron Globes chased away

"In some districts, burning iron globes were seen in the air, some of which, while flying, burnt some farms and houses. But in some places they were repelled by opposing them with crucifixes, episcopal blessing and holy water."

Source: *Flodoardi Annales*, in *Monumenta Germaniae Historiae*, trans. G. H. Pertz, Tome III (Hanover, 1839).

71.

9 September 967, Japan, exact location unknown ⓘ
Triangular formation

Numerous objects in *triangular formation flying under the rain clouds*, trajectory east-west. This description, if it is reliable, excludes the meteoric interpretation.

Source: *Brothers* III, 1, 1964. No original source given.

72.

989, Constantinople (Istanbul), Turkey
Erratic "comet"

"The star appeared in the west after sunset; it rose in the evening and *had no fixed place in the sky. It spread bright rays, visible from a great distance, and kept moving, appearing further north or further south, and once when it rose changed its place in the sky, making sudden and fast movements*. The people who saw the comet (sic) were stunned, in awe, and believe that such strange movements are an evil omen. And just as people expected, something happened: in the evening of the day when they usually celebrated the memory of Velikomuchenik (a martyr of early Christianity), a tremendous earthquake brought down the towers of Byzantium..."

It seems to us today that an object that "changes its place in the sky, making sudden and fast movements," cannot be a comet if the description is accurate. However Gary Kronk's *Cometography* indicates that Halley's Comet was visible in the night sky during July and August of 989, based on Chinese accounts, so some confusion is possible.

Source: *Istoria* ("History"), a 10th century manuscript by Byzantian writer Lev Diakon. Quoted from a modern Russian edition: Lev Diakon, *Istoria*, trans. M. Kopylenko (Moscow: G. Litavrin, 1988), 91.

73.

3 August 989, Japan, exact location unknown ⓘ
Three bright objects meet in the sky

"The three objects became bright, in extraordinary fashion, and met at the same point of their trajectory."

Source: Christian Piens, *Les Ovni du passé* (Paris: Marabout, 1977), 41.

74.

Circa 998, Budapest, Hungary
King Stephen's aerial trips

King Stephen (Istvan), who lived from 975 to 1038, and was crowned King of Hungary in 997, was said to be lifted to the sky with some frequency. His biographer, Chartruiz, Bishop of Hungary, revealed that this sometimes happened spiritually and at other times physically.

On one occasion, as detailed in Chartruis' *Life of St. Stephen, King of Hungary*: "While praying in his tent, he was lifted into the air by the hands of angels."

Source: Ebenezer Cobham Brewer, *A Dictionary of Miracles* (London: Chatto & Windus, 1901), 217.

75.

Circa 999, Abbey of Saint-Léger, Côte d'Or, France ⓘ
Bedroom visitation

Rodulphus Glaber, a monk and chronicler, writes "Not so long ago such (visions) happened to me, by the favor of God. At the time I was staying in the monastery of the martyr Saint-Léger, also named *Abbaye de Champeaux*. I saw one night, before Matines, a hideous little monster of vaguely human form appear at the foot of my bed. It seemed to be, as much as I could discern, of medium size with a frail neck, a thin face, very black eyes, a wrinkled and narrow forehead, a goatee, straight and pointed ears, straight and dirty hair, dog teeth, a sharp occiput, its breast swollen, a bump on the back, hanging buttocks and dirty clothes, with its whole body appearing to shake.

"He grabbed the edge of the bed in which I was lying and shook it with terrible violence and said: 'You will not stay here any longer.' And at once I woke up terrified, and suddenly I saw the figure I just described. It was gnashing its teeth while repeating the same thing: 'You will not stay here any longer.' I got up from bed at once and ran to the monastery, where I kneeled in front of the altar of the very Holy Father Benedict, extremely terrified. And I began to recall the offenses and serious sins I had committed by being impudent or negligent."

Source: Rudolphi Glabri, *Historiarum Libri Quinque ab anno incarnationis DCCC usque ad annum MXLIV*, book V, chapter I, paragraph 2.

According to Ernest Petit ("Raoul Glaber," in *Revue Historique*, XLVIII, 1892) Glaber stayed in Saint-Léger between 997 and 1005 AD. This is the first of his visions. The five books of his *Historiae* contain other such experiences as well as many anecdotic accounts of superstitions around the year 1000.

Epilogue to Part I-A

What can we say about the above sightings? They range in credibility and significance from curious events where a natural explanation is improbable (but not entirely impossible, if some of the elements of the observation were reported mistakenly) to extraordinary stories that have evoked paranormal, or even mystical interpretations among the people of the time. All of them made enough of an impact for a record to have been kept by the witnesses and later chroniclers. The very fact that they have come down to us through so much troubled history is quite remarkable.

These reports do not constitute "evidence" for physical visitation by non-human creatures. All we can say is that they are consistent with modern descriptions of unidentified phenomena and the secondary effects surrounding them. In fact, we could have stopped our work at the year 1000 and we would have presented a fair cross-section of phenomena gathered by modern authors under the label of ufology, including abductions and hard traces.

These ancient records show how powerful the concept of such intervention into human affairs can be: most of our religious texts today can be traced to such events, and to the philosophical movements they triggered.

About the year 1000 many things changed on our planet. Large towns became real cities; in Europe, the feudal system stabilized society. Stone castles and monasteries would become genuine centers of learning while commerce expanded, bringing faraway lands in more frequent contact with Europe. Even the Crusades, the source of so much pillage and bloodshed, would soon play a role in creating an infrastructure for the exchange of knowledge, the rudiments of international banking and the management of complex projects. The nature of the reports will be even more intriguing and detailed in the following sections.

PART I-B

Chronology: 1000 to 1500 AD

The Second Millennium opened with intense religious fervor: the world was in terrible fear of cosmic upheaval, the Last Judgment and the end of the world, but a new spirit of exploration also appeared: Viking Leif Ericson (c. 980-1020), the son of Eric the Red, discovered America while Christianity reached Iceland and Greenland. The Chinese perfected their invention of gunpowder. Normans extended their influence to England after the battle of Hastings (1066). Conflict between different faiths intensified, leading in 1096 to the first of eight murderous Crusades that would force the blending of two great civilizations and help introduce new philosophical ideas, including Hermeticism, into European kingdoms barely emerging from the Dark Ages.

The rudiments of science arrived in Europe from the Middle East, with primitive astronomical instruments, early tables of star positions, and knowledge of Greek medicine and philosophy transmitted by Arabic scholars. The first cathedrals were built, the gothic style appeared at *Saint-Germain-des-Prés* in Paris, and the first account of the use of a mariner's compass was noted (in 1125). During the eleventh and twelfth centuries, a new invention started its slow spread into Western Europe from Spain, with the adoption of paper as a replacement for parchment. Far less expensive, paper greatly accelerated the spread of knowledge. The twelfth century would also see the founding of Cambridge University in England, the compilation of the Edda mythologies in Scandinavia, and the teachings of Albertus Magnus. Early in the thirteenth century Fibonacci introduced Arabic numerals into Europe, and the great University movements expanded in all countries, from Brussels and Salisbury to Salamanca, Siena, Toulouse and Vicenza, supported by great scholars like Roger Bacon (1214-1294). Libraries appeared everywhere, preserving ancient knowledge and contemporary chronicles.

Travelers became increasingly ambitious, encouraged by Marco Polo's voyage to China from 1271 to 1295. Knowledge about the world began circulating more widely, while the Crusades ended (in 1291) with the Knights of St. John of Jerusalem settling in Cyprus. Things took a disastrous turn in the mid-fourteenth century when the Black Death devastated Europe, killing a third of the population of England (1347). Early in the fifteenth century the Chinese compiled the first Encyclopedia (in 22,937 volumes!), a civil war began in France, Joan of Arc led the French armies against England, and Portuguese navigators found the first Negroes near Cap Blanc in western Africa, starting the slave trade again.

Everything suddenly accelerated in the last years of the fifteenth century: Leonardo da Vinci made his famed scientific discoveries, Copernicus studied at Cracow, the first terrestrial globe was constructed in Nuremberg, Johan Gutenberg used metal plates for printing and the king and queen of Spain, against the advice of their committee of experts, financed the voyage of an Italian navigator named Christopher Columbus. The world had changed.

76.

25 April 1001, Foggia, Italy
Strange flashes and a luminous lady

The count of Aviemore, tired after a day of hunting, decided to spend the night in a rustic hut. In the middle of the night he was awakened by servants and friends frightened by strange flashes, who urged the hunters to flee with them, fearing a forest fire. He decided not to run away with his companions but to cautiously study the strange phenomenon. Heading for the place where the flashes came from, the Count realized that there was no fire or burning trees, but a strange light. Among the flashes he saw a beautiful lady, whom he took to be the Virgin Mary.

A farmer named Nicholas, nicknamed Strazzacappa, who was going to work, saw the vision as well and reportedly heard a request from the apparition for a place of worship to be erected there. The case received publicity and a small chapel built at the spot became a center of pilgrimage. After a few years the Verginiani, led by William of Vercelli, settled there. When they merged with the Cistercians, the now famous monastery passed to the care of these monks. The church was elevated to the dignity of basilica by Pope Paul VI on 31 May 1978.

Source: Marino Gamba, *Apparizioni mariane nel corso di due millenni* (Udine: Ediz. Il Segno, 1999).

77.

Circa 1010, Ostium, Italy
Five-year old child abducted

Peter Damian, Cardinal-Bishop of the Italian city of Ostium (1007-1072), recorded what would be regarded today as a typical abduction involving the five-year-old son of a nobleman: "One night he was carried out of the monastery into a locked mill, where he was found in the morning. And when he was questioned, he said that he had been carried by strangers to a great feast and bidden to eat; and afterwards he was put into the mill through the roof."

What we see here is an early instance of a thread that will become increasingly important as the chronology develops, focusing on alleged interaction between human witnesses and creatures of another order. While a simplistic Christian interpretation classifies them as "demons," more sophisticated scholars recognized they did not fit easily within the biblical definitions of good and evil. In the Moslem world they would be recognized as the Djinni. In the later medieval world they will become the Fairies, the elves, the Elementals of the Alchemical tradition, the "Good Neighbors" of the Celtic world. The parallels are obvious between the beliefs in such beings and contemporary abduction stories made popular by television.

"Great feasts" are a staple of fairy folklore. Abductees were usually "bidden to eat" when the fairies whisked them off to their hidden palaces, just as people often claim to be given pills or liquids to swallow in today's accounts. Even being pulled, pushed or dragged through the roof has its parallel in modern UFO lore.

Source: *Malleus Maleficarum*, written in 1486 by Heinrich Kramer and James Sprenger, translated with an introduction, bibliography and notes by Montague Summers (London: Bracken Books, 1996), 105.

The works of S. Peter Damian, which have been more than once collected, may be found in Migne, *Patres Latini*, CXLIV-CXLV.

78. 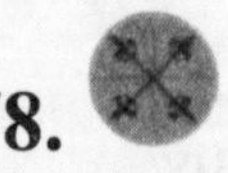

7 July 1015, Kyoto, Western Japan
Objects emerge from 'mother stars'

The Director General of Saemonfu [the Royal Guard] said that he had witnessed two stars meeting at night. "*The circumstances were as follows: Both stars flew slowly towards each other and the moment they were 10 meters or so from each other, there came little stars rushing out of each big star*, coming towards the other big star, and soon returned to their respective mother star, then the two mother stars flew away swiftly. After this meeting, clouds appeared and covered the sky. I hear that people in ancient times also witnessed such a phenomenon, but recently it was so rare that I was impressed not a little."

Source: Masaru Mori, "The Female Alien in a Hollow Vessel," *Fortean Times*, 48 (1987): 48; *Inforespace* 23:35.

79.

Autumn 1023, France: A ballet of stars

"There were seen in the southern part of the sky in the Sign of the Lion, two stars that fought each other all Autumn; the largest and most luminous of the two came from the east, the smallest one from the west, the small one rushed furiously and fearfully at the biggest one which didn't allow the speck to approach, but he struck her with his mane of light, repulsing her far towards the east."

Source: Adémar de Chabannes, *Chronicon*, book 3, ch. 62, in J. Chavanon, *Adémar de Chabannes, Chronique* (Paris: A. Picard, 1897).

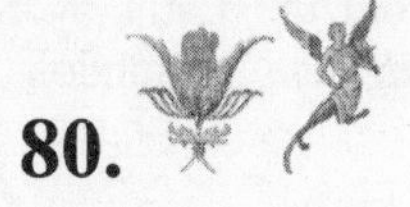

80.

1036, Taichang, China ⓘ
Bedroom visitation, abduction

A cloud carrying a female from the sky is said to have come down to the bedroom of Wang's daughter and flown away with her. Chinese writer Sheng Gua reports: "Under the reign of Jinyou (1034-1038) a scholar from Taichang named Wang Lun saw (goddess) Zigu flying down into his daughter's bedroom. This goddess knew how to write and was very pretty. A cloud floated under her feet, and she moved fast without effort. Zigu asked Wang Lun's daughter: 'Do you want to travel with me?' She agreed with a sign of her head. At once, clouds formed in the courtyard and the girl was lifted, but the clouds could not carry her. Zigu said at once: 'There is dust on your shoes, take them off before coming up.' The girl did as she was told and she rose in the clouds that lifted her to the sky."

Source: Shi Bo, *La Chine et les Extraterrestres*, op.cit., 27.

81.

1045, England: A "witch" gets abducted ⓘ

"When Henrie the third of that name was Emperour of Rome, in England a certain southsaying Witch was caried away by the Divel, whyche being drawen after him uppon his horsse with a horrible crye, he caryed away up into the ayre, the cry of whiche old woman was heard for certaine houres almost foure miles in that Countrey."

This constitutes only one of hundreds of similar stories about witches carried away by paranormal means or by non-human beings, usually thought to be demonic.

Source: Lycosthenes, *Prodigiorum ac Ostentorum Chronicon* (Basel, 1557). Translation from the Latin by Stephen Batman, *The Doome, warning all men to judgment...* (London, 1581).

82.

Ca. 1050, Vinland (Newfoundland): Woman in black ⓘ

The Greenlanders Saga includes a report about a woman named Gudrid who was sitting near the doorway beside the cradle of her son Snorri when "a shadow fell across the door and a woman entered dressed in a black close-fitting dress. She was rather short, wore a band round her head and had light-brown hair; she was pale and had such large eyes that their equal had never been seen in a human head."

The entity walked over to where Gudrid was sitting and said: "What is your name?"

"My name is Gudrid, but what is your name?"

"My name is Gudrid," she replied.

"Then Gudrid the housewife held out her hand, that she should sit by her. But it happened at the same moment, that Gudrid heard a great crack, and was then the woman lost to sight, and at the same time one Skraling was killed by a house carle of Karlsefne's, because he would have taken their weapons. And went they now away as usual, and their clothes lay there behind, and their wares; no man had seen this woman, but Gudrid alone."

This episode, an early instance of the meme of a "Woman-in-black," took place in the days of Thorfinn Karlsefni, the son of Thord Horsehead, the son of Snorri Thordason of Hofdi. Karlsefni was a companion of Leif Eirikson at Brattahlid. The two authors have disagreed about this case, since it could be considered a ghost story rather than a UFO case, but numerous modern claims of alien visitation fall in the same category and follow the identical model.

Source: Helge Ingstad, *Westward to Vinland* (London: Jonathan Cape, 1969).

83.

14 April 1054, Rome, Italy
A bright circle in the midday sky

In their paper "Do We Need to Redate the Birth of the Crab Nebula?" astronomers Guidoboni, Marmo, and Polcaro quote from the *Tractatus de Ecclesia S. Petri Aldeburgensi*, written by a monk or a clerk of the church of St. Peter in the town of Oudenburg, in present-day Belgium, regarding aerial phenomena observed at the time of the death of Pope Leo IX. They argue that the event described was a supernova, which is possible but unlikely.

"The most blessed Pope Leo, after the beginning of the construction of the aforementioned church of St. Peter, in the following year, on the 18th day before the first of May (i.e., 14th April 1054), a Monday, around midday, happily departed this world. And at the same time and hour as his leaving of the flesh, not only in Rome, where his body lies, but also all over the world appeared to men a circle in the sky of extraordinary brightness which lasted for about half an hour. Perhaps the Lord wished to say that he [the Pope] was worthy to receive a crown in Heaven between those who love Him."

The supernova that gave rise to the Crab Nebula was first seen by Chinese astronomers who noted a "guest star" in the constellation Taurus on July 4, 1054, fully three months after the Rome sighting. Simon Mitton lists 5 independent preserved Far-East records of this event (one of 75 authentic guest stars – novae and supernovae, excluding comets – systematically recorded by Chinese astronomers between 532 BC and 1064 AD). This star became about 4 times brighter than Venus in its brightest light, or about magnitude -6, and was visible in daylight for 23 days. It was probably also recorded by Anasazi Indian artists (in present-day Arizona and New Mexico), as findings in Navaho Canyon and White Mesa as well as in the Chaco Canyon National Park (New Mexico) indicate.

The astronomers note that the English translation of the Latin terms "circulus" and "corona" is not perfect, because they do not convey the original sense of "disc" that the Flemish writer expressed in his text. "The fact that *corona* was conceived as a bright disk (or shield) makes us understand that also *circulus* in this context must mean the same object. In conclusion, the Flemish chronicler saw a bright disk in the sky, and not a halo. Furthermore, we can observe that in this document, the author describes the phenomenon in neutral terms, unaffected by any set of beliefs: the disk-like shape, the intense brightness and the duration of the phenomenon are all elements common to very different cultures. The author separates the description of the phenomenon from his cautious symbolic interpretation, showing a clear awareness of the different levels of discourse." Note that the text of the *Tractatus* does not give the correct date for the Pope's death, which was 19 April 1054.

Source: E. Guibodoni, C. Marmo and V.F. Polcaro, "Do we Need to Redate the Birth of the Carb Nebula?" *Memorie della Societa Astronomica Italiana* 65 (1994): 624.

84.

Circa 1059, Fanliang, China
The bright pearl in the lake

Sheng Gua, a Chinese scholar of the Song Dynasty, recorded an interesting sighting in Chapter 369 of his *Stories on the Bank of a Stream of Dreams*:

"In the middle of the reign of emperor Jia You [1056-1063], at Yangzhou, in the Jiangsu province, an enormous pearl was seen especially in gloomy weather. At first it appeared in the marsh of the Tianchang district, passed by the lake of Bishe and disappeared finally in the Xinkai lake. The inhabitants of that region and travelers saw it frequently over a period of ten years. I have a friend who lives on the edge of the lake. One evening, he looked through the window and saw the luminous pearl near his house. He half-opened his door and the light entered, illuminating the room with its brightness. *The pearl was round, with a gold-colored ring around it. Suddenly, it enlarged considerably and became bigger than a table. In its centre, the luminary was white and silvery, and the intensity was such that it could not be looked at straight on.*"

The light it emitted even reached trees that were some 5 kilometers away and as a result these cast their shadow on the ground; the faraway sky was all alight. Finally, the round luminous object began to move at a breathtaking speed and landed on the water between the waves, like a rising sun.

As the pearl often made its appearance in the town of Fanliang in Yangzhou, the inhabitants, who had seen it frequently, built a wayside pavilion and named it "The Pearl Pavilion." Inquisitive people often came from afar by boat, waiting for a chance to see the unpredictable pearl.

Source: Shi Bo, *La Chine et les Extraterrestres*, op.cit., 26. The case is also mentioned by Paul Dong in *China's Major Mysteries: Paranormal Phenomena and the Unexplained in the People's Republic of China* (China Books, 2000), 69-71. Dong quotes from an article in Peking's *Guang Ming Daily* of February 18th 1979, "Could It Be That a Visitor from Outer Space Visited China Long Long Ago?" written by Professor Zhang Longqiao of the Chinese department of Peking Teachers College. The actual account comes from the book "Meng Qi Bi Tan" ("Essays of the Meng Hall") by Shen Kua of the Song Dynasty (960-1127).

85.

1067, Northumbria, England
Fiery sign revolves, moves up and down

"In this year, truly, several people saw a sign; in appearance it was fire: it flamed and burned fiercely in the air; it came near to the earth, and for a little time quite illuminated it; *afterwards it revolved and ascended up on high*, then descended into the bottom of the sea; *in several places it burned woods and plains*. No man knew with certainty what this divined, nor what this sign signified. In the country of the Northumbrians this fire showed itself; and in two seasons of one year were these demonstrations."

The original account, in Gaimar's *History of the English* (in *Chronicles and Memorials of Great Britain and Ireland during the Middle Ages*. London: Her Majesty's Stationery Office, 1889, Kraus reprint, 1966) runs thus:

> Many folks saw a sign
> In likeness of fire it was,
> In the air it greatly flamed and burned:
> Towards the earth it approached,
> For a little it quite lighted up.
> Then it revolved above,
> Then fell into the deep sea.
> In many places it burnt woods and plains.

Source: C. E. Britton, *A Meteorological Chronology to A.D. 1450* (London: H.M.S.O., 1937), 44. Britton comments: "Anglo-Saxon Chronicle gives the date of the return from Normandy as December 6 but does not mention the auroral appearances." Also mentioned by Geoffrey Gaimar in *L'Estoire des Engles solum la Translacion Maistre Geffri Gaimar*, a 12th century manuscript.

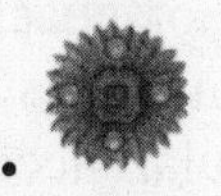

86.

December 1071, Zhengjiang, China ⓘ
Light rising from the river

Scholar Su Dongpo saw a big light emerge from the Yangtse River, scaring away the mountain birds.

Source: Shi Bo, *La Chine et les Extraterrestres*, op.cit., 26.

87.

July 1085, Estella, Navarra, Spain: A great star, and the Holy Virgin

Estella, in Navarra, is another place named after an unusual aerial sighting. The Virgin and a great star are said to have appeared to a group of shepherds on the mountain. The consequent worship of the area brought in pilgrims by the hundreds and King Sancho Ramirez built a sanctuary there. A sign on its capilla reads:

Ésta es la estrella	*This is the star*
Que bajó del Cielo	*That came down from the Sky*
a Estella	*to Estella*
Para reparo de ella.	*To observe it.*

Source: Javier Sierra and Jesús Callejo, *La España Extraña* (Madrid: Editorial EDAF, 1997), 131-2.

88.

1092, Drutsk and Polotsk, Ukraine
First reference to the Devil's Hunt

A common theme in ancient folklore refers to mysterious sounds in the sky reminding terrified people of the passage of dozens of men on horseback riding at full speed, with their dogs and servants, leaving enormous destruction behind. In this particular account the phenomenon first appeared in Drutsk, as a great sign "like a very large circle in the middle of the sky." That summer the weather was very dry, with numerous forest fires and many deaths. In Polotsk people heard great noises in the night, seemingly of devils galloping along the streets. Later they manifested during the day on horseback, but the only visible part was the hooves of their horses.

Another version of the text (Radziwill's) suggests that "the people of Polotsk are devoured by the dead," showing ambiguity between the deceased and demons. It is related in the *Povest' vremennykh let,* usually referred as the *Nestor's Chronicle* or *Chronicle of Bygone Years.* The following text was extracted by Yannis Deliyannis from the Laurentian codex (Лаврентьевский список) which includes the oldest version of the *Povest' vremmenykh let*. Apart from the manifestation of demons in the streets of

Polotsk, the reference to the appearance of a 'great circle' in the sky is of particular interest.

Fig. 7: An illumination from the Radziwill Chronicle.

The text reads: "Year 6600 (note: since the creation of the world in 5500 BC) This year there was a very peculiar prodigy in Polotsk. At night, a great noise was heard in the street: demons ran like men and if someone went out of his house, he was hurt right away by an invisible demon with a deadly wound. No one dared to leave his house. Then the demons manifested themselves on horses in plain day: they could not be seen themselves but only the hooves of their horses. They also hurt people in Polotsk and in the neighbourhood. So it was said: "There are ghosts killing citizens in Polotsk". These apparitions began in Droutchesk. Around this time a sign appeared in the heavens. A great circle was seen in the middle of the sky."

Source: Claude Lecouteux, *Chasses fantastiques et cohortes de la nuit au Moyen Age* (Paris: Imago, 1999), 31-32, quoting *La Chronique des Temps Passés*. See also, in Russian, Повесть временных лет (*Povest' vremennykh let*) as published in ПОЛНОЕ СОБРАНИЕ РУССКИХ ЛЕТОПИСЕЙ, *Том первый* ЛАВРЕНТЬЕВСКАЯ ЛЕТОПИСЬ, Ленинград: Издательство Академии Наук СССР, 1926-1928.

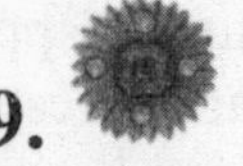

89.

July 1096, Japan, exact location unknown ⓘ
A necklace of ten lights in the sky

Ten flying objects combined to form a necklace in the sky in the northwest. In the absence of an original quote, it is impossible to analyze this event further.

Source: Takao Ikeda, *Nihon nu ufo* (Tokyo: Tairiku shobo, 1974).

90.

Circa 15 September 1098, Antioch, Turkey
Scintillating globe

In the *Historia Francorum qui Ceperint Jerusalem* of Raymond d'Aguiliers, Count of Toulouse, we read that during the First Crusade: "very many things were revealed to us through our brethren; and we beheld a marvelous sign in the sky. For during the night there stood over the city a very large star, which, after a short time, divided into three parts and fell in the camp of the Turks."

Alfred of Aachen writes: "In the silence of the night, when benevolent sleep restores men's strength, all Christians on guard duty were struck by a marvelous sight in the sky. It seemed that all the stars were concentrated in a dense group, in a space the size of about three *arpents*, fiery and bright as coals in a furnace, and gathered as a globe, scintillating. And after burning for a long time, they thinned out and formed the likeness of a crown, exactly above the city; and after remaining for a long time gathered in a circle without separating, they broke the chain at a point on that circle, and all followed the same path."

Source: August C. Krey, *The First Crusade: The Accounts of Eyewitnesses and Participants* (Princeton, 1921); Albert d'Aix, *Alberti Aquensis Historia Hierosolymitana* in *Recueil des Historiens des*

Croisades. Historiens Occidentaux. RHC. OCC0 Tome IV, 265-715. Translation by Yannis Deliyannis.

91.

Circa 1100, Germany
Prodigies herald the coming Crusade

"The signs in the sun and the wonders which appeared, both in the air and on the earth, aroused many who had previously been indifferent...A few years ago a priest of honorable reputation, by the name of Suigger, about the ninth hour of the day beheld two knights, who met one another in the air and fought long, until one, who carried a great cross with which he struck the other, finally overcame his enemy...Some who were watching horses in the fields reported that they had seen the image of a city in the air and had observed how various troops from different directions, both on horseback and on foot, were hastening thither.

"Many, moreover, displayed, either on their clothing, or upon their forehead, or elsewhere on their body, the sign of the cross, which had been divinely imprinted, and they believed themselves on this account to have been destined to the service of God."

Source: Ekkehard of Aurach. *On the Opening of the First Crusade* (1101).

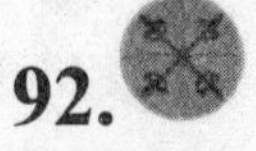

92.

11 February 1110, Pechorsky Monastery, Russia
A fiery pillar

There was an omen in the Pechorsky monastery: "On February 11th there appeared a fiery pillar that reached from the ground to the sky, and lightning lit all earth, and thunder rattled at the first hour of night, and everyone saw it. The pillar first stood over the stone trapeznitsa (monastery dining room), blocking the sight of the cross, and, after a short while, moved to the church and stood over Feodosiev's (Theodosius) tomb; it then went to the top of the church, turning its face to the east, and afterwards made itself invisible." The record reads: "It wasn't an usual fiery column, but the apparition of an angel, because angels often appeared as a fiery column or a flame."

Source: Nestor, *Russian Primary Chronicle* (Cambridge, MA: Harvard University Press, 1930), 296-297. The date is sometimes given as 1111.

93.

1130, Bohemia, Czechoslovakia ⓘ
Flying "serpent" in the sky

"A sign (or "a monster") resembling a flying serpent" is said to have flown over Bohemia, and was recorded by two separate historians. This could have been a natural phenomenon.

Source: Czech magazine *Vecerni Praha*, quoted in *The Washington Post*, August 2, 1967. The original sources are *Canonici Wissegradensis Continuatio Cosmae*, in Monumenta Germaniae Historica (MGH) SS 9, 136 and *Annales Gradicences* in MGH SS 17, 650.

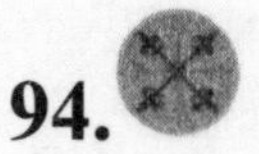

94.

12 August 1133, Japan, exact location unknown ⓘ
Close encounter

A large silvery object is reported to have come down close to the ground. We have failed to locate an actual quote, so we give this case with reservations.

Source: Morihiro Saito, *Nihon-Tenmonshiriyou*, chapter 7.

95.

1142, Bohemia, Czechoslovakia: Flying "dragon" ⓘ

Similar to the case from 1130, a phenomenon described as a "flying dragon" flew over Bohemia.

Source: Czech magazine *Vecerni Praha*, quoted in *The Washington Post*, August 2, 1967. Original: *Monachi Sazavensis, cont.Cosmae (a. 932-1162)*, in MGH SS 9, 159.

96.

1155, Rome, Italy
Three lights and a cross in the sky ⓘ

During the coronation of Federico (Frederick) Barbarossa by Pope Eugene III "there appeared in the sky three lights, and a cross formed by stars."

Source: B. Capone, "Luci dallo Spazio" in *Il Giornale dei Misteri*, Dec. 1972, which gives the year as 1152. While Barbarossa was declared king of the Holy Roman Empire in that year, the coronation did not actually take place until 1155 because of widespread unrest within the lands he supposedly controlled, and the disloyalty of his rival Henry the Lion. We hope that future researchers will be able to trace a more precise reference.

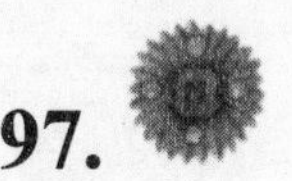

97.

1161, Thann, Alsace, France ⓘ
Three lights in the sky

Three lights or luminous objects were observed by the Lord of Engelburg over the village of Thann, in Alsace, in 1161. A servant of Ubald, bishop of Ombrie, had stolen a relic from the Saint's body, hiding it in his walking stick, which he planted in the ground next to a pine tree. Three aerial lights were seen coming over the top of the tree. The next morning the servant found his stick immobilized and was unable to pick it up. This impressed people so much that they built a chapel to commemorate the 'miracle.'

Each year in Thann, on the 30th of June, three fir-trees (in reference to the three lights) are cremated in front of the main church in celebration of this foundation legend. The celebration is known as the "crémation des trois sapins" and still occurs today.

Source: Johannes Andreas Schenck, *Sanctus Theobaldus* (Freiburg, 1628).

98.

25 December 1167, England, location unknown
Two objects

Two "stars" appear on Christmas Day. The actual quote is from Nicholas Trivetus (*Annales*): "At the watch night (*vigilia*) of the Lord's Nativity, two fiery stars appeared in the western sky. One was large, the other small. At first, they appeared joined together. Afterwards, they were for a long time separated distinctly." It is probable, but not certain, that the sighting was made in England.

Source: Nicholas Trivetus (1258-1328), *Annales sex regum Angliae*. Trivetus was not contemporary with the event, so he must have copied it from an older chronicle.

99.

1169, China, location unknown
Wheels fall off as two dragons fly away

In the history of the Song Dynasty it is written that in the fifth year of the K'ien Tao (now known as Qiandao) era, which corresponds with 1169 AD, dragons were seen battling in the sky during a thunderstorm:
"Two dragons fled and *pearls like carriage wheels fell down on the ground*, where herds' boys found them."

Source: Dr. M. W. De Visser, *The Dragon in China and Japan* (Amsterdam: Johannes Müller, 1913), 48. Visser quotes from "the Wu ki."

100.

1171, Teruel, Aragon, Spain
The King observes a mystery

Alfonso II and his men observed a wandering bull and a mysterious, star-like luminous object hovering above. As described in a current historical brochure about the town, "Tradition says that in the XII century, during the Reconquest of Spain, King Alfonso II, after taking several important positions, continued along the banks of the River Martin and upon reaching what is now Teruel, he split his army up, leaving part of his warriors in the Cella Plains with orders to remain on the defensive, and he then proceeded to confront the rebels in the mountains of Prades. This is the point where history and legend blend together. The warriors disobeyed the king's orders and ran after a bull that was being followed by a star from heaven because they had seen it in premonitory dreams: a sign, according to them, which marked the place where a new town was to be established. In this way they took the fortress of Teruel planting their banner in the conquered fortress."

This fact is still represented today on the shield of Teruel, with a bull and a star above it.

Source: Javier Sierra and Jesus Callejo, *La España Extraña* (Madrid: Ediciones EDAF, 1997), 122-4.

101.

18 June 1178, Canterbury, England: cosmic catastrophe

Gervase of Canterbury wrote that about an hour after sunset five witnesses watched as the upper horn of the bright new moon suddenly split in two. From the midpoint of this division a flaming torch sprang up, spewing out fire, hot coals and sparks. The moon "writhed [and] throbbed like a wounded snake." This happened a dozen times or more, "turning the moon blackish along its whole length."

When a geologist suggested in 1976 that Gervase's account referred to the meteor impact that created the 22 kilometer lunar crater called Giordano Bruno, the theory was widely accepted. However, as reported in several scientific journals in 2001, new calculations show that such an event would have resulted in a fierce, week-long meteor storm on Earth with 100 million of tons of ejecta raining down on our planet. Of course, this did not take place in the twelfth century AD, or archives all over the world would have recorded it! This begs the question "What did Gervase's contemporaries really see?" Did they observe the dramatic entrance of a comet into the Earth's atmosphere – or something even stranger? Were they even looking at the moon?

Source: University of Arizona news release dated 19 April 2001. The BBC website posted a report on May 1st, 2001: "Historic lunar impact questioned" (http://news.bbc.co.uk/2/hi/science/nature/1304985.stm).

102.

27 October 1180, Kii Sanchi, Nara, Japan ⓘ
Glowing vessel

A glowing "earthenware vessel" (a saucer?) maneuvers in the sky between the mountains of Kyushu, flies off to the northeast towards Mount Fukuhara. It changes course abruptly, turns south and disappears with a luminous trail.

Source: Sobeps (Société Belge pour l'Etude des Phénomènes Spatiaux), *Inforespace* 23 ; *Brothers* III, 1 (1964).

103.

1182, Friesland, Holland
Four suns, armed men in the sky

The Chronicler Winsemius (1622) reports, in his *Croniek van Vriesland* that four Suns and a score of armed men were seen in the sky and a bloody rain fell.

Source: M.D. Teenstra, *Volksverhalen en Legenden van vroegere en latere dagen* (Geertsema: Groningen, 1843), 117.

104.

1185, Mount Nyoigadake, Japan ⓘ
A luminous wheel enters the sea

First there were red beams behind the mountain, and then an object like a luminous wheel flew over, and entered the sea. The witnesses were fishermen.

Source: *Brothers Magazine* III, 1, 1964.

105.

9 August 1189, Dunstaple, Bedfordshire, England
Marvelous vision

Numerous amazed observers see the sky "open up" as a huge cross hovers till midnight.

The British monastic chronicler William of Newburgh (1136-1198) noted the sightings of several prodigies in the sky in his *Historia rerum anglicarum*, a philosophical commentary dealing with his own times. In chapter four of Book VI, we read:

"Nor ought I to pass over in silence a most amazing and fearful prodigy, which about this time was seen in England by many, who to this day are witnesses of it to those who did not see it. There is upon the public road which goes to London a town, by no means insignificant, called Dunstaple.

"There, as certain persons happened to be looking up at the sky in the afternoon, they saw in the clear atmosphere the form of the banner of the Lord, conspicuous by its milky whiteness, and joined to it the figure of a man crucified, such as is painted in the church in remembrance of the passion of the Lord, and for the devotion of the faithful (…)

"When this fearful sight had thus been visible for some time, and the countenances and minds of those who were curiously watching it were kept in suspense, the form of the cross was seen to recede from the person who seemed affixed to it, so that an intermediate space of air could be observed between them; and soon afterwards this marvelous vision disappeared; but the effect remained, after the cause of this prodigy was removed."

Other sources: the case is also mentioned by writers Thomas Wykes (with a date of 1191), Wilhelmus Parvus, and Walther of Hermingford (the latter two give a year of 1189).

106.

June 1193, London, England
Bright white ball of light, hovering

"On the 7th of the Ides of June, at 6 o'Clock, a thick black Cloud rose in the Air, the Sun shining clear all round about. In the middle of the Cloud was an Opening, out of which proceeded a bright Whiteness, which hung in a Ball under the black Cloud over the Side of the Thames, and the Bishop of Norwich's Palace."

Source: Thomas Short, *A general chronological history of the air, weather, seasons, meteors, &c. in sundry places and different times* (1749).

107.

Late December 1200, Yorkshire, England
Five Moons in formation

"In the third year of John, King of England, there were seen in Yorkshire *five Moons; one in the East, the second in the West, the third in the North, the fourth in the South, and the fifth (as it were) set in the midst of the other, having many Blazing Stars* about it, and went five or six times encompassing the other, as it were the space of one hour, and shortly after vanished away."

Source: William Knight, "*Mementos to the World, or, An historical collection of divers wonderful comets and prodigious signs in heaven...*" (printed by T. Haly, for T. Passinger, London, 1681).

Note: King John reigned from April 6, 1199 until his death in 1216. The third year of his reign would therefore correspond with 1202 AD. However, another writer, Roger de Hoveden, states it occurred "a little before the Nativity of the Lord," 1200. As Hoveden himself died in 1201, we think the sighting probably took place in King John's second year of reign.

108.

14 September 1224, Mount Alverne, Italy ⓘ
Mystical light, visitation

Brother Leo saw a ball of light suspended above St. Francis of Assisi while he conversed with an invisible being. "He heard voices which made questions and answers; and he remarked that Francis, who was prostrate, often repeated these words: 'Who are you, o my God? And my dear Lord? And who am I? a worm, and Thy unworthy servant.' He also saw him put his hand out three times into his bosom, and each time stretch it out to the flame. The light disappeared, the conversation ceased."

This kind of narrative bridges the gap between lights and objects that fly through the atmosphere, reports of luminous orbs at ground level and "earth lights" over special spots. Here we have a ball of light (observed by an external witness) over a man who appears in communication with it, hence the relevance to our study.

Source: Father Candide Chalippe, *The Life and Legends of Saint Francis of Assisi* (Teddington: Echo Library, 2007), 191-2.

109.

3 May 1232, Caravaca, Spain ⓘ
Lights, and an "angel" brings a cross

Luminous phenomena attend a double-armed cross apparently brought down by an entity assumed to be an angel. The religious context surrounding the observation has enabled it to survive as a legend, told by multiple authors. Traditionally, the most authoritative of them is considered to be the 13th century Franciscan Juan Gil (Egidio) de Zamora. He relates that the cross was brought inside the church by two angels. The current whereabouts of the artifact are equally uncertain. The cross that can be seen today in the sanctuary in Caravaca is a copy, the original having been stolen in 1934, probably by a cult.

Source: Clara Tahoces, "Caravaca, ¡Qué Cruz!" *Más Allá* 127 (September 1999).

110.

2 October 1235, Japan: circling lights in the sky

About 8 P.M., by clear sky, a fortune teller named Suketoshi Abe, consultant to Shogun (Warlord) Yoritsune

Fujiwara, reported to his palace that mysterious sources of light had been seen swinging and circling in the southwest. These lights moved in loops until the early hours of the morning. Yoritsune ordered an investigation and his astrology consultants, who were skilled in astronomy, conducted the study: "It is only the wind making the stars sway," they reported after hearing the statements of Suketoshi Abe.

With arrogance worthy of our modern academic experts, they even suggested that he should write a letter of apology. A high government official, Yasutoki Houjo, denied their request.

Source: This case is mentioned in the Japanese magazine *Brothers* (No. I) and by one of us (Vallee) in *Anatomy of a Phenomenon* (1965) with an incorrect date. The original source is the book *Azumakagami*, edited in 1605 (see *Shinjinbutuouraisha*, vol. 4, 1977). Azumakagami means "Mirror of the East." It was a chronicle covering the period 1180 to 1266, and was compiled after 1266 under the directive of the Hojo regent. It is usually written in two words: Azuma Agami.

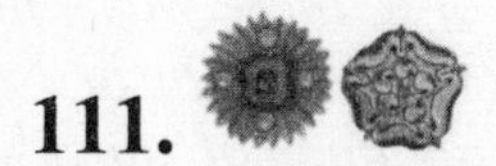

111.

1237, El Puig, Valencia: A fleet of seven lights

According to one record, seven mysterious lights in the night sky were seen on four Saturdays in a row. They appeared to be falling from the sky and entering the earth at a particular spot. Quoting from Tirso de Molina's *Historia de la Orden de la Merced*, the chronicle in which the story was originally written:

"The sentries and custodians of the castle [at El Puig] observed that every Saturday, at midnight, a fleet of luminous stars, seven in number, consecutively descended upon the summit nearest the said fortress, in the same place where our monastery now lies." When the guards informed their masters, Pedro Nolasco (1189-1256) and the mayor, supposing that the phenomenon was trying to announce something important, went up to the site together and carefully excavated the spot.

Whether by some amazing coincidence or divine providence it did not take too long to find a hidden treasure: a bell, and below that a sculpted image of the Virgin Mary. Nolasco thanked the angels for the wondrous gifts and a little time afterwards constructed an altar at the spot.

Source: Tirso De Molina, *Historia de la Orden de la Merced* (1637). Today the monastery has a website: www.monasteriodelpuig.es.tl.

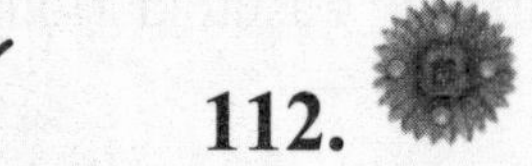

112.

24 July 1239, France
A great light, ascending

"On July 24, 1239, at dusk, but not when the stars came out, while the air was clear, serene and shining, a great star appeared. *It was like a torch, rising from the south, and flying on both sides of it, there was emitted in the height of the sky a very great light. It turned quickly towards the north in the aerie region, not quickly, nor, indeed, with speed, but exactly as if it wished to ascend to a place high in the air."*

This sequence of motion is not typical of a natural phenomenon, and it certainly was not a "star."

Source: Matthew Paris, *Chronica Majora* (London: Longman, 1880), vol. 3, 566.

113.

1252, Padua, Italy: Flying light, seen for an hour

This event is described in the *Cronaca di Rolandino da Padova*, where a witness reports:

"A certain great star, like a comet, but it was not a comet because it did not have a tail and it was a portentous thing because it looked almost as large as the moon, and it moved faster than the moon, but as fast as falling stars, and indeed it was not the moon. It was observable for an hour and then it vanished."

This object, as described, was not a comet or a meteor.

Source: U. Dall'Olmo, "Meteors, meteor showers and meteorites in the Middle Ages: From European medieval sources," *Journal for the History of Astronomy* 9 (1978).

114.

14 October 1253, England: A battle of stars

Nicholas of Findern reported to Burton Abbey that "About the hour of vespers, the sky being clear, suddenly a large bright star appeared out of a black cloud with two smaller stars in the vicinity. A battle royal soon commenced, the small stars charging the great star again and again, so that it began to diminish in size, and sparks of fire fell from the combatants. This continued for a considerable time, and at last, the spectators, stupefied, by fear and wonder, and ignorant of what it might portend, fled."

Source: *Annales de Burton*, in H. R. Luard, ed., *Annales Monastici* vol 1 (London: HMSO, 1864).

115.

12 September 1271, Japan
Saved from execution by a flying sphere!

At midnight one of Japan's greatest saints, Nichiren Shonin (1222-1282), was being escorted to the beach to be executed. Just before the fatal moment, a brilliant sphere as large as the moon flew over, illuminating the landscape. The authorities were so frightened by the apparition that they changed their minds about putting Shonin to death. Instead, they exiled him to Sado Island, though this did not prevent his teachings from spreading. A branch of his teachings, the Sokka-Gakkei, has millions of adherents throughout the world today.

Source: Rev. Ryuei Michael McCormick, *Lotus Seeds: The Essence of Nichiren Shu Buddhis* (Nichiren Buddhist Temple of San Jose, 2000).

116.

1273, Naples, Italy ⓘ
A light enters the bedroom of a sick man

The biography of St. Thomas Aquinas (ca. 1225 to 7 March 1274) states that on the year before his death he returned to Naples, staying in that city for a few weeks during an illness. While he was there two monks saw a light described as a big star coming through the window. It rested for a moment on the head of the sick man and disappeared again, just as it came.

The link with ufology here is very much open to debate, yet abduction researchers have been looking into stories of this kind with increasing interest.

Source: Antonio Borrelli, "San Tommaso d'Aquino Sacerdote e dottore della Chiesa," citing the *Life* of Saint Thomas Aquinas. www.santiebeati.it/dettaglio/22550.

117.

3 June 1277, China, location unknown
Strange event seen at dawn

An unknown object was immortalized in a poem by Liou Ying, a Chinese poet of the Yuan Dynasty. The title of the poem, which can be found in chapter three of *The Yuan Literature Collection*, was simply "Event Seen at Dawn."

"I rise at dawn and, through the window, I see a very bright star that crosses the Milky Way. Now I see three luminous objects appear in the southern sky, of which two fly away and disappear suddenly from my sight. The one which remains possesses five unequalled lights beneath it, and above its upper part I see something in the form of a dome. The unknown object begins to move in a zigzag, like a dead leaf. At the same time, some fiery thing falls from the sky. A short time afterwards, the sun rises but its brightness is dulled by the luminous object that moves quickly in a northerly direction. In the western sky, a green cloud is suddenly disturbed by another unknown object, oval in shape, flat, that descends quickly. This object is more than three metres long, and is surrounded by flames. It rises again shortly after its descent.

"In view of this splendid and amazing spectacle, I rush to the village to alert the inhabitants. When my friends come out of their houses, the flying machine has disappeared. After the event, I reflect on it very much but do not find a reasonable explanation. I have the impression I have come out of a long dream. I hasten to write down all that I have seen at the time so that whoever understands these events can give me an explanation."

Source: Shi Bo, *La Chine et les Extraterrestres*, op.cit., 37.

118.

Circa 1284, Parma, Saint Ruffino, Italy
A duel of stars

On 6 August 1284 the naval battle of the Meloria, between the forces of Genoa and Pisa, took place. It is said in the *Chronicon Parmesan*, of the Franciscan monk Salimbene de Adam (1221-1287):

"It should be known that this battle and massacre between the Genovesi and Pisani had already been foretold and announced long before it happened. In the town of Saint Ruffino, in the diocese of Parma, some women peeled [washed?] the linen at night: and they saw two great stars meeting in the sky. They drew away from each other and still collided again, and chased one another, and more than once…"

Source: Giuseppe Scalia, ed., *Cronica Fratris Salimbene de Adam ordinis minorum* (Laterza, 1966).

119.

3 August 1294, Japan, exact location unknown ⓘ
Red shining object

During a parade, a red shining object appeared, coming from the direction of a shrine. It resembled the Moon, and flew north.

Source: Morihiro Saito, *Nihon-Tenmonshiriyou*. Chapter 7, "Meteor, the Messenger from Space."

120.

May 1295, I Hing, China
Two flying Dragons fall into a Lake

A strange phenomenon was witnessed in the fifth month of the year yih-wei, which corresponds with 1295:

"In a short time a heavy wind came riding on the water, reaching a height of more than a chang (ten ch'ih or feet). Then there fell from the sky more than ten fire balls, having the size of houses of ten divisions. The two dragons immediately ascended (to the sky), for Heaven, afraid that they might cause calamity, sent out sacred fire to drive them away."

The 14th-century chronicler of this incident, Cheu Mih, adds that he had personally observed the results of another 'dragonfall.' Seeing the scorched paddy fields of the Peach garden of The Ts'ing, he interviewed one of the villagers. "Yesterday noon a big dragon fell from the sky," he was told. "Immediately he was burned by terrestrial fire and flew away. For what the dragons fear is fire."

Source: M. W. De Visser, *The Dragon in China and Japan* (Amsterdam: Johannes Müller, 1913), 48-49.

121.

8 September 1296, Loreto, Italy
Globes of light, and an elliptical object

Before dawn, mysterious globes of light appeared repeatedly in the sky of Loreto, falling, stopping and disappearing suddenly. The phenomenon was witnessed by a hermit, Paul Selva, who wrote a famous letter to Charles II dated June 1297. The phenomenon appeared as a body of elliptical shape. A writer named Mantovano who obtained the information from a record dating back to 1300, notes: "He saw a light in the shape of a very bright comet measuring twelve feet in length and six in width, coming down from heaven in the direction of the church and after it approached, vanished at the site." The object, obviously, was not a comet.

Sources: G. Garrat, *Loreto, nuova Nazaret* (Recanati, 1894); O. Torsellini, *Laurentana istoria*, trans. B. Quatrini (Bologna, 1894).

122.

24 December 1299, Tier (Trèves), Germany
Globes of light, and an elliptical object

The Chronicle of the Archbishops of Trier, the *Gesta Trevirensium Archiepiscoporum*, makes an interesting reference to an object in the sky. The term they employed, *cometa*, could actually refer to virtually any luminous body in the sky, not necessarily to a comet as we define it today. In fact, this particular "comet" behaved very strangely.

It was just after midnight. The sky was unusually misty and a foggy frost covered the land.

"Inside the darkness itself, a comet the size of the moon appeared as if hanging in the air, tinted by an ardent redness and which disappeared after an hour. And again, in-between a small interval, two comets appeared simultaneously a short distance from one another, exhibiting the same size and color as earlier; but they disappeared immediately. A third time, after a short hour, [another] one appeared, in all respects visible in the size and color of its predecessors, and which also vanished immediately."

Source: *Gesta Trevirensium Archiepiscoporum*, in E. Martene and U. Durand, *Veterum Scriptorum et Monumentorum... amplissima collection*, vol. IV (Paris, 1729, Col. 370).

123.

1320, Durham, Saint Leonard, England ⓘ
Luminous phenomenon over a the burial site

Upon the death of the Abbott of Saint Gregory monastery, an unknown object ("a great light") lit up the sky over his burial site in Saint Leonard. Later it came lower, moved away and disappeared. The symbolic meaning of this event leaves its connection with ufology open to debate.

Source: Robert de Graytanes, *Historia Dunelmensis*. As published in *Historiae Dunelmensis Scriptores tres* (London-Edinburgh: Publications of the Surtees Society, 1839).

124.

4 November 1322, Uxbridge, England
A pillar with a red flame

"In the first hour of the night there was seen in the sky over Uxbridge a pillar of fire the size of a small boat, pallid, and livid in color. It rose from the south, crossed the sky with a slow and grave motion, and went north. Out of the front of the pillar, a fervent red flame burst forth with great beams of light. Its speed increased, and it flew thro' the air... Many beholders saw it in collision, and there came blows as of a fearful combat, and sounds of crashes were heard at a distance." We note that an object moving with a "slow and grave motion" could have been neither a comet nor a meteor.

Source: *Flores Historiarum* attributed to Robert of Reading, *Rerum Brittannicarum Medii Aevi Scriptores* 95, v. 3: 210-211.

125.

About 1347, Florence, Italy: Low-flying cigar-shaped objects at the time of the Black Plague

Writer Gianfranco degli Esposti mentions that "reports relating to the period of the famous Black Plague, between 1347 and the 1350, speak of strange cigar-shaped objects slowly crossing the sky, sometimes at low altitude, dispersing in their passage a disturbing mist."

He attributes the Black Plague to these objects because "immediately after the appearance of these shocking events, the epidemic exploded in that area."

In Florence a huge mass of vapors appeared in the sky, coming from the north. It spread throughout the land. In the East in the same year, many animals fell from the sky. Their decomposing animal carcasses were said to make the air fetid and to cause the spread of the infamous illness that was fatal in India, Asia, and Britain. In Florence alone it killed 60,000 people.

Source: Gianfranco Degli Esposti, "Travi di fuoco e segni divini: paura nei cieli del Medioevo" (www.edicolaweb.net/ufost16r.htm); Lycosthenes, *Prodigiorum ac ostentorum Chronicon* (Basel, 1557).

126.

20 July 1349, Japan, exact location unknown ⓘ
Two shining objects clash

Two shining objects appeared from the southeast and northwest. They had a terrible clash as they appeared to maneuver acrobatically, emitting flashes.

Source: Morihiro Saito, *Nihon-Tenmonshiriyou.* Chapter 8, "The Messenger from Space."

127.

5 Feb. 1355, Suzhou (Pingjiang) Jiangsu prov., China
Enigma

Big noise in the sky, vision of a large black cloud with flames and lights, loaded with troops. Physical destruction.

Source: Shi Bo, *La Chine et les Extraterrestres*, op.cit., 30, citing writer Tao Zhongyi.

128.

Summer 1360, England and France
Armies and towers in the sky

"And in the summertime of this year in flat and deserted places in England and France, and often visible to many, there suddenly appeared two towers, from which two armies went out, one of which was crowned with a warlike sign, and the other was clothed in black. And a second time the warriors overcame the blacks, and returned to their tower, and the whole vanished."

Source: The *Chronicon Angliae*, covering 1328 to 1388, is attributed to Thomas Walsingham (d. 1422). C. E. Britton, *Meteorological Chronology to A.D. 1450 (*London: H.M.S.O., 1937), 144.

129.

1361, Yamaguti prefecture, Western Japan ⓘ
Drum-like object emerges from the sea

A drum-shaped object, six meters in diameter, is said to have emerged from the sea. It flew overhead, going west. We have not traced an exact reference to this case but its abundant use in databases and on websites has influenced our decision to include it here.

130.

Circa February 1382, Paris, France
Roaming, flashing globe

Before the Maillets uprising, a fiery flashing globe was seen for a period of eight days, "roaming from door to door above the city of Paris, without there being any wind agitation nor lightning or noise of thunder, and on the contrary, the weather never ceased to be serene."

Source: *Chronique du Religieux Saint Denys contenant le règne de Charles VI de 1380 à 1422.* Tome II (Paris, 1840).

131.

1384, Caravaca, Spain ⓘ
Two lights watch over holy relics

Strange aerial lights were frequently associated with miracles in Medieval Spain. In 1384, while the Caravaca cross (see case 109 above) was being transported from the village of Caravaca in Murcia to the village of Lorca y Totan, two lights in the sky accompanied the cross-bearers throughout the journey. They did not disappear until the object was in place. There are several stories about the Caravaca cross, which had more than its fair share of magical adventures. There are legends about how it 'teleported' from one place to another and how it attracted luminous phenomena on more than one occasion.

Source: Clara Tahoces, "Caravaca, ¡Qué Cruz!" *Más Allá* 127 (September 1999).

132.

15 July 1385, London and Dover, England
Three lights join as one

On July 15th 1385 "at London and likewise at Dover, there appeared after sunset a kind of fire in the shape of a head in the south part of the heavens, stretching out to the northern quarter, which flew away, dividing itself into three parts, and travelled in the air like a bird of the woods in flight. At length they joined as one and suddenly disappeared."

Source: C.E. Britton, *A Meteorological Chronology to A.D. 1450* (London: H.M.S.O., 1937), 149; also noted by John Malvern, a monk of Worcester, who certainly contributed to the *Polychronicon* (begun by Ranulph Higden, a monk of Chester), but continued the chronicle only as far as 1377.

133.

14 October 1387, Leicester and Derbyshire, England
Revolving wheel in the sky

"A certain appearance in the likeness of a fire was seen in many parts of the kingdom of England, now in one form, now in another, nearly on a single night, yet in various places, throughout the months of November and December (…) and some appeared in the form of a burning revolving wheel, others again in the form of a round barrel of flame emitting fire from above, yet others in the shape of a long fiery beam, and it thus appeared in one form or another through a great deal of the winter, especially in the counties of Leicester and Northants."

Source: "Chronicon Henrici Knighton, vel Cnitthon, monachi Leycestrensis," or the Chronicon of Henry Knighton (d. 1396). The book covers 1337-1396, and after Knighton's death was continued by another scribe. See also C.E. Britton, *A Meteorological Chronology to A.D. 1450* (London: H.M.S.O., 1937), 150. Note that the date might be November.

134.

1390, Bologna, Italy ⓘ
Unknown creatures flying aboard a fiery object

"One tradition states that in 1390 the guardian of the Asinelli Tower saw a "ball of fire" that rotated over of the roofs, and inside were seen the faces of some devils who were trying to see outside."

Source: "Quegli Ufo sopra le Due torri," *Il Domani di Bologna*, 21 October 2006, 10.

135.

26 January 1390, Messina, Sicily, Italy
Light descending, ascending

Mongitore writes "A similar appearance was seen at two hours of the night in Messina, as you saw fall from the sky above the Cross, at the top of the dome of the Monastery of the Fathers of St. Salvadore. The monks were astonished at this view, but it was not certain how long the light was seen, so the fear was brief; as having lasted half an hour, (then it) went back up to Heaven..."

We retain this case because a "light" visible for half an hour going up in the sky is unlikely to be a meteor.

Source: Antonino Mongitore, *Della Sicilia ricercata nelle cose più memorabilia* (Palermo, 1742).

136.

Winter 1394, England: Another wheel-shaped object

According to Raphael Holinshed's landmark chronicle of British history, a wheel- or barrel-shaped object appeared in several areas of England:

"A certain thing appeared in the likeness of fire in many parts of England every night. *This fiery apparition, oftentimes when anybody went alone, it would go with him, and would stand still when he stood still... To some it appeared in the likeness of a turning wheel burning; to others as round in the likeness of a barrel, flashing out flames of fire at the head; to others in the likeness of a long burning lance."*

Whatever it was that caused such an impression on the people of England, it does not seem meteoritic in nature.

Source: Raphael Holinshed, *Chronicles of England, Scotland and Ireland* (London: J. Johnson, 1808) vol. II, 829. Raphael Holinshed, though not the book's sole author, is thought to have helped inspire William Shakespeare to write at least two of his plays. Both Macbeth and King Lear were based on material contained in Holinshed's book.

137.

2 September 1394, Forlì, Italy: Huge celestial object

At the second hour of the night, men walking in the main square of Forlì saw an enormous "asub" (celestial object) fly over very slowly. Duration: "the time of two Pater-nosters." It left a smell like burning wood. Some witnesses described it as motionless in the sky for a while.

Source: F. Guarini, *I Terremoti a Forlì* (Croppi, Forlì, 1880), 142.

138.

1395, Languedoc, France: Aerial combat

"In the land of Languedoc, *a big star and five small ones were seen in the sky*. These, as it seemed, attacked and sought to fight the big one, which they followed for half an hour. Also a voice was heard in the sky, shouting. Then a man was seen, who seemed to be made of copper, holding a spear in his hand, and throwing fire. He grabbed the big star and hit it; after which, nothing more was seen."

Source: *L'Histoire de Charles VI, Roy de France, et des choses mémorables advenues durant quarante-deux années de son règne, depuis 1380 jusqu'à 1422*, by Jean Juvénal des Ursins, évèque de Reims. Published by Michaud and Poujoulat in *Nouvelle Collection des Mémoires pour servir à l'histoire de France depuis le XIIIe siècle jusqu'à la fin du XVIIIe*, Tome II (1836).

139.

16 September 1408, Rome, Italy: Flying formation ⓘ

Three "stars" were seen to fly over Rome. The incident was described by Antonio Di Pietro, canon of the Vatican, in his *Diarium Romanum* (Diary of Rome from 1404 to 1417), now conserved in the Vatican Archive. Di Pietro narrates that on that evening he was going to supper with friends near where the ancient Basilica of Saint Peter stands today.

"Suddenly after sunset…we saw...a beautiful star that, coming from the sky of Tarrione, headed towards Castel Sant' Angel with two other small, splendid bright stars. And we were all very surprised by that spectacle."

The sighting may have been of an unusual meteor train.

Sources: Antonio di Pietro, *Diarium Romanum* (Diary of Rome from 1404 to 1417), preserved at the Archivio Capitolare Vaticano. This Latin manuscript was found in the Library of Modena by L. A. Muratori, who inserted the text in Volume XXIV (ed. 1734) of the *Rerum Italicarum Scriptores*. See also *Coelum* astronomy magazine,

No. 5-6, May-June 1977, article "Gleanings from science fiction medieval texts" by Umberto Dall'Olmo, 107. Credit: Umberto Cordier.

140.

2 July 1420, Castle Godego, Treviso, Italy ⓘ
The lady in the light

In the evening a Hungarian merchant, Peter Tagliamento, was leading his herd of cattle to Bassano del Grappa. As he came to an area of thick brush, close to Castel di Godego, he realized he had lost the way. All around him were only shadows, the woods, and deep silence. In despair, Peter started praying and suddenly he saw a great light. Still trying to realize where he was and what was happening, Peter saw a young woman of great beauty, who told him how to get to the road towards Bassano, but requested that a chapel be built at that place. She planted a cross in the earth as proof of her visit.

Peter found his herd and reached the leaders of the community of Godego to fulfill the mandate he had received. At first no one believed him, but they found the cross planted in the woods. This convinced them and they decided to erect a chapel, where people came in solemn processions.

Source: Marino Gamba, *Apparizioni mariane nel corso di due millenni* (Udine: Ediz. Il Segno, 1999).

141.

3 March 1428, Forlì, Italy: Celestial object

Another case of a celestial object ("asub") in Forlì: At 1:30 A.M. a fiery lamp was observed for about two hours. The city archives also mention "a very high flame in the shape of a tower, and a column of apparent fire rising in the air."

Source: Filippo Guarini, *I Terremoti a Forlì* (Croppi, Forlì, 1880), 12-13 and 143.

142.

5 January 1433, Nice, France: Luminous globe

A luminous globe appears, seen several hours. "On January 5th, 1433," writes Abbé Joseph Bonifacy, "a luminous globe appeared in the air for several hours."

Source: G. Tarade, *Soucoupes volantes et civilisations d'outre-espace* (Paris: J'Ai Lu, 1969). Also *Cielo e Terra*, April 1972, 9. We have been unable to verify the text by Bonifacy, which is only available in manuscript form.

143.

June 1444, Bibbiena, Arezzo, Italy
Unexplained golden globes of light

Over three months multiple witnesses saw globes of light, golden in color, both inside and outside a church. The story by Don Massimo, a Benedictine monk, mentions that "turning to the church he and his companions saw a globe as thick as a printing press."

Mr. Lorenzo Piovano of Bibbiena stated that he saw more lights day and night, moving around the church and leaving a smell of remarkable sweetness. Don Massimo is careful to add that the mayor and others who ran into the church saw nothing, but they did notice the smell.

Sources: Don Massimo's manuscript of "Miraculous facts that occurred near Bibbiena, etc." inserted in the Moreno Frullani Collection No. 29, 56, in the Riccardiana library in Florence. Also see: http://www.mariadinazareth.it/apparizione%20bibbiena.htm

144.

29 May 1453, Constantinople: Light from the sky

"Every night [during the siege by the Turks] a fire descended from the sky, stood over the City, and enveloped her with light all night long. At first the Christians read this light as a sign of God's wrath and the coming destruction of the city, but initial success against the Turks led to the reinterpretation that God had sided with the Christians and that they would prevail.

"Thus the sultan and his entire retinue became visibly depressed...and were considering lifting the siege...On the night before their scheduled departure the heavenly sign descended in its customary manner but did not envelop our City as it had before...[N]ow it seemed to be far away, then scattered quickly, and vanished at once. The sultan and his court were immediately filled with joy."

Source: Makarios Melissinos, "Chronicle of the Siege of Constantinople" in George Phrantzes, *Fall of the Byzantine Empire* (Amherst, MA: University of Massachusetts Press, 1980), 97-136.

145.

Late December 1456, Piacenza, Italy
Four unknowns

In the *Annali Piacentini* of Antonio da Ripalta, we read of the apparition "of four wonderful stars that proceeded directly from the east to the west and were positioned almost in the sign of a cross."

Source: U. Dall'Olmo, "Meteors, meteor showers and meteorites in the Middle Ages," *Journal for the History of Astronomy* 9, 1978.

146.

7 March 1458, Kyoto, Japan: Five stars circle the moon

Five "stars" appeared to circle the moon, changed colors three times and vanished suddenly.

Source: *The Taiheiki: A Chronicle of Medieval Japan*, trans. Helen Craig McCullough (North Clarendon, VT: Tuttle Publishing, 2004).

147.

October 1461, Metz, France: Many lights, seen twice

"Between Saint Remy's day (October 1st) and All Saints' Day (November 1st) numerous and marvelous signs like great firebrands the length of four fathoms and a foot large were seen in the air. "It lasted for half of a half quarter of an hour and was seen twice. Some people also said they had seen by night the like of a battle, and heard a great uproar and noise."

Two significant meteor showers happen in October: the Draconids (between the 8th and 10th day of the month) and the Orionids (around the 21st). The sightings might have been caused by these events, but meteors would not account for the report of "great uproar and noise."

Source: Jean-François Hughenin, *Les Chroniques de la ville de Metz recueillies, mises en ordre et publiées pour la première fois* (Metz 1838), 297.

148.

1 November 1461, Arras, France: Hovering object

Jacques Duclercq, legal adviser to Philippe III, writes: "On this day of Our Lord, All Saints Day, there appeared in the sky an object as bright as burning steel, as long and wide as half of the moon. *It was stationary for fifteen minutes. Suddenly, the strange object began to spiral upwards and then it spun around and rolled over like a loose watch spring, after which it disappeared in the sky.*"

Source: *Mémoires de Jacques du Clercq, sur le Règne de Philippe le Bon, Duc de Bourgogne, publiés pour la première fois par le Baron de Reiffenberg*, Tome III (2nd ed., Bruxelles, 1836), 189.

149. 

19 February 1465, Italy: Great ship in the air

From the *Notabilia Temporum* of Angelo de Tummulillis: "There appeared many signs in the air in the same month, always in the morning, at daybreak. At the first hour of the 19th of this month a kind of great flaming ship appeared in the air towards the north and it appeared again on the 20th and 21st, not at the same time but later."

Source: U. Dall'Olmo, op. cit.

150. 

8 March 1468, Mount Kasuga, Japan: Dark object ⓘ

In the middle of the night a dark object took off from Mount Kasuga flying west towards the bay of Osaka, with a sound like a spinning wheel. Its size was estimated as 9 by 6 feet.

Source: *Brothers Magazine* I, 1, no full quotation found.

151. 

27 September 1477, Japan, location unknown ⓘ
Object, unknown substance

A luminous object crossed the sky. A cotton-like substance fell for the next six hours.

Source: Case summary in *Brothers* I, 1, but no original source given.

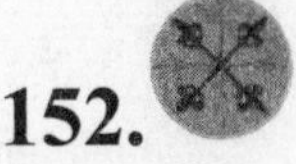

152.

1478, Milan, Italy: Two flying objects during a war ⓘ

Two unexplained flying objects are observed during a battle.

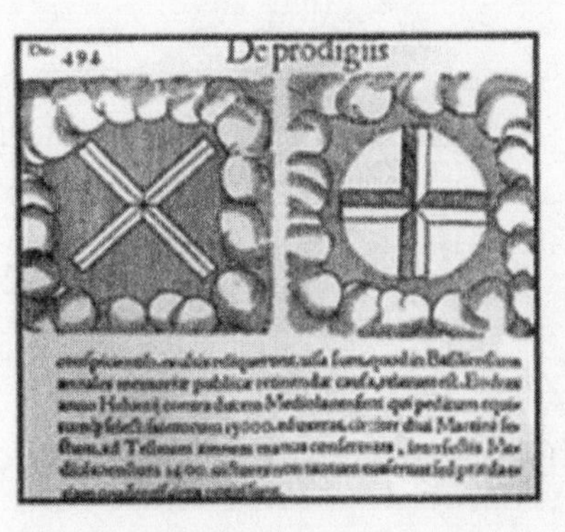
494 De prodigiis

Fig. 8: Illustration from Lycosthenes

An engraving of the scene accompanies the text in Lycosthenes' edition but there is no guarantee it was made especially for the book because images such as these were recycled from publication to publication.

Source: Lycosthenes, *Prodigiorum ac ostentorum Chronicon* (Basel, 1557).

153.

1479, Arabia: Pointed object in the sky

A remarkable engraving highlights this observation. Lycosthenes notes that this object, which he calls a "comet," was seen in Arabia, "in the manner of a sharply pointed beam." The illustration shows that whatever was observed does not to conform to our knowledge of comets.

However we should note that the illustration looks similar to drawings of the first multi-stage rockets built around the same time by Conrad Haas. We therefore doubt it was

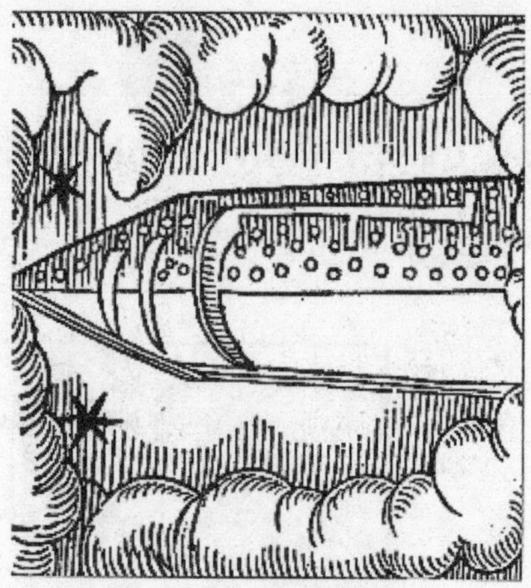

Fig. 9: Arabian "rocket"

drawn in Arabia. Usually such drawings would have been provided by the printers, often taken from very different contexts.

154.

18 October 1482, Albisola, Savona Province, Italy
A dazzling object stops a battle

The facts date back to when the inhabitants of two neighboring villages, Albisola and Stella, decided to fight over territorial issues. A bloody attack took place. The sky was totally clear of clouds, when, an hour after midday, a small white cloud came from the east, so bright it seemed to collect all the rays of the sun. Dazzled by the light, the fighters heard a sweet clear voice repeating three times the word "peace," and then everything disappeared. All were astonished, their eyes looking up at the sky, and they abandoned their weapons.

Sources: Giovanni Bernardo Poggi in the manuscript of his *Memoirs*. Details can be found in the magazine *Maria Ausiliatrice*, September 2005, and at this webpage:
http://www.mariadinazareth.it/Prodigi%20luminosi/Madonna%20della%20Pace.htm

155.

August 1487, Forlì, Italy: Floating cartwheel in the sky

"In that same year, on a morning (two hours before day) *a big star seemingly appeared, coming from the mountain and going toward Ravenna. It looked quite like a butterfly flying in the air.* I saw it and so did a lot of peasants who had put either wood or wheat onto their carts, and also those coming to Forlì. I was in the country and had risen to come to Forlì in the cool hours. It lasted a very short time. Some people say they had seen it when on the mountain, half an hour before." Others saw it as a huge 'cartwheel' floating in the sky.

Source: Leone Cobelli, an Italian historian, in *Cronache Forlivesi dalla Fondazione della Città all'anno 1498* (Bologna, 1874).

156.

1491, Vladimir, Russia: A Figure rises in the air

The apparition in the air of the Saint Grand Prince Alexander Nevsky. "In the year 6999 (of the old Russian calendar) in the great city of Vladimir there was an awe-inspiring apparition and frightful and terrible sign of the wrath of the Lord. Thus the Lord punishes us and leads us from sin toward repentance.

"On a certain day after morning Mass many saw the following appearance above the stone church of Our Lady and the residence of the abbot: just above the place where the remnants of the Saint prince Alexander repose, on the very dome of that church, they saw a strange sign.

"*It was as if a light cloud or thin smoke spread around, white as a pure frost and bright as the sun. Then the people saw the likeness of the Holy Prince on a white horse rising up toward the sky. The people who saw it were very frightened and began to toll the bells all over the city.*"

Source: A. N. Vyssotsky, "Astronomical Records in the Russian Chronicles from 1000 to 1600 A.D." Historical Notes and Papers 22, *Meddelande Fran Lunds Astronomiska Observatorium,* Ser. II., 126, (Sweden, 1949), 45.

157.

13 August 1491, Milan, Italy: Summoning the Aliens

Seven "men" appeared before Philosopher Facius Cardan (Fazio Cardano) in his study. According to his son Jerome the story left by his father, a mathematically-gifted lawyer and friend of Leonardo da Vinci, read as follows:

"When I had completed the customary rites, at about the 20th hour of the day, seven men duly appeared to me clothed in silken garments resembling Greek togas, and wearing, as it were, shining shoes. The undergarments beneath their glistening and ruddy breastplates seemed to be wrought of crimson and were of extraordinary glory and beauty. Nevertheless all were not dressed in this fashion, but only two who seemed of nobler rank than the others. The taller of them who was of ruddy complexion was attended by two companions, and the second, who was fairer and of shorter stature, by three. Thus in all there were seven. They were about forty years of age, but they did not appear to be above thirty. When asked who they were, they said they were men composed, as it were, of air, and subject to birth and death. It was true that their lives were much longer than ours, and might even reach to three hundred years' duration. Questioned on the immortality of our soul, they affirmed that nothing survived which is peculiar to the individual…

"When my father asked them if they did not reveal treasures to men if they knew where they were, they answered that it was forbidden by a peculiar law under the heaviest penalties for anyone to communicate this knowledge to men. They remained with my father for over

Fig.10: Jerome Cardan (Girolamo Cardano, 1501-1576)

three hours. But when he questioned them as to the cause of the universe they were not agreed. The tallest of them denied that God had made the world from eternity. On the contrary, the other added that God created it from moment to moment, so that should He desist for an instant the world would perish."

Source: Jerome Cardan, *De Subtilitate Rerum Libri XXI* (Nuremberg, in-folio 1550), XIX.

158.

11 October 1492, Guanahany, Bahamas
The light seen by Columbus

Two hours before midnight Christopher Columbus and a crew member saw a light alternately going higher and lower. The actual passage reads as follows:

"The land was first seen by a sailor called Rodrigo de Triana, although the Admiral at ten o'clock that evening standing on the quarter-deck saw a light, but so small a body that he could not affirm it to be land; calling to Pero Gutierrez, broom of the King's wardrobe, he told him he saw a light, and bid him look that way, which he did and saw it; he did the same to Rodrigo Sanchez of Segovia, whom the King and Queen had sent with the squadron as comptroller, but he was unable to see it from his situation.

"The Admiral again perceived it once or twice, appearing like the light of a wax candle moving up and down, which some thought an indication of land. But the Admiral held it for certain that land was near; for which reason, after they had said the Salve which the seamen are accustomed to repeat and chant after their fashion, the Admiral directed them to keep a strict watch upon the forecastle and look out diligently for land, and to him who should first discover it he promised a silken jacket, besides the reward which the King and Queen had offered, which was an annuity of ten thousand maravedis."

Source: *Personal Narrative of the First Voyage of Columbus to America: From a Manuscript Recently Discovered in Spain,* trans. Samuel Kettell (Boston: T. B. Wait and Son, 1827), 32-33.

159.

1494, Apulia, Italy: Three suns at night ⓘ

"Upon the coming of the little King Charles the VIIIth in Naples...in Apulia during the night three suns appeared in the middle of the sky which was all around covered with clouds, accompanied by many lightnings and horrible thunders."

Source: Francesco Guicciardini, *Storia d'Italia* (Turin: Giulio Einaudi, 1971).

160.

20 September 1498, Japan, location unknown ⓘ
Umbrella-shaped object

A bright object resembling an umbrella crossed the sky with a rumbling sound.

Source: Brothers I,1, Dainihonjisinsiriyo Nihon-jisinsiriyo; Takao Ikeda, *UFOs over Japan.*

161.

1499, South Atlantic, off Africa
The slow-moving light seen by Pedro Cabral

A phenomenon difficult to explain as a meteor occurred when Pedro Alvares Cabral left Portugal on an expedition of 13 vessels and a crew of 1,200 men. The expedition was plagued with incidents. However, had it not been for one such near-disaster they would never have headed west and gone down in history as the first men to reach Brazil in the year 1500. As they were sailing around Africa they saw a luminous object in the southern sky. It only remained in sight for 8 minutes, moving slowly towards the Cape of Good Hope. Shortly after, a hurricane arose. Six ships sank or ran aground. The remaining seven went on to the Americas. These vessels made it back to Lisbon bringing with them a fortune in spices and news about the discovery of Brazil and Madagascar.

Source: W. B. Greenlee, ed., *The Voyage of Pedro Alvares Cabral to Brazil and India* (London: Hakluyt Society, 1937).

Epilogue to Part I-B

By the close of the 15th century the use of printing had changed everything in terms of the generation and distribution of knowledge. German inventor Johann Gutenberg (1397-1468), who may have been aware of earlier Chinese and Korean printing methods, had developed molds that allowed for the mass production of individual pieces of metal type. Printing presses soon began to appear all over Europe.

Books were scarce, copied by monks or (after the 13th century) in commercial scriptoria, written by hand. While it might take someone a year or more to hand copy a Bible, with the Gutenberg press it was possible to create several hundred copies a year. Printed works were not immediately popular: some nobles refused to keep printed books in their libraries, fearing that would depreciate their valuable manuscripts. Much of the Islamic world, where calligraphic traditions were extremely important, also resisted. In spite of this, Gutenberg's printing press spread rapidly. Within thirty years of its invention in 1453, towns across Europe had the equipment.

For the purposes of our study, it is important to note that printing, which made an impact only comparable in modern times to that of the Internet, led to information spreading more quickly, within a more literate citizenry, so that more reports of unusual events survived. On the negative side, it also spread disinformation and misinformation, just as the Internet does today. Publishers shamelessly exploited people's fears by trumpeting strange events, while stories of portents and signs in the sky were cynically invented to support political or religious objectives.

Printing was expensive. It became a source of significant profits, two facts that combined to spread sensational news broadsheets of dubious validity, creating incentives to compile information about unusual incidents. Chroniclers correlated such visions with current affairs and future predictions. As we study the records of unusual sightings in the sixteenth century and beyond, increasingly sharp analysis is required to take these social distortion effects into consideration.

Another important factor appears in the late fifteenth century with European scholars' novel obsession with witchcraft, putting the topic of unusual phenomena (and ordinary folks' interaction with them) in a new and dangerous light. The most authoritative and influential treatise on the subject of witchcraft was indisputably the *Malleus Maleficarum*, or "The Witch's Hammer," written in 1486 by two erudite Dominican friars. It served as the official witch-hunter's handbook for nearly two centuries, the maximum authority used by inquisitors, magistrates and priests to justify the brutal torture and execution of alleged witches in every European country. The text was reprinted at least sixteen times in German, eleven times in French, twice in Italian and went through more than half a dozen editions in English. It became the principal source of inspiration for every work published after it.

The authors of the *Malleus Maleficarum*, Jacob Sprenger (1436-95), Dean of the Theology Faculty at the University of Cologne, and the prior Heinrich Kramer (1430-1505), divided their treatise into three parts. The first part discussed the need for governing authorities to comprehend the true diabolical nature of witchcraft in all its aspects: the threat posed to Catholicism, pacts with the Devil, problems caused by lascivious demons, and so on. The second part deals with the three kinds of *maleficia* (dark magic) and how they may be "successfully annulled and dissolved." The third part considers methods to hold a witchcraft trial and the punishment that best suits each crime. Here we can find advice on what punishment should be given "in the Case of one Accused upon a Light Suspicion" and about "the Method of passing Sentence upon one who has been Accused by another Witch, who has been or is to be Burned at the Stake."

Of particular interest to us is the issue of physical contact with beings assumed to be demons, a form of interaction the two scholars call "transvection". Of all the issues dealt with in the *Malleus Maleficarum*, the most prominent were (1) whether humans could feasibly procreate with demons and bear their children, and (2) whether people were taken physically by demonic beings and transported to secret locations, or if it was all in the mind.

In other words, five hundred years ago they were debating the exact same issues as ufologists today. It may seem a horrid, unfair thought, but it is difficult to read ancient books such as the *Malleus Maleficarum* or Rémy's later *De Demonolatriae* (1595) without coming away with the impression that today's leading abduction researchers, who abuse witnesses with dubious hypnotic techniques to extract information, would have enjoyed a successful collaboration with the chief inquisitors of yore.

PART I-C

Sixteenth-Century Chronology

The sixteenth century is marked, first and foremost, by extraordinary expansion of the knowledge of the world, thanks to numerous expeditions to the Americas. In 1519 Magellan leaves for the first voyage around the world; he sails into the Pacific Ocean, previously unknown to Europeans. As a result, the commercial prominence of Mediterranean cities decreases, to the advantage of ports like Lisbon and the premier colonial empires, Portugal and Spain. In Mexico and Peru, explorers find thriving civilizations and many opportunities for enrichment that help transform European society.

Parallel to the expansion of geographic knowledge, navigation, and trade, the world undergoes a deep transformation of ideas under the influence of the humanist philosophy that feeds the Renaissance, blending with mystical notions that refuse to disappear, while early scientists like Copernicus (1473-1543) and Galileo (1564-1642) use critical observation and the experimental method to build new theories of the world that conflict with traditional teachings.

It is the time of Leonardo da Vinci, Michelangelo, and Raphael in art, and of Martin Luther in religion. In 1520 the Reformation shakes the foundations of the Catholic church, which reacts with renewed commitment to its mystical basis: When Ignatius of Loyola founds the Jesuit Order in

1534, first as a tool against the Moslem religion, and eventually against Protestantism, he is said to have been threatened by an assassin, who fled when an angel came down from the sky and confronted him!

Fig. 11: Ignatius of Loyola saved by an angel

Thus reports of unusual phenomena gradually become caught between increasing rational interest in all natural effects and lingering temptation to attribute them to celestial powers, in the phraseology of traditional religion– a polarity that has survived to the present day.

162.

1501, between Urbino and Gubbio, central Italy ⓘ
Abortions caused by a horrible object in the sky

Professor Carlo Pedretti, specialist of Leonardo da Vinci Studies at the University of California, has published an article about relations between the "monstrous" and the Renaissance. He stated that the Florentine physician Antonio Benivieni (1443-1502), who was interested in monsters from the pathological point of view, mentions a wonder that took place between Urbino and Gubbio in 1501: "a horrible appearance in the stormy sky that caused many abortions – what we today call the appearance of a UFO."

Source: Antonio Benivieni, *De abditis nonnullis ac mirandis morborum et sanationem causis*, G. Weber, ed., in *Academia Toscana di Scienze e Lettere "La Columbaria,"* 142 (Florence: Leo S. Olschki, 1994), 190; Carlo Pedretti, *La Nazione*, 28 July 1979.

163.

29 September 1504, Tirano, Val Poschiavo, Italy
Abducted by a lady of light

At dawn a man named Mario Omodei, who had gone into his garden, was suddenly surrounded by a dazzling light and heard a voice that called him by name. At the same time he felt lifted from the ground and found himself carried away to a land owned by a man named Alojsio Quadrio. Here, in an even more dazzling vision, an apparition he took to be Mary instructed him to make public the fact that she wanted a temple dedicated to her. Indeed it was completed in 1513 and consecrated in 1528.

A priest named Simone Cabasso wrote in 1601 about the adventure of Omodei: "it seemed that the mountains (...)

were illuminated by an unusual light (...) He clearly felt lifted from the earth, and transported to a garden, and was taken down to the ground." The luminous apparition looked like a 14-year old girl.

Sources: Gamba, Marino. *Apparizioni mariane nel corso di due millenni.* Udine: Ediz. Il Segno (1999); and Cabasso, Simone, *Miracoli della Madonna di Tirano.* Vicenza: Ed. Pietro Gioannini (1601).

164.

19 March 1509, Villefranche-du-Rouergue
Vehicle interference: Light beings free up a chariot

A man named Collongis (or Collonges) who was driving a cart across a fork in a shallow river tried in vain to free it when it became stuck. Having prayed fervently, he saw a being in a blinding light coming from the East, accompanied by twelve other figures. He took them to be the Virgin Mary and the Apostles. They stepped on thirteen stones local people used to cross the river and disappeared in the West. As they vanished, Collongis found that his cart was free from the mud. An investigation by the bishop of Rodez led to the founding of the Chapel of the Thirteen Stones on July 1st, 1510.

Sources: Bernard, Gilles and Guy Cavagnac, *Villefranche-du-Rouergue, histoire et génie du lieu*, Ed. Privat (1991), 82, and Chiron, Yves. *Enquête sur les Apparitions de la Vierge.* Paris: Ed. Perrin (1955), 65-66. (Credit: Franck Marie)

165.

1513, Rome, Italy: Michelangelo's flying triangle

The celebrated sculptor Michelangelo Buonarroti (1475-1564) observed a triangular light with three tails of different colors. He painted a picture of it but this has not survived.

According to Benedictine chronicler Benedetto Lushino's book *Vulnera Diligentis* (second book, chapter XXII) Michelangelo saw a "triangular sign" one calm night.

Fig. 12: Michelangelo

It resembled a star with three tails, one silvery, the second one red, and the third fiery and bifurcated.

Source: Giovanni Papini, *La vita di Michelangiolo nella vita del suo tiempo* (Milano: Garzanti, 1949), 198-200.

166.

8 November 1517, Moldavia, Romania ⓘ
An object resembling a face

According to the 17th century Moldavian chronicler, Grigore Ureche (ca. 1590-1617), "a great blue sign shining like the face of a man" appeared in the sky. After some time without moving, it "hid itself in the sky again."

Source: Grigore Ureche, *Letopisetul Tarii Moldovei*. As published by E. Picot, *Chronique de Moldavie depuis le milieu du XIVe Siècle jusqu'à l'an 1594* (Paris: E. Leroux, 1878).

167.

April 1518, Yucatán, Mexico: A "star" with rays of light

Spanish conquistador Juan de Grijalva (ca.1489-1527) wrote: "On this day, in the evening, we witnessed a big miracle, and it was that there appeared a star above the ship after sunset, and it moved away, emitting rays of light continuously until it was above the town or large village, and it left a trace in the air that lasted for three long hours, and we also saw other very clear signs, by which we understood that God wanted us to populate that land…" The village was Coatzalcoalco.

Source: "Itinerario de la armada del rey catolico a la isla de Yucatàn, en la India, el año 1518, en la que fue por Comandante y Capitàn General Juan de Grijalva. Escrito para su Alteza por el Capitàn mayor de la dicha armada." In Joaquin Garcia Icazbalceta, *Colleccion de documentos para la historia de México*, Volume 1. México (Libraria de J. M. Andrade, Portal de Agustinos N. 3, 1858), 302.

168.

1520, Hereford, England: Fiery circle flies up

A case with thermal effects was recorded by Lycosthenes in his *Prodigiorum ac Ostentorum Chronicon*:

"In 1520 AD in England, at Hereford, a colossal beam of fire was seen in the sky. Approaching the earth, it burned many things with its heat. After this, it ascended into the sky again and was seen to change its shape into a circle of fire."

Source: Lycosthenes, op. cit., 527.

169.

1521, Cuenca, Spain: Contact with a flying Alien ⓘ

Dr. Eugenio Torralba was in contact with "Zekiel," a being who taught him many secrets, and flew him to Rome. Torralba received such fame by virtue of his new-found knowledge and medicines that even Cervantes mentioned him in Don Quijote: "Remember the true story of Dr. Torralba," says Quijote, "who was taken by the devils through the air...and in twelve hours arrived in Rome..." In 1525 he became the personal doctor of the widowed queen of Portugal, Leonor.

Zekiel (or "Zequiel"), however, proved to be Torralba's curse as well as a blessing. The Inquisition didn't take long to find out about the good doctor's dealings with the entity, and it was soon revealed that Torralba had been physically transported to faraway places by magical means. Torralba was arrested in 1528 and cruelly tortured, despite his insistence that he had never entered into a pact with the creature nor gone against the Catholic faith at any time. He was sentenced to prison on March 6th 1531, but was soon released and allowed to continue his medical practices on the condition that he never again had any contact with Zekiel.

Source: Marcelino Menendez y Pelayo, *Historia de los Heterodoxos Espanoles* (Madrid: La Editorial Católica, 1978).

170.

1523, Changsu, Jiangsu Province, China
Flying ships, carrying men

The Chinese book *Stories in a Summerhouse of Flowers*, written by Qiu Fuzuo, includes an account of an encounter that took place in 1523, in the second year of Emperor Jianjing's reign. At this time there lived a teacher called Lü Yu in the village of Yujiu.

"One day when it was raining without stopping the teacher observed two ships sailing over the woolly clouds above the ruins, in front of his house. On these two ships that measured more than ten fathoms [over sixty feet], two tall men were busying themselves, each one twelve feet tall and wearing a red hat and multicoloured clothes. They held a pole in their hands. The ships moved very quickly.

"In the home of the teacher Lü Yu that day there happened to be a score of scholars who, alerted by Lü Yu, came out of the house and stood next to him to observe the phenomenon. Then, the men in multicolored clothing passed their hand over the scholars' mouths; their mouths at once became black and as a result none of them could speak. At that moment, they saw a man, escorted like a mandarin, dressed like an old scholar, emerge on one of the ships accompanied by a bonze. A long time after this, the ships flew away, as if carried by the clouds, and descended again a kilometer away, in a cemetery. The ships set off again; the scholars felt their mouths return to normal. But five days later, Ju Lu died, though nobody knows why."

Source: Shi Bo, *La Chine et les Extraterrestres*, op.cit., 42.

171.

1526, Rome, Italy: Demonic transportation

The Italian inquisitor Paulus Grillandus, whose *Tractatus de Hereticis et Sortilegiis* had almost as much impact as the *Malleus Maleficarum*, wrote that a countryman in Rome saw his wife take all her clothes off and go out of the house.

The next morning he asked his wife where she had been all night. At first she refused to tell him, but when he started to become more aggressive she told him she went to a witch gathering. He demanded that she take him with her the next time, and not long after this they were both "transported" by two he-goats. However, she warned her husband not to pronounce the word "God" during his time with the demons, to which he agreed. The man saw many famous people at the meeting, all of whom declared their devotion to the Devil in a ceremony. There was a dance and a banquet. The man noticed that the food on the table lacked salt. Of course, salt has purifying qualities associated with warding off evil spirits and was therefore shunned by demons and fairies alike. The man was unaware of this fact, much to his misfortune. He asked for the salt and, when he thought he had it in his hand, exclaimed, "Thank God, the salt has come!"

Suddenly, everything disappeared before his eyes. Men, women, tables and dishes evaporated and everything went dark. He found himself naked in the countryside, in the cold night. At dawn he met some shepherds who informed him he was near Benevento, some 100 miles from Rome. They gave him something to eat and clothes to wear, and eventually he found his way home, begging for money on the way. When he reached Rome, starving and exhausted, the first thing he did was to report his wife, who was forced to confess and promptly burnt at the stake.

Medieval demonologists, similar to today's abductionists, could be divided into two broad groups: the skeptics and the "true believers." The sceptics regarded the whole subject of transvection as a mental illusion that gave a person the *sensation* of being lifted bodily by devils and taken through the sky. There was no need or precedent for complex psychological theories as such illusions were generally attributed to dark Satanic forces. In fact, it was heretical to think otherwise: "The act of riding abroad may be merely illusory, since the devil has extraordinary power over the minds of those who have given themselves up to him, so that what they do in pure imagination, they believe they have actually and really done in the body."

Some researchers have speculated on the possibility that some, if not all, abductions and encounters occur in altered states of consciousness and not in the physical world as we know it.

Source: *Malleus Maleficarum: The Classic Study of Witchcraft*, Part I, Question I, 7, trans. Montague Summers (London: Bracken Books, 1996).

172.

1528, Utrecht, Netherland: Yellow object in the sky

"Cruel and strange observation" of a yellow object in the sky, flying over during the siege of the city. The inhabitants of Utrecht panicked, while attackers took it as a sign of impending success. Lycosthenes writes: "At the time the city of Utrecht was heavily besieged, a terrible sign was seen in the sky which threw the town inhabitants into dismay and the enemies into the hope of capturing the town. For a sign in the sky, resembling a cross of a yellowish color (and of terrible aspect) appeared over the town. And because it was the symbol of Burgundy, they believed on both sides that the town would shortly belong to the Burgundians."

Source: *Terribile visione in cielo durante l'assendio di Utrecht* (1528), and Lycosthenes, *Prodigiorum ac ostentorum chronicon*, 536.

173.

9 October 1528. Westrie, N. Germany: Horrible object

A bizarre sighting was chronicled by Pierre Boaistuau in his *Histoires Prodigieuses*. Ambroise Paré describes a blood-red 'comet' that appeared over Westrie. It so terrified the populace that some reportedly died of panic and others became ill. The 'comet' emerged from the east and was seen for an hour and a quarter, disappearing finally towards midday – which implies, by the way, that it could not have been a comet unless by "midday" was meant "the southern direction."

At the top of the object people described an arm that held a great sword, the blade pointing downwards. There were three stars towards the tip of the sword, the one right on the end being the brightest. On both sides of the 'comet' were a great number of axes, blades and bloody swords and repulsive, bearded human faces.

Source: Ambroise Paré and all subsequent authors drew from a booklet by Peter Creutzer: *Auslegung Peter Creutzers, etwan des weytberhümbten Astrologi M. Jo. Liechtenbegers (sic) discipels über den erschrecklichen Cometen... erschynen am xi. Tag Weynmonats des MCCCCxxvii. Jars*...(1527).

PRODIGIEVSES. 68

QVelque temps apres que ceſte prodigieuſe planette fut apparue, toutes les parties de l'Europe furēt preſque baignées de ſang humain, tāt de l'incurſion des Turcs q̃ des autres plaies que receut l'Italie par le ſeigneur de Bourbon, lors qu'il miſt Rome à ſac, & que luy meſme y laiſſa la vie, Petrus Creuſerus & Ioannes Liechtber excellens aſtrologiens interpreterent par écrit la ſignification

Planette hideuſe qui apparut l'an que Bourbon miſt Rome à ſac.

Planette interpretée.

Fig.13: "Horrible object" seen in Westrie

174.

August 1533, Peru: Another mysterious "non-comet"

Garcilaso de la Vega, the Incan, writes in Chapter 23 of his work *Historia General del Perú*, that Tupac Huallpa's fear of his own death was exacerbated by the sighting of a great greenish black 'comet' in August 1533. It was an unusual comet, "a little narrower than the body of a man and longer than a pike" (spear-headed medieval weapon), and had been seen by many witnesses on several occasions at night. This made Huallpa particularly depressed because a similar object had been observed a few days before the death of his father, Huayna Capac. This comet was evidently not of the ordinary kind, at least in the opinion of Huallpa, who was accustomed to heavenly phenomena. Besides, comets are not "greenish-black!"

Tupac Huallpa was executed on 29 August 1533.

Source: Inca Gracilaso de la Vega, *Historia General del Perú* (1617), Book I, Chapter XXXIV.

175.

31 May 1536, Monte Stella, Brescia, Lombardy, Italy
Apparition, with a message ⓘ

Antonio de' Antoni, a poor deaf-mute shepherd of Gardone Val Trompia, was reciting the rosary while his flock was grazing. Suddenly there came before his eyes a light more intense than the Sun, in the middle of which the Virgin appeared with Jesus in her arms. Dressed in a simple way, she wanted the shepherds to build a temple there.

Antonio went off and described the event to everyone he knew, and when he returned the site was "illuminated by the beams of an overhanging star." Pope Paul III gave the place a consecrated status, and construction was completed in 1539.

Source: Marino Gamba, *Apparizioni mariane nel corso di due millenni* (Udine: Ediz. Il Segno, 1999).

176.

1537, near Florence, Italy
Benvenuto Cellini's "enormous splendor"

In his autobiography well-known artist Benvenuto Cellini relates the following:

Fig. 14: Benvenuto Cellini

"On horseback, we were coming back from Rome. Suddenly people cried 'Oh God what is that great thing we see over Florence.' It was a great object of fire, twinkling and emitting enormous splendor..."

Source: Benvenuto Cellini, *Vita*, 1558-1566, Book I, 89.

177. 

16 January 1538, Franconia, Thüringen, Germany
Disk, melted metal

A large "star" was seen in the sky and came down, emitting balls of fire that melted metal objects. Scholar Simon Goulart lists the case in his chronicles as follows:

"It was seen in Franconia, between Bamberg and Thuringia Forest, a star of marvelous magnitude, which came lower gradually, and became a great white circle from which whirlwinds and clumps of fire emerged. When they fell onto the earth they melted the tips of spears, irons and horses'bits without hurting either men or buildings."

Source: Simon Goulart, *Trésors d'Histoires Admirables et Mémorables de notre Temps* (1600). Genève: P. Marceau (1610), 53-54. Jobus Fincelius, *Wunderzeiche, Warhafftige Beschreybung und gründlich verzeichnuss schröcklicher Wunderzeichen und Geschichten* (Jhena: Rödinger, 1556); Lycosthenes, op. cit., 563.

178.

15 May 1544, Nay, Béarn, France: Crashing object

An object shaped like a fiery sword (variously described as "three fireballs") hovered above the town, then fell and crushed a house with a frightening noise.

Source: Pierre Boaistuau, *Histoires Prodigieuses* (1560), vol. II, 148.

179.

1546, Caranza near La Spezia, Italy: Disk changes color

A manuscript by the chronicler, Father Antonio Cesena, found in the public library at La Spezia, tells of farmers reporting "a strange disk, changing from yellow to red, with red fireballs shining beneath it". It was seen in two separate areas including the small village of Caranza, near Passo del Bocco "from time to time."

Cesena interpreted the sighting as a portent of the death later that year of Count Luigi Fieschi, the governor of Varese Ligure.

Source: Antonio Cesena, *Relatione dell'origine e sucessi della terra di Varese* (1558). The original manuscript is lost but a copy made in 1683 is still held by the Bibioteca della Società Economica di Chiavari (ms. Z VI 29).

180.

24 April 1547, Halberstadt, Saxony, Germany: Black sphere

A black ball-shaped object was seen, apparently "emerging from the middle of the moon" and flying fast towards the North.

Source: Simon Goulart, *Trésors d'Histoires Admirables et Mémorables de notre Temps* (1600) ; Jobus Fincelius, *Wunderzeiche, Warhafftige Beschreybung und gründlich verzeichnuss schröcklicher Wunderzeichen und Geschichten.* (Jhena: Rödinger, 1556); Lycosthenes, op. cit., 595.

181.

13 November 1547, Near Rome, Italy
Strange objects fly over

A rod and a cross appeared in the sky at 3 P.M., with a bird-like object above them. The weather was clear and the sky was calm. The objects were seen for three days.

The event is depicted in a German broadsheet in the Johann Jacob Wick's collection, held by the Zürich Zentralbibliothek.

Source: Jobus Fincelius, *Wunderzeiche, Warhafftige Beschreybung und gründlich verzeichnuss schröcklicher Wunderzeichen und Geschichten* (Jhena: Rödinger, 1556); *Erschreckliche unerhorte warhafftige Gesichten so gesehen ist zu Rhom...* (Strassburg: Jakob Frölich, 1547), ZB PAS II 12/29.

182.

15 December 1547, Hamburg, Germany
Heat-generating globe

"The sailors of Hamburg saw in the air, at midnight, a glistening globe fiery like the Sun, rolling towards the northern part. Its rays were so hot that passengers could not remain inside the ships, but were forced to hide and take cover, thinking that their vessels were about to burn."

Source: Simon Goulart, *Trésors d'Histoires Admirables et Mémorables de notre Temps* (1600); Fincelius, Jobus, op. cit.; Lycosthenes, op. cit., 595.

183. 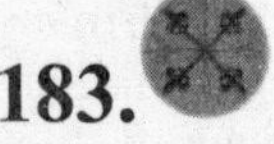

28 June 1548, Oettingen, Bavaria
Flying vehicles, red flames

The sky became darker and about twenty flying "vehicles" were seen coming and going above the houses, along with red flames. The witness says he saw the phenomenon on two occasions: on 28 June and on 26 July 1548.

Source: Bruno Weber, *Wunderzeichen und winkeldrucker*, 1543-1586, Urs Graf Verlag (Zürich: Dietikon, 1972), 93.

184. 

19 June 1550, near Trebnitz, Saxony, Germany
Bloody rain, split sun

The people of Saxonia, near Wittemberg, beheld a strange sight, according to Boaistuau in *Histoires Prodigieuses*. A great cross appeared in the sky, surrounded by two large armies that made a lot of noise while they fought. Blood fell to the ground like rain and the sun split in two, one piece of which seemed to drop to the earth.

Boaistuau drew this story from Lycosthenes, who in turn took it from Fincelius. Lycosthenes made an error in the date and location, but we were fortunate in finding a contemporary broadsheet from 1550 that depicts and describes the phenomenon just as Fincelius wrote.

Source: Jobus Fincelius, op. cit.; *Ein new streydtbars / grausam / glaubhafftigs wonderzeychen / so dieses Fünfftzig (...) Junij / am himel gesehen worden ist* (Nürnberg: Stephan Hamer, 1550?), GNM Nürnbergt. HB 2795/1204.

185.

15 October 1550, Biubiu River, Chile
The Lady from the Comet

Pedro de Valdivia (1500-1554), a conquistador who went to America to seek fame and fortune, fought for Francisco Pizarro in the Battle of Salinas, and later headed the conquest of Chile in 1540, founding Santiago del Nuevo Extremo (nowadays Santiago) the following year, Concepción in 1550 and the city of Valdivia in 1552, before his death at the hands of the Araucan Indians.

Valdivia left very few written documents, but one of these, an "instruction" addressed to his representatives at the Court, dated October 15th 1550, mentions two mysterious figures that appeared to the Mapuche Indians shortly before an attack. The beings, a beautiful woman and an old man on a horse, both dressed in white, had come to warn the Indians that they would perish if they tried to retaliate. When the first visitor, the woman, had disappeared, the devil himself intervened to reiterate the message!

Valdivia made two reports, one of which, the most complete of the two, he sent to the king Charles V on the same day. This is what the conquistador wrote:

"And it seems our God wants to use His immortality for his divine cult to be honored in it and for the devil to come out from where he has been worshipped for so long; thus, according to the native Indians, the day they came upon this fort of ours, at the same time as those that rode on horseback assaulted them, there fell in the middle of the squadrons an old man on a white horse, and he told them:

"'Flee all of you, these Christians will slaughter you,' and their fright was so great that they began to flee. They [the Indians] told more: that three days before this, [when the Indians were] passing the Biubiu river to overcome us, a comet fell among them, on Saturday at midday, which was seen by many Christians at our fort as it traveled with greater brightness than other comets, and from which, once fallen, a beautiful woman came out, also dressed in white, who told them: 'Serve the Christians, don't go against them because they are very brave and will kill you all.'

And when she went, the devil came, their chief, and he told them to gather a large multitude of people, and that he would come with them, because, on seeing so many of us together, we would drop dead with fear; and thus they proceeded with their journey."

Source: Pedro de Valdivia. *Cartas* (1552) in *Crónicas del Reino de Chile* (Madrid: Atlas, 1960).

186.

Circa 1551, Morbecque, France ⓘ
Sex with the Devil: A flying contactee condemned to die

A woman named Jacquemine Deickens, the wife of Hirache, was accused of sorcery after a being (thought to be the Devil) appeared to her as she was milking the cows.

She was said to have known him carnally, and received a mark on her back, below the left shoulder blade, which proved her guilt. Every three or four weeks she flew out of her house to meet with other witches at the crossroads in front of the house of Mr. Pierre Depours, there to dance and partake of a feast with many demons.

She was tried and executed in 1557.

Source: Claude Seignolle, *Les Evangiles du Diable selon la croyance populaire* (Paris: Maisonneuve & Larose, 1964), 245.

187.

1551, near Waldstadt, Germany
Woman taken up by the devil and dropped from the sky

A woman who had uttered some blasphemies during a drinking party was taken up in the air by the devil "in the presence of everyone." The witnesses rushed out to watch where she was carried. They saw her hovering up in the sky outside the village, after which she dropped and was found dead in the middle of a field.

Source: Dr. Jean Wier, *Histoires, disputes et discours des illusions et impostures des diables, des magiciens, infâmes, sorciers et empoisonneurs, le tout compris en 5 livres*. Translated from the Latin, ca. 1577.

188.

3 January 1551, Lisbon, Portugal: Flying red cylinders ⓘ

Red cylinders in the sky are rumored to have terrified the population. We note that another reference speaks of a "fiery meteor" seen on 28 January at the time of a great earthquake. In the absence of a precise source we cannot say if confusion exists between the two reports.

Source: H. Wilkins, *Flying Saucers on the Attack* (New York: Ace Books, 1967), 183. We have found no original reference for this event.

189.

March 1551, Magdeburg, Germany A phenomenon scares an Emperor ⓘ

Three suns were seen in the sky. Emperor Charles Quint decides to halt the siege of the city. In the absence of more information, these may have been parhelia. Historians report that "The emperor Charles V laid siege to it; but was prevailed upon to withdraw his army for a great sum of money…"

Source: Rev. Alban Butler, "Life of Saint Norbert" in *The Lives or the Fathers, Martyrs and Other Principal Saints*, vol. 6 (New York: D. & J. Sadlier, & Company, 1864).

190. 

13 January 1553, Porco, Peru
An unexplained "comet" is taken as an omen

Nicolás de Martínez Arzanz y Vela, author of *Historia de la Villa Imperial de Potosí* (1705), writes in Chapter 2:

"As Don Sebastian and his allies were getting the people and the weapons ready to carry out the revolt in this Town, and Captain Francisco Hernández Girón in his division of Chaqui…was also preparing for his…there appeared in the sky, above Porco, three suns and two moons in the middle of a great ring, and within it two blue and red arches. On the same day there appeared above this rich Imperial Hill and Town another two arches of various colors and a notable comet as red as blood. Enrico Martínez, His Majesty's Cosmologist in the Peruvian Kingdom, says the following (…)

"'On Friday January 13th 1553, fifty two days before General Hinojosa was killed, at seven o'clock in the morning there appeared in the sky, in Porco, the large ring that passes through the middle of the natural sun and through the other Suns and Moons; it was stretching towards the west, and was entirely white, a span in thickness; this ring seemed to be half a league in diameter. The natural sun was a little red, almost like blood, and the two at its side very red, just like blood, so much so that the brightness and fire caused those who saw it to avert their eyes. The two Moons at the front were like white Moons, slightly red; the two Arches that appeared were blue and red, as they usually appear; the small arch was wider than the blue one.'"

So far it seems we are dealing with an unusual, but perfectly natural, atmospheric phenomenon. More interesting is the so-called "comet" that was seen in connection with it:

"The comet that appeared outside the ring was very fiery and blood-red, with a formidable curly head and the tail was similar. This comet was seen in Porco and in all the surrounding areas. The comet was seen for seven days at dawn over the rich land of Potosí, with another two arches, one very white which looked like polished silver, and the other was above this one and was almost blood-red and was as bright as fire…"

This description does not match a cometary object, and at 7 A.M. this could hardly have been an aurora borealis. The social reaction to the phenomenon is even more interesting: "The astonished Indians, covering their faces and spitting in the air, cried: *Aucca, Aucca, maiccan Apuhuañuncca.* These words signified some bad event, abominable action or frightful ruin, which is all conveyed by the word *Aucca*, a name they also give to any visible or invisible enemy (…)

The death of the great Apuc followed that of Francisco Hernández Girón, who, with the income from his villages, was to appear like the Sun three times in the bloody victories he had; and twice like the Moon, in the waning of his fortune with his imprisonment and death."

The complex phenomena described here could only be accounted for by a combination of very unusual atmospheric circumstances.

191.

1 February 1554, Salon de Provence, France
Sighting report by Nostradamus

Dr. Nostradamus and a thousand other witnesses have seen a big "bright burning rod" that changed its flight path, between 7 and 8 P.M. It was in the sky for two hours, displaying a swinging motion. Here is the actual report:

TO THE ILLUSTRIOUS, HIGH-BORN AND
ALMIGHTY LORD CLAUDE,
DUKE OF TENDE,

Knight of the order of Regents and of the King and Honorary Citizen of Provence, Michael de Nostre Dame, his humble and obedient Servant bids greeting and good fortune.

"Gracious Lord,

"According to reports received, on the first day of February in this year of 1554, a most terrifying and horrible sight was seen on [....] towards evening, apparently between 7 and 8, which I am told was seen as far as Marseille. Then it was also seen at nearby St. Chamas by the sea, such that near the moon (which at that time was near its first quarter) a great fire did come from the east and make its way towards the west. This fire, being very great, did by all accounts look like a great burning staff or torch, gave out from itself a wondrous brightness, and flames did spurt from it like a glowing iron being worked by a smith. And such fire did sparkle greatly, glowing aloft like silver over an immense distance like Jacob's road in the sky, known as the 'Galaxy' [i.e., the Milky Way], and raced overhead very fast like an arrow with a great roaring and crackling thunderous din' and as though it were being blown hither and thither by the [raging and roaring?] of a mighty wind.

"Then slowly, over the course of 20 minutes, it turned until we saw it passing over the region of Arles via that we call the 'stony road' which the poets do call immensum fragorem [i.e. the Crau]. Then it turned towards the south, high over the sea, and the fiery stream that it created retained its fiery color for a long time, and cast fiery sparks all around it, like rain falling from heaven.

"This sight was much more terrifying than human tongue could say or describe. And I thought that it might have come from a mountain known as a volcano. But on the 14th of this month I was called to [Bry?], where I asked diligently of many people whether they also had seen it, but

Ein Erschre[illegible]lich vnd Wunderbarlich Zeychen/so am Sambstag für Judica den zehenden tag Martij zwischen siben vnnd acht vhrn in der Stadt Schalon in Franckreych / von vielen leuten gesehen worden.

Dem durchleuchtigen Hochgebornen vnd Großmechtigen Herrn Hern Clodio/ Grauen von Tende /Rittern des ordens Regenten/ vnd des Königes in der Prouintz Stathaltern /Entbeutet Michael De Nostre [illegible]me/sein vnterthenigér gehorsamer Diener/ seinen grus / vnd alle wolfart.

GNediger Herr /nach beschehener musterung/ so gewesen am ersten tag Februarij dises 1554.jars / ist allhie zu Schalon ein sehr erschrecklich/ vnd entsetzlichs gesicht am 10 Martij vngeuehrlich zwischen 7 vnd 8 vhrn gegen dem abendt gesehen worden/ welchs/meins erachtens/biß gen Marsiliam geraicht hat/ Dann es auch zu Sanct Chamas bey nahend dem Meer gesehen worden/ also das bey dem Monn/ welcher dieselbe zeyt nahend dem ersten viertheyl gewesen/ ein groß Fewer vom Auffgang kommen ist/ vnd sich gegen dem Nidergang hat gewendet. Dises Fewer sehr groß/ vnd aller massen wie ein grosse brinnende stangen oder fackel gestalt/ hat einen wunderbarlichen schein von sich geben/ vnd dauon seyen die flammen gesprungen/ wie von einem glüenden Eysen/ das der Schmid arbeytet/ Vnd hat solch Fewer vil funcken in die höhe/ glintzend wie das Silber / vnmessiger leng auffgeworffen/ Gleich der Jacob [illegible], den am himmel/ Galaxia genant/ sehr geschwind wie ein pfeyl/ mit einem grossen rauschen vnd prasseln / welches die Poeten immensum fragorem nennen/ vnd gleicher gestalt wie die bletter vnd bawme von einem gewaltigen Winde hin vnd wider getrieben/ für vberglauffen. Hat lang/ vnd biß in 20 Minuten gewehret/ biß wir es gesehen/ gehend vber die gegend Arla / sunsten den steinigen weg genant/ da hat es sich gewendet gegen Mittag/ hoch ins Meer/ vnd der Fewrige streime den es machet / behielt lange zeyt seine Fewrigefarbe / vnd warff geringes vmb sich die Fewrigen funcken/ wie der plitz/ so vom Himmel fellet. Dieses gesicht ist viel erschrecklicher gewesen/ dann das es Menschliche zunge möcht außsprechen / oder beschrieben werden/ Vnd mich beduncket das es von einem Berge jesseyt Air/ sanct Wilkum genant herkäme/ Aber den 14 dises Monats / bin ich nach Air erfordert worden / daselbst bey vielen Bürgern fleyssige nachfrag gehabt/ ob solch gesicht auch gesehen/ aber alda nichts erfaren/ Sonder von dannen erst bey zweyen meylen ist erschinen vnd der Herr derselben ort hat es selb gesehen/ vnd begert das ich sein geuatter alda sein/ solchs auch sehen vnd außlegen möchte. Zwen tag nach dem man seiner acht genommen/ ist der Balbirer von Sanct Chamas zu mir kommen / vnnd angezeygt/ das ers vnd andere Bürger daselbst auch gesehen / vnd es gestalt gewesen in form eines halben bogens/ vnd gewehret hat biß zu dem Spanischen Meer/ Vnd wo es nider/ wie es in der höhe gewesen/ hette es alles vorprent vnd zu puluer verzeret/ do es für vber gangen. Dann am Himmel vnd in der weyten/ ist es bey einem Pisanischen lauff oder stadio breyt gewesen/ dauon für vnd für das Fewer gesprühet vnd gefallen. Vnd souil ich dauon judiciren kan/ nach gelegenheyt diser gegent vnd Climatis/ ist es new vnd sehr frembd zuhören / vnd were viel besser/ das es nicht erschinen. Dann dises gesicht oder Comet ein gewisse anzeygung gibt/ das dieser gegent der Prouentz vnd andern Flecken am Meer / ein vnuerhoffter vnd vnuersehener vnfall begegnen soll/ durch Krieg/ Fewer/ Hunger/ Pestilentz oder andere frembde Kranckheyten/ oder sonst von frembden Nationen beschwert vnd vnderdruckt werden. Dises Zeychen haben mer dann tausent menschen gesehen/ vnd bin dasselbig zuuerzeychnen gebetten worden/ vnd Ewer Herrligkeyt zu zuschreyben / so es/ meins erachtens selbs gesehen/ vnd gehört/ wie es beschehen ist/ Vnd bitte Jhesum vnsern Herrn das er E. H. Ehr / lang leben vnd wolfart reichlichen mehren vnd erweytern wölle. Datum in Franckreych zu Schalon in der Prouentz 19 Martij / Anno 1554.

Ewer Herrligkeyt.

Vntertheniger vnnd gehorsamer Diener.

Michael De Nostre Dames

Aus Frantzösischer Sprach Transferirt/ vnd gedruckt zu Nürnberg bey M. Joachim Heller.

Fig. 15: Object seen in Salon de Provence

not all of them had experienced it. But it did appear only seven miles from there, and the Lord of that same place had seen it, and desired that I should be his witness that he had seen and wished to record it. Two days after the fire had been seen, the Prefect of St. Chamas came to me and indicated that he and other townspeople had seen the same thing, and that it had taken the shape of a half-rainbow stretching as far as the Spanish Main. And if it had been low down rather than high up, it would have burnt up everything and reduced it to ashes as it went by.

"They also said that its breadth in the sky was around a [Pisan?] running distance or stadium [about 200 yards], from which the fire sprayed and fell. And so far as I can judge in the circumstances, it is [...] very strange to hear, and it would be much better had it not appeared. For this apparition or comet gives certain indication that this Ruler of Provence and other stretches by the sea shall encounter unexpected and unforeseen calamity through war, fire, famine, pestilence or other strange diseases, or otherwise shall be attacked and subjugated by foreign nations.

"This omen was seen by more than a thousand people, and I have been bidden to confirm this and write to your Eminence about it, insofar as I have in my own estimation seen and heard how it happened. And I pray Jesus Our Lord that he may grant Your High Eminence long life, and that he may richly multiply and extend your good fortune.

"Given in France, at Salon-de-Provence, this 19th of March in the year 1554. Your Eminence's most humble and obedient servant, Michael de Nostre Dame."

Source: Translated from the French and printed in Nuremberg by M. Joachim Heller (with the illustration given here), cited in Mathias Miles, *Siebenbürgischer Würgengel* (1670). The above translation is from: www.propheties.it/Nostradamus/inedites/inedites3.htm.

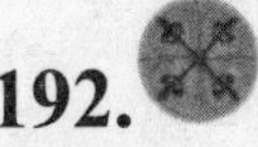

192.

13 June 1554, Iena, Germany
Spheres and disks

A large number of spheres and disks flew over the city of Iena. They had sudden variations of speed and turned to a red color as they flew north.

Source: Mathias Miles, *Siebenbürgischer Würgenengel* (Hermannstadt, 1670).

193.

5 March 1555, Buendía, Cuenca, Spain ⓘ
Hovering cross

On Tuesday March 5th 1555, in Buendía, in the Spanish province of Cuenca, many people saw an enormous object in the shape of a cross in the sky. It was stationary, floating next to a new calvary (a life-size representation of a crucifix on raised ground) erected by the Brotherhood of the True Cross.

Seven other women saw it there. The Inquisition of Cuenca dispatched somebody to investigate the occurrence, but nineteen people described the sighting to notaries.

We include this case because of its interest as an unidentified aerial phenomenon, though the religious significance of the case should not be ignored.

Source: William A. Christian, *Local Religion in Sixteenth-Century Spain* (Princeton University Press, 1989), 186-7.

194.

14 April 1561, Nuremberg, Germany
Vertical cylinders

At sunrise many spheres and disks, red, blue and black

Fig. 16: The sighting at Nuremberg

were seen to come out of two vertical cylinders. They flew across the face of the Sun in an apparent "aerial fight". In a contemporary engraving some of the spheres appear to have landed on a hill to the right of the city, where much smoke is rising, while an elongated shape resembling a great black spear is seen in a horizontal position.

"Beyond balls of a red color, bluish or black, and circular disks, two large pipes were seen…within which small and big pipes were found three balls, also four and more.

"All these elements started fighting against one another." The fight seems to have lasted about an hour, then "as mentioned above, from the Sun and the sky, it fell onto the earth as if everything was burning, and with great smoke everything got consumed."

Source: This pamphlet is preserved in the Wickiana collection of the Central Library in Zurich: *Erscheinung am Himmel über Nürnberg am 14. April 1561.* Zürich Zentralbibliothek [ZB PAS II 12:60].

195.

1 March 1564, near Brussels, Belgium
Aerial bombardment

In Gilbert's *Annalen* for 1806 is an account of a fearful phenomenon seen between Mechel and Brussels. The sky was clear at first, but about 9 o'clock became fiery, throwing down a reflection upon the earth so that everything became yellowish. In the meantime there appeared in the sky figures of three men in royal robes with crowns upon their heads, remaining visible for nearly three-fourths of an hour, when they gradually drew near together and in the course of another 15 minutes disappeared. Then frightful stones fell, large and small, some of which were five or six pounds in weight. So far as known none of this material has found its way into collections.

Source: George P. Merrill and William F. Foshag, *Minerals from earth and sky,* Volume 3 (New York: Smithsonian Institution Series, Inc., 1938), 12-13.

196.

7 August 1566, Basel, Switzerland: Aerial Combat

Many black spheres in apparent aerial combat. Several turn red and disintegrate.
"At sunrise were seen in the air numerous large black balls that flew at high speed towards the Sun, then turned around, hitting one another as if they were fighting. Many of them became red and fiery, and later they consumed themselves and were extinguished."

A contemporary engraving shows the spheres in the sky above the "Munster" cathedral with the Antistitium.

Fig. 17: The sighting at Basel

Source: Samuel Coccius (Koch), *Seltzame gestalt so in diesem M.D.LXVI. Jar, / gegen auffgang und nidergang, under dreyen malen am Himmel / ist gesehen worden, zu Basel auff den xxvij. und xxviij. Höwmonat / und volgends auf den vij Augsten* (Basel: Samuel Apiarium, 1566), ZB PAS II 6/5.

197.

7 April 1567, Basel, Switzerland: A black sphere

A black sphere appears in the sky and covers the face of the Sun. It was seen all day long.

Source: Samuel Coccius (Koch), *Wunderbare aber Warhaffte Gesicht vii erscheinung in Wolcken des Himmels auff den andern tag Menens in diesem lauffenden acht und sechtzigsten Jar. [...] Sampt angehenckter geschicht / inn dem vergangnen LXVII. Jar auff den vii. tag / Aprellens ausz dem lufft geoffenbaret / bende vorhin niemalen / aber jetz. under zur warnung im truck auszgangen* (Basel: Samuel Apiarius, 1558), Ms. F. 18.

198.

26 September 1568, Tournai, Belgium ⓘ
Great circles of fire

“Marvelous signs in the sky were seen from the seventh to the twelfth hour in the evening. At first, great circles of fire were seen with rays emerging like suns dragging water (?), afterwards a black cloud was seen and after that, great lights appeared. That being gone, men on horses were seen fighting each other and it seemed as if several musketeers were skirmishing against one another. Sparkles of fire were seen which illuminated the ground with a terrible shine.”

A possible interpretation of this case would invoke an aurora borealis, but not enough is precisely known to make that determination, so we keep the event with reservation.

Source: Alex Pinchart, *Mémoires de Nicolas Soldoyer*, as published in *Mémoires de Pasquier de la Barre et de Nicolas Soldoyer, pour servir à l’histoire de Tournai* (Brussels, 1865), vol. II, 304.

199.

20 July 1571, Prague, Czechoslovakia
Mysterious round “chariot”

“About midnight there was a great wind over Prague that made such a rumbling noise that it sounded like an earthquake. The people woke up with a start and hurried to their windows. Looking towards the cattle fair (today Charles Square) they saw a marching army coming along Spalena Street.

“The soldiers held their weapons in their hands and witnesses found their appearance somewhat unnerving. Behind the soldiers came something resembling a large round ‘chariot’ drawn by oxen.

"The object, which made a loud noise, was apparently made of metal *and had no wheels*. Eight large human figures marched behind the vehicle. They looked frightful because they had no faces but wore enormous spurs on their feet, adding to the noise.

"Once they had crossed the square, a great fire appeared on the ground in front of the Church of the Sacred Heart. On one side of the fire there were a large number of boxes, and on the other there were barrels. These barrels looked as if they could have been used to transport gunpowder. The big chariot arrived near the fire and all the boxes and the barrels were thrown on it. Then again a frightful wind arose at the same time as a kind of rain of fire and all this horrifying vision disappeared. However, a luminous object could be made out in the air, a circle of fire that persisted until dawn. That year there was a great famine and many people died."

Source: J. Beckovský, *Poselkyně starých příběhův českých. Díl druhý. Od roku 1526-1715. Sepsal Jan Beckovský, kněz řádu Křižovníků s čevenou hvězdou. K vydání upravil Dr. Antonín Rezek, docent rakouského dějepisu na universitě pražské.* V Praze, nákladem Dědictví sv. Prokopa, písmem knihtiskárny B. Stýblovy 1879-1880.

200.

20 September 1571, Lepanto, Italy
A flaming column guides the fleet

On the night of September 20th 1571 a fiery object was seen over Lepanto. The official historian of the papal fleet of Rome, Alberto Guglielmotti, recorded the event in a report based on statements given by two witnesses, Sereno and Caracciolo. In his summary, Father Guglielmotti wrote that the "sign in the sky…was considered by everyone to be a miracle."

"It was a clear, starry night with a cool wind coming from the north. Suddenly, a colossal fire in the shape of a shining, flaming column was seen by everyone to cross the sky over a long period of time, filling all the witnesses with great admiration... All the witnesses regarded this as a good omen and sensed they were on the verge of a great victory. They believed this column of fire was showing them the way, guiding the Christian fleet in the sea in the same way the people of Israel were guided across the desert in biblical times.

On October 7th, Selim II, the Sultan of Turkey, was defeated at sea by the Christian fleet, just off the shore of Lepanto.

Source: Padre Alberto Guglielmotti, *La Guerra dei pirati e la marina pontificia dal 1500 al 1560* (Florence: Le Monnier, 1876).

201.

16 February 1572, Constantinople, Turkey ⓘ
Crosses in the air

The people of Constantinople were amazed to see cross-shaped objects flying above their city.

It was said that for the consolation of the miserable Christians held captive in Constantinople, and for the confusion of the Turks and Jews living in that city, God placed three crosses in the sky above three Turkish mosques (Piali Baja, Capassi and Saint Sophia). These appeared three days in a row, from Thursday to Saturday.

The crosses were seen by everyone in the area. They floated in the air, high above the roofs of the mosques. Their color changed continuously. The Christians rejoiced and gazed at the objects in wonder, while among the Turks there was only confusion and uproar.

Finally, the Turks met with the Jews and asked them what they thought was happening. The Jews replied that the Christians were great magicians, and were trying to frighten them with their magic arts. The Turkish soldiers decided to retaliate more violently than ever, crushing the

Christian troops in armed combat. They even started to shoot at the crosses. However, whenever a missile was about to hit them, the crosses vanished momentarily, only to reappear immediately afterwards.

Source: W*arhaffte Zeitung vnd beschreibung der Stadt Constantinopel dreyer Creutz gesicht. Auff S. Sophia / Patriarcha / vnd Andrea Kirchen gesehen worden seind / Dreytag auff jeder besonder / vnd allmal von einer Kirchen auff die ander sich erzeigt / geschehen den 16. Februarij des 72. jars. Auch ist warhafftig vnden hernach gesetztes schreiben von Constantinopel aus / von einem Ritter Grio Malluj genandt / Bebstlicher Heiligkeit für warhafftige zeitung zugeschrieben worden / den 10. Martij. im 72. Jar* (Augsburg: Hans Rogel, 1572), SUB Göttingen 4° H TURC.712.

202.

15 November 1572, Romerswil, Switzerland
A farmer's abduction

Hans Buchmann, a 50-year-old Swiss farmer from Romerswil, had gone to Sempach, a nearby village. When he failed to return, his wife sent out their two sons to look for him. The boys found their father's hat, coat and gloves. They also found his saber and its sheath, lying on the path. This frightened them and they suspected that Klaus Buchmann, their father's cousin, who had for years been an enemy to the family, could have murdered him. The authorities had Buchmann's property searched, in vain.

Four weeks later, the family received news about Hans Buchmann's whereabouts: he was in Milan! On 2 February 1573, two and a half months after he disappeared, he came back. His wife and children were astonished to see that he did not have a single hair on his head, his face or his chin. His face was so swollen that they didn't recognize him at first. When the authorities learnt that the man had returned they interrogated him, as so much trouble had been caused to cousin Klaus. The town chronicler, Renward Cysat (1545-1614), was present at the interrogation.

Buchmann explained that on the day he disappeared he carried money to pay Hans Schürmann, the owner of the Romerswil inn, to whom he owed sixteen florins. Schürmann was not at home so he decided to go to Sempach on other business matters. There he stayed until dawn, drinking something but very little, and then set off for home. As he was passing through the forest he suddenly heard a strange noise. At first he thought it was the buzz of a swarm of bees, but then he realized it sounded more like music. He felt afraid, and was no longer sure where he was nor what was happening. He unsheathed his sword and swiped at the air around him, losing his hat, gloves and coat in the process. Before losing consciousness he could feel that he was being lifted up into the air. He was taken to another country. He was disoriented and confused, with no idea where he was. He felt pain and swellings in his face and around his head.

Two weeks after his abduction he found himself in Milan, with no idea how he had got there. He was weak because he had not eaten or drunk anything in days, but he was determined to find his way home. Hans Buchmann neither knew the city nor spoke the language, and had no way of communicating his situation to anyone until he came across a guard of German origin who took pity on him.

Source: Cysat, who knew Buchmann personally, was sure the man had been kidnapped by a fairy of some kind. In his *Collectanea Chronica und Denkwürdige Sachen pro Chronica Lucernensi et Helvetiae*, where the story is described in detail, he mentions several other cases of fairy abduction.

203.

29 January 1574, Japan, exact location unknown ⓘ
Flying umbrella

A large object shaped like an umbrella flew over, illuminating the sky and the ground. It was seen twice.

Source: *Brothers* I, 1. We lack any earlier reference.

204.

21 December 1576, Mount Kasuga, Japan ⓘ
Wheel in the sky

A wheel-shaped object flew for an hour over the castle on Mount Kasuga. No historical source is given for this tantalizing case, which would deserve further study by researchers familiar with Japanese chronicles.

Source: *Brothers* magazine, again with no reference.

205.

5 December 1577, near Tübingen, Germany
Flying black hats

"Numerous black clouds appeared around the Sun, similar to those we see during major storms; shortly thereafter, other clouds of blood and fire emerged from the Sun, and yet others yellow as saffron.

"From these clouds came luminous effects like big, high and broad hats, and the earth itself appeared yellow, bloody, and covered with high and broad hats that took various colors such as red, blue, green, and most of them black.

"Everyone can easily understand the meaning of this miracle, and know that God wants men to repent and make penance. May the all-powerful God help all men to recognize Him. Amen".

Source: Pierre Boaistuau, *Histoires Prodigieuses* (2nd edition, 1594). Also described in a German broadsheet printed in early 1578: *Schröckliche Newe Zeitung / von dem Wunderzeichen / welches den kurtzverschinenen fünfften deß Christmonats / zu Alttorff inn dem Land Würtenberg ist gesehen worden* (Strassburg: Bernhard Jobin, 1578), ZB PAS II 15/1. This belongs to J.J. Wick's collection (Wickiana).

206.

21 December 1578, Geneva, Switzerland
Signs and prodigies

"Marvellous and terrifying discourse of the signs and prodigies that appeared over the city of Geneva the 21st day of December 1578" is the title of a brochure published by G. Stadius, mathematician of the Duke of Savoy and noted astronomer. It describes strange phenomena, including "a comet surmounted by a small cross."

Source: *Discours merveilleux et espouventable des signes et prodiges qui sont apparuz au ciel sur la ville de Genefve le XXI. jour de décembre mil V. cens LXXVIII* [par B. Du Coudre, avec réponse de G. Stadius] (Paris: J. Pinart, 1579), Bibliothèque Nationale de France, BN MP-3321.

207.

18 February 1579, Paris, France: Flying intruder

A "great and wondrous flying serpent or dragon" appeared, according to a leaflet ("*canard*") of the time.

It was seen by many over Paris from two o'clock in the afternoon until evening.

Source: A leaflet at the Library of Amiens entitled *Du serpent ou dragon volant, grand et merveilleux, apparu et veu par un chacun sur la ville de Paris, le mercredi XVIII febvrier 1579, depuis deux heures après midi, jusques au soir* (1579).

208.

7 February 1580, Straits of Magellan
Red, fiery flying shield

In *Viajes al Estrecho de Magallanes* by Pedro Sarmiento de Gamboa (ca.1530-1592) we read of this navigator's travels to the Strait of Magellan in 1579 and 1581. On Thursday,

February 7th 1580 at 1:00 A.M., he wrote that towards the south-southeast:

"We saw a round thing appear, red like fire, like a shield, that rose up on the air or on the wind. It became longer as it went over a mountain and, in the form of a lance high above the mount, its shape became like a half-moon between red and white in color."

The three simple shapes accompanying the text are a circle, an oval and a half-moon.

Sarmiento de Gamboa compares the object to an "adarga." This was a round, oval or heart-shaped shield used in the time of Don Quijote.

209.

1586, Grangemuir, Scotland: Healer abducted by fairies

A woman named Alison Pearson confessed that she had met with the "Good Neighbors," who had given her a salve that could cure every disease. She had seen a man clad in green, who was accompanied by many men and women making merry with good cheer and music, and she was carried away by them. She was tried and put to death in 1586, although she had treated the Archbishop of Saint Andrews, who stated he had received benefit from it.

Source: James Grant, *The Mysteries of all Nations: Rise and Progress of Superstition, laws against and trials of witches, ancient and modern delusions* (Edinburgh: Reid & Son, 1880), 517.

210.

1586, Beauvais sous Matha near Tors, France
Flying Hat-shaped object

A brown "hat" with horrible red colors was observed flying close to the steeple shortly before sunset.

Fig. 18: Agrippa d'Aubigné

The witnesses were the great French poet Théodore Agrippa d'Aubigné (1552-1630) and the Marquis de Tors.

"The Marquis, lord of that place, took his guest to a garden, shortly before sunset, and they saw a round cloud come down over the hamlet of Beauvais-sous-Matha, with a color that was horrid to see, for which one is forced to use a Latin word: *subfusca* (dark brown).

"This cloud resembled a hat with an ear in the middle, the color of the throat of an Indian rooster (…) This hat with its sinister sign came into the steeple and melted there."

Source: Agrippa d'Aubigné, *Histoire Universelle* (1626), III, iv, ch. 3.

211.

12 January 1589, Saint-Denis, France
Sky phenomena

A text published in French around 1599 makes the following report:

"We have seen at night two large clouds between Paris and St. Denis, which radiated great light, and they moved towards one another, joined and then separated again, and a large number of sagettes (arrows) and spears of fire came out, which lasted a long time in combat, then after having been fighting well, they retreated, then began to travel, and passed over the City of Paris, and went straight on southwards.

"Then on Friday 13th of the month of January, we have also seen in the Sky a great Crescent and one Star above it, like a Comet, which was bright all day, and people were amazed. Christians prayed for God to save us from the menace. Amen"

Source: Anon., *Signes merveilleux aparuz sur la ville & Chasteau de Bloys, en la presence du Roy: & l'assistance du peuple. Ensemble les signes & Comete aparuz pres Paris, le douziesme de Janvier, 1589 comme voyez par ce present portraict* (1599?), Bibliothèque Municipale de Blois, n° Inv.: LI 13.

212.

1590, Scotland, location unknown ⓘ
Tubular object

In 1590 Scottish peasants informed the shire reeve (the king's representative) that a large tubular object had been seen hovering over their town. It hung motionless in the sky for several minutes before it vanished.

Despite abundant references online and in print, an original source has not been located.

213.

1592, Pinner, Middlesex, England ⓘ
Transported away!

Farm worker Richard Burt was confronted by a being he described as "a large black cat" and was transported magically to Harrow.

Source: *A Most Wicked Work of a Wretched Witch* (1593).

214.

15 October 1595, Targoviste, Wallachia, Romania
Hovering object

When prince Michel the Brave besieged the city of Targoviste, the capital of Wallachia, temporarily occupied by Turks, "a large comet appeared" above the military camp and rested for two hours (according to an Italian report of the facts, redacted in Prague). After three days the Turks were defeated. No such comet is mentioned in astronomical records.

Source: Calin N. Turcu, *Enciclopedia observatiilor O.Z.N. din Romania (1517-1994)* (Bucharest: Ed. Emanuel, 1994), 3.

Epilogue to Part I-C

The end of the sixteenth century finds France devastated by fanaticism and a religious civil war between Catholics and Protestants, only resolved in 1598 by the Edict of Nantes, through which King Henry IV establishes for the first time the dual principles of freedom of conscience and freedom of religion. Spain and France are both exhausted, while England dominates the seas and becomes a great commercial and industrial power, extending its colonial empire to America with the rise of Virginia.

It is the end of the Elizabethan period. Shakespeare and Cervantes reign over literature.

Science now progresses in great strides: in the last decade Galileo publishes his observations on falling bodies and (in 1593) invents the thermometer. Botanical gardens are established at the University of Montpellier, and the first manuals of veterinary science appear.

When 1600 comes around, astronomers Tycho Brahe and Johannes Kepler are working together in Prague, while Dutch opticians have invented the telescope, Kircher has built the first magic lantern and William Gilbert has published the first scientific treatise on magnetism and electricity; The Western world has entered a new era.

ART I-D

Seventeenth-Century Chronology

Spurred on by strategic and scientific interest in navigation, astronomy underwent unprecedented growth during the seventeenth century. Experimental and theoretical publications flourished under the pen of Galileo, Huygens, Cassini, and numerous observers of the Moon and planets using the newly-invented telescopes. Similar progress revolutionized physics, mathematics and medicine, often in spite of the dictates of the Church.

This movement towards better understanding of nature and man's relationship to it, long repressed by religious ideology, found its expression in the "Invisible College" and culminated in the creation of the Royal Society in London in 1660, while Harvard College in the colony of Massachusetts was awarded its charter in 1650.

Similar forces were at play in Asia, where Chinese naturalist Chen Yuan-Lung published his treatise on "New Inventions," and in Japan where Seki Kowa, "the Arithmetical Sage," anticipated many of the discoveries of Western mathematics. He was the first person to study determinants in 1683, ten years before Leibniz used determinants to solve simultaneous equations.

Political aspirations created turmoil in the background, particularly in England with the parliamentarian revolution led by Cromwell, the restoration of the Stuart monarchy and the further upheaval leading William of Orange to the throne. France fared better, dominating European culture and politics for most of the century, until the disastrous Revocation of the Edict of Nantes that forbade Protestantism and drove leading Huguenot families out of the country: hundreds of thousands fled to Switzerland, Germany, Holland and Great Britain, ruining entire provinces and decimating French industry at the end of the reign of Louis XIV. Huguenots took the art of clock making to Geneva, the steamboat to England and paper making to Holland.

News of extraordinary phenomena was greeted with keen interest, either for their "philosophical" value or as omens of mystical importance. Antiquarians and Chroniclers collected such reports and compiled information from various countries, including North and South America. We even begin to find reports of unusual aerial sightings in the pages of the early scientific journals, like the *Philosophical Transactions* of the Royal Society, often in terms that seem surprisingly open and free compared to the staid, self-censored, dogmatic, and often arrogant scientific literature of today.

215.

23 January 1603, Besançon, France
Self-propelled cloud

"In the year 1603, being in Besançon for the duties of my charge as Visitor to Sainte Claire monastery, it happened that on a Thursday, the 23th day of January, between 7 and 8 P.M., we were told that all the people were assembling in the streets, terrified. I went out, and like the others I saw a great light in the air over the cathedral, covering the whole of Mount Saint Etienne with a round-shaped, heavy cloud, reddish in color, while all the air was clear and the sky so devoid of fog that the stars were seen shining brilliantly.

"This light remained quasi-motionless over Mount Saint Etienne, and from there we saw it coming so low that it nearly touched the houses and lit up the nearby streets, but with a motion so slow that it was hardly noticeable, and it halted for at least a quarter of an hour over Saint Vincent Abbey, where some pieces of relics of two glorious Saints are kept. Then, escaping over the *Grande place* of Chammar to the Doubs river, it went away through the *Grande rue* that goes to the bridge, and straight to the cathedral where it vanished, but as we said before, with such a slow motion that its travel lasted until 9:30 at night, which is to say at least two hours."

Source: Révérend Jacques Fodéré, *Narration historique et topographique des convens de l'ordre de St-François* (Lyon: Pierre Rigaud, 1619), 10-11.

216.

May 1606, Kyoto, near Nijo Castle, Japan ⓘ
Hovering red wheel

Numerous witnesses, including Samurais, see balls of fire kept flying over Kyoto and one night, a red wheel had come over and hovered above Nijo castle. We have not located an original Japanese source for this interesting case.

Source: Michel Bougard, *La chronique des OVNI* (Paris: Delarge, 1977), 86.

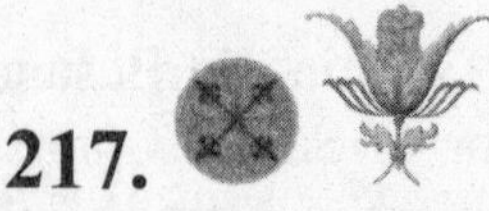

217.

1608, between Angoulême and Cognac, France
Flying warriors

"The day was calm and clear, and in an instant a large number of small, thick clouds appeared. They came down to the ground and turned into warriors. Their number was estimated between 10,000 and 12,000, all handsome and tall, covered with blue armor, aligned behind deployed red and blue banners (…)

"This sight was such that peasants and even the nobility took alarm. They assembled in large number to observe these soldiers' progress; they noticed that when they came near a thick wood, to maintain their good order, they rose above it, only touching the leaves of the trees with the bottom of their feet, eventually walking on the ground again to a forest where they disappeared.

"I have written this based on a manuscript report by the late M. Prévost, curate of Lussac les Eglises."

Source: *Chronique de Pierre Robert*, cited by A. Catinat, *Chartes, Chroniques et Mémoriaux* (Lyon, 1874).

218.

1 August 1608, Genoa harbour, Italy
Fighting creatures from the sea

Two human figures holding what looked like flying snakes were seen fighting over the sea. Only their torso was visible above the waves. Their cries were so "horrible" some of the witnesses were sick with fear. They were seen repeatedly for a couple of weeks, and about 800 cannon shots failed to scare them away.

It should be noted that this event is often confused with a series of unrelated weather phenomena between Nice and Lambesc in France, where "bloody rains" were reported.

Source: Anon., *Discours au Vray des terribles et espouvantables signes...* (Troyes: Odard Aulmont, 1608).

219.

15 August 1608, Genoa harbour, Italy
Three coaches drawn by fiery creatures

"Over the sea off the harbor of Genoa, there appeared three coaches, each drawn by six fiery figures resembling

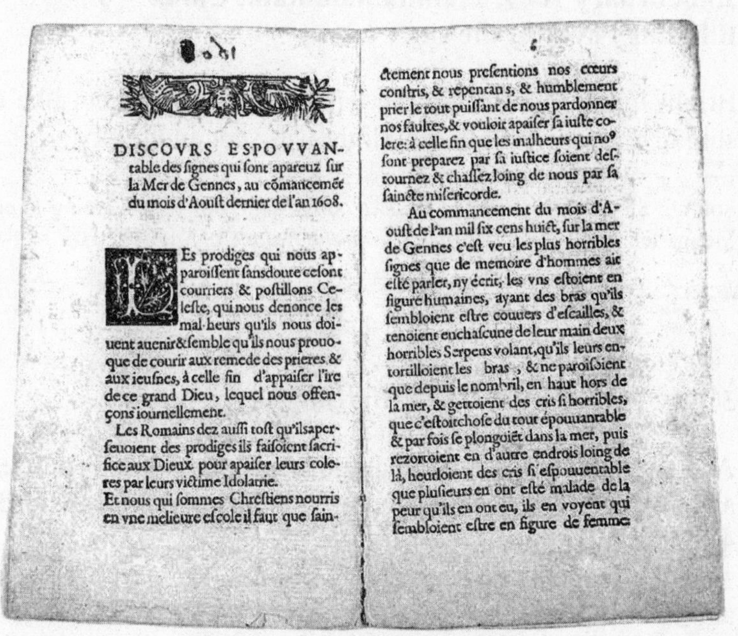

DISCOVRS ESPOVVANTable des ſignes qui ſont apareuz ſur la Mer de Gennes, au cõmancemẽt du mois d'Aouſt dernier de l'an 1608.

LEs prodiges qui nous apparoiſſent ſansdoute ceſont courriers & poſtillons Celeſte, qui nous denonce les mal-heurs qu'ils nous doiuent aucnir & ſemble qu'ils nous prouoque de courir aux remede des prieres & aux ieuſnes, à celle fin d'appaiſer l'ire de ce grand Dieu, lequel nous offençons iournellement.

Les Romains dez auſſi toſt qu'ils aperſeuoient des prodiges ils faiſoient ſacrifice aux Dieux pour apaiſer leurs coleres par leurs victime Idolatrie.

Et nous qui ſommes Chreſtiens nourris en vne melieure eſcole il faut que ſainctement nous preſentions nos cœurs conſtris, & repentans, & humblement prier le tout puiſſant de nous pardonner nos faultes, & vouloir apaiſer ſa iuſte colere: à celle fin que les malheurs qui no⁹ ſont preparez par ſa iuſtice ſoient deſtournez & chaſſez loing de nous par ſa ſaincte miſericorde.

Au commancement du mois d'Aouſt de l'an mil ſix cens huict, ſur la mer de Gennes c'eſt veu les plus horribles ſignes que de memoire d'hommes ait eſté parler, ny écrit, les vns eſtoient en figure humaines, ayant des bras qu'ils ſembloient eſtre couuers d'eſcailles, & tenoient en chaſcune de leur main deux horribles Serpens volant, qu'ils leurs entortilloient les bras, & ne paroiſoient que depuis le nombril, en haut hors de la mer, & gettoient des cris ſi horribles, que c'eſtoit choſe du tout épouuantable & par fois ſe plongoiẽt dans la mer, puis rezortoient en d'autre endrois loing de là, heurloient des cris ſi eſpouuentable que pluſieurs en ont eſté malade de la peur qu'ils en ont eu, ils en voyent qui ſembloient eſtre en figure de femme

Fig. 19: Discours Espouvantable…

dragons. With the said coaches were the aforesaid signs that still had their serpents and went on screaming their horrible cries and came close to Genoa, so that the spectators, or at least most of them, fled in fear of such a prodigy. However, when they had made three times a trip along the harbor and uttered such powerful screams that they resounded across the mountains around, they got lost over the sea and no news of them has been heard since.

"This caused great damage to the citizens of Genoa, as among them the son of Sr. Gasparino de Loro and also the brother of Sr. Anthonio Bagatello. Several women were also afflicted and had such fear that it caused their death."

Source: Anon., *Discours au Vray des terribles et espouvantables signes...* (Troyes: Odard Aulmont, 1608).

220.

15 February 1609, Tiannin mountain, China
Blinding "eye" in the sky

Bright lights illuminated the temple walls: an object like a ship or an eye with blinding light was seen in the sky.

Source: Shi Bo, *La Chine et les Extraterrestres*, op.cit., 43, citing Feng Mengzhen, *Collection of Stories from the Palace of Snow.*

221.

3 July 1612, Switzerland: Battling sky armies

According to this broadsheet, some terrible and wondrous signs were seen in the heavens in Switzerland, on the 3rd, 4th, 5th, and 6th of July, 1612.

They included three suns, three rainbows, a white cross, and two battling armies.

The text in this broadsheet contains rhymed strophes. It describes the events rather superficially and repetitively addresses its Christian readers 'young and old' to repent their sins because wondrous signs from heavens are indicators of bad times, war, and menace.

Fig. 20: Sky armies over Switzerland

Source: *Beschreibung der am 3.4.5. vnd 6. Julii dises 1612. Jars erschienen vnd grausamen erschröcklichen Wunderzeichen am Himmel* (Basel: Johann Schröter, 1612). With 22 strophes from Christian Fischer. Herzog August Bibliothek, HAB 38.25 Aug. 2, fol. 799.

222.

1613, Perth, Scotland: Abducted by fairies

Isobel Haldane testified that she was carried out of her bed, "whether by God or the Devil, she knew not," to a hillside which opened before her. She went in, and stayed three days in "the fairy kingdom." Eventually she was brought out by "a man with a grey beard" who taught her to cure diseases and foretell the future.

Far from being rare, such reports of abductions by non-human creatures became commonplace during the 17th century, eventually giving rise to accusations of commerce with the Devil. Unfortunately many of these stories did not include a date or a place. This one is an exception.

Source: Robert Pitcairn, *Criminal trials in Scotland* (Edinburgh, 1833).

223.

4 March 1614, Kinki, West Japan ⓘ
Four-sided object

A single object was seen in the sky, described as a four-sided figure. Unfortunately, we have not been able to uncover a quotation from an original source.

Source: Quincy catalog typescript, and the "*Nihon-Kaikimonogatari.*"

224.

1619, Flüelen, Lake Lucerne, Switzerland
Fiery dragon

Herr Christophorus Schere, prefect of Uri County, saw a bright, long object, fiery in color, near Flüelen, flying along Lake Uri: "As I was contemplating the serene sky by night, I saw a very bright dragon flying across from a cave in a great rock in the mount called Pilatus toward another cave, known as Flue, on the opposite side of the lake.

"Its wings were agitated with much celerity; its body was long as well as its tail and neck. Its head was that of a serpent with teeth, and when it was flying, sparkles were coming out of it like the ones thrown by an incandescent iron when struck by smiths on an anvil. At first, I thought it was a meteor, but after observing more closely, (I saw) it was truly a dragon from the recognizable motion of the members. This I write to you with respect, that the existence of dragons in nature is not to be doubted any more."

Source: Athanassius Kircher, *Mundus Subterraneus* (Amsterdam, 1665), Lib.VIII, 93-94.

225.

1619, Prague, Czechoslovakia: Flying Globes

A succession of fiery globes is observed. Some of them split into several parts or other globes. The report reads:

"A strange and prodigious thing was seen in a village that is 6 leagues from Prague, the capital city of Bohemia. Never had we seen such a spectacular or frightful sign before. The inhabitants of the village were on guard as the country is full of soldiers, because of the partialities and differences that exist in the empire today. The village priest was with them at about 10:00 in the evening. He was praying, looking up at the sky, when suddenly he stopped, astounded by what he saw. He could see a globe that resembled the moon, but fiery. It divided into two parts, and one of the parts divided into four smaller globes.

"The most amazing thing was that one of the globes disappeared, and in its place we saw a bloody crucifix. These things stayed [in the sky] for a short time, and then disappeared gradually, finally vanishing into a big hole. Then we just saw a great globe which resembled the moon, as we had witnessed at the beginning. This whole process was repeated three or four times, and then everything disappeared."

Source: An 8-page pamphlet titled *Signes Prodigieux d'un Globe de Feu apparu en Allemagne, Capable d'Espouventer toute la Chrestienté, y ayant esté veu un Crucifix sanglant* (Prodigious Signs of a Globe of Fire that appeared in Germany, Able to Frighten all the Believers, and a Bloody Crucifix) (Paris: Pierre Bertault, 1619).

226.

1 February 1620, Quimper-Corentin, France
Green flying creature

Many witnesses: thunder falls on the cathedral. A green "demon" is seen inside the fire.

"On Saturday a great disaster took place in the town of Quimper-Corentin; namely that a beautiful and tall pyramid (note: bell tower) covered with lead, being atop the nave of the great church, and over the cross of that said nave, was burnt by the lightning and fire from the sky, from the top down to the said nave, without any way to remedy it.

"And to know the beginning and the end, it is that about seven and a half to eight in the morning, there was a clap of thunder and terrible lightning, and at that instant was seen a horrible and frightening demon, taking advantage of a great downpour of hail, seizing the said pyramid from the top under the cross, being the said demon of green color, having a long tail of the same color. No fire or smoke appeared on the said pyramid, until about one in the afternoon, when smoke started coming out from the top of it, and lasted a quarter of an hour, and from the same place fire appeared, while it ran higher and lower, so that it became so large and frightening that it was feared the whole church would burn, and not only the church but the whole town.

"All the treasures of the church were taken outside; neighbors (of the church) had their goods transported as far as they could, in fear of the fire. There were more than 400 men to extinguish the fire, and they could not do anything to stop it. Processions went around the church and other churches, all in prayers. Finally, for all resolution, holy relics were placed on the nave of the said church, near and before the fire. Gentlemen of the Chapter (in absence of Monsignor the Bishop) began conjuring this evil demon, which everyone could see clearly in the fire, sometimes green, yellow, and blue. (They) threw *Agnus Dei* into it, and nearly a hundred and fifty barrels of water, forty or fifty carts of manure, yet the fire went on burning.

"For an ultimate resolution a loaf of rye bread worth four *sols* was thrown into it, within which a consecrated host had been placed, then holy water with the milk of a wet nurse of good morals, and all that was thrown into the fire; at once the demon was forced to leave the fire and before getting out it made such trouble that we all seemed to be burned, and he left at six hours and a half on the said day, without doing any damage (thank God) except for the total ruin of the said pyramid, which is of the consequence of twelve thousand *écus* at least.

"This evil being out, the fire was conquered. And shortly afterwards, the loaf of rye bread was found still intact, without any damage, except that the crust was somewhat blackened. And about eight or nine and a half, after the fire was out, the bell rang to assemble the people, to give graces to God. The gentlemen of the Chapter, with the choir and musicians, sang the *Te Deum* and a *Stabat Mater*, in the chapel of the Trinity, at nine in the evening."

Source: Lengley-Dufresnoy Vol. I, Part 2, 109, citing *La Vision Publique d'un Horrible et très Epouvantable Démon, sur l'Eglise Cathédrale de Quimper-Corentin, en Bretagne, le Premier Jour de ce mois de Février 1620. Lequel Démon consuma une pyramide par le feu, et y survint un grand tonnerre et feu du Ciel* (A Paris, chez Abraham Saugrain, en l'Isle du Palais, jouxte la copie imprimée à Rennes par Jean Durand, rue Saint Thomas, près les Carmes, 1620).

227.

9 April 1620, Geneva, Switzerland
Flying hats and men in black

"Two suns were seen, one red and the other one yellow, hitting against each other (...) Shortly afterwards there appeared a longish cloud, the size of an arm, coming from the direction of the sun, which stopped near the sun, and from that cloud came a large number of people dressed in black, armed like men of war. Then arrived other clouds, yellow as saffron, from which emerged some 'reverberations' (?) resembling tall, wide hats, and the earth was seen all yellow and bloody. The sun became double and it all ended with a rain of blood."

Source: *Effroyable bataille aperçue sur la ville de Genesvre le dimanche des Rameaux dernier...* (brochure published in 1620) cited in *Les Soucoupes chez Heidi* (GREPI, 1977).

228.

13 October 1621, Nîmes, France
Fiery chariots, a great sun

"Over the city of Nîmes, about 9 to 10 P.M., over the amphitheater, was seen something like a great sun, very resplendent, which was surrounded by a number of other luminous torches.

"It seemed to want to move straight towards the Roman Tower, over which appeared something like fiery chariots surrounded by very bright stars."

Source: *Les Signes Effroyables Nouvellement Apparus...*, Cited by Veronica Magazine (Gouiran & Lamblard, 1976). Also see Michel Bougard, *La chronique des OVNI* (Paris: Delarge Ed., 1977), 92-93.

229.

12 May 1624, Anhalt, Germany ⓘ
Chariots in the sky

From six to eight o'clock in the evening a multitude of men and chariots were observed, emerging from the clouds over Gierstedt (Bierstedt), Anhalt, in Germany.

Source: L. Brinckmair, *The Warnings of Germany* (London: John Norton, 1638), 18-19.

230.

1634, Wiltshire, England
Dancing elves, a paralyzed witness

Mr. Hart was paralyzed and assaulted by a group of dancing elves at night. He woke up in a fairy ring.

The curate of a Wiltshire 'Latin Schoole', Mr. Hart was assaulted by a group of elves one night in 1633 or 1634. Whilst walking over the 'downes,' he saw "an innumerable quantitie of pigmies or very small people" dancing in a typical fairy ring "and making all maner of small odd noyses." Mr. Hart, "being very greatly amaz'd, and yet not being able, as he sayes, to run away from them, being, as he supposes, kept there in a kind of enchantment," fell to the ground in a daze. The "little creatures" surrounded their prey and "pinch'd him all over, and made a sorte of quick humming noyse all the time..." Hart awoke to find himself in the centre of a ring pressed into the grass—a fairy ring. "This relation I had from him myselfe, a few days after he was so tormented," writes Aubrey.

Source: K. Briggs, *A Dictionary of Fairies* (London: Penguin Books, 1976).

231.

Circa 1635, Port-Louis, Brittany, France
A Procession of sky beings

A 60-year old man named Jean Le Guen, who lived in Riantec near Port-Louis, asserted that he had observed a procession of beings he took to be "angels" in the sky. They were going from Port-Louis to Caudan.

Source: *The Diary of Jesuit Father Julien Maunoir*, written in 1672, recording a statement about the case by the Lord of Lestour. Published as *Miracles et Sabbats. Journal du Père Maunoir, missions en*

Bretagne (1631-1650) presented by Eric Lebec (Paris: Editions de Paris, 1997), 85.

232.

28 May 1637, Between Chartres and Paris, France
Three unexplained "stars"

Travelers marveled at three large 'stars' surrounded by smaller ones, with a long streak of other 'stars.' In the absence of a better description, including the duration and trajectory of the phenomena, the skeptic may well decide that the travelers in question simply saw a series of bright meteors.

Source: *Les prodigieux Signes nouvellement apparus au ciel en plusieurs lieux et notamment aux environs de la ville de Paris. Avec l'explication d'iceux, lesquels sont très-favorables pour la France* (Paris: C. Morlot, 1637), BN 8-LK7-7793.

233.

March 1638, Muddy River near Charlton, Massachusetts: Missing time among the Puritans

Puritan James Everell and two others were stunned as they saw a luminous mass that hovered and returned over a three-hour period. Their boat was pulled upstream by the phenomenon.

The settling of the first Puritan colony in Boston was chronicled by Governor John Winthrop, who arrived in Massachusetts Bay in 1630 with one thousand English emigrants. A historian himself, Winthrop kept a record of the colony's first years in the New World. His journal is far from being a mere collection of unlikely anecdotes or village gossip. It is quite significant, therefore, that he regarded two spectacular sightings of unexplained phenomena as being sufficiently important to be recorded for posterity.

The first sighting took place in March 1638. A member of the Puritan Church, James Everell, "a sober, discreet man," was crossing the Muddy River one evening in a small boat with two companions. Suddenly a great luminous mass appeared in the sky above the river. It seemed to dart back and forth over the water. When it remained motionless, it "flamed up" and seemed to measure three yards square. When it moved, it "contracted into the figure of a swine" and flew away towards Charlton.

It did this repeatedly over a period of two or three hours, always returning briefly to the same spot above the water before shooting off again.

When the light had finally vanished, Everell and his friends stood up and were surprised to learn that the boat was now further upstream than it should have been, as if it had been pushed, pulled or carried by an unknown force. In fact they had been carried against the tide to their original starting point, one mile away.

Why the light would be swine-shaped is a mystery not even the Puritan colonists could interpret, though it should be noted Everell was a leather dresser by trade, and he could have sought a familiar shape in an otherwise amorphous light.

It is curious that the men observed that "two or three hours" passed during the spectacle. Can we believe they sat watching the phenomenon for such a long time?

The mysterious repositioning of the boat could suggest that they were unaware of part of their experience. Some researchers would interpret this as a possible alien abduction if it happened today.

Any speculation at this late date is merely conjecture, but it is interesting to note that at least a superficial resemblance exists between this case and recent claims in the American abduction literature.

Source: John Winthrop, *The History of New England from 1630 to 1649* (Boston: Little, Brown and Company, 1853), 349-350.

234.

April 1639, Yuan, Fengxian, Shansi province, China
Flying star at funeral

A red, white, yellow and blue "star" flew over a funeral, circling the village for a long time. The villagers were presenting their condolences to the family of Yuan Yingta, a minister of war under the Ming dynasty who had sacrificed himself on the battlefield while resisting the Man army. Suddenly a luminous object like a star, red, white, yellow and blue in color, flew over the funeral procession. This brilliant thing did not touch the ground, but it flew around the village for a long time, then rose up in the sky again. Its light was visible five kilometers away.

Source: Shi Bo, *La Chine et les Extraterrestres*, op.cit., 45, citing scholar Lou Ao, *Histoire Locale du District Fengxian.*

235.

July1639, Santiago, Spain: ships in the sky

A short pamphlet published in Seville in 1639 titled *An Account of the Prodigious Visions of Armies of Men, Standards, Flags, Vessels, and Other Things, that Visibly have been Seen over a Long Time, near the Town of Santiago in Galicia, in the Fields of Lerida, since June 24th to this Present Year of 1639*, reported that "in Santiago three ships appeared in the air with the sound of drums and many people." Unfortunately no more details are given.

Source: *Relacion certissima de las prodigiosas visiones... que visiblemente se han visto largo tiempo, cerca de la ciudad de Santiago de Galizia...* (Seville, 1639).

236.

9 May 1641, Braga, Portugal: Flying disk, entities

A flat disk ("wafer") appeared in the sky along with two human figures designated as "angels." There were numerous witnesses, and the original text makes fascinating reading:

"Year of the birth of Our Lord Jesus Christ 1641, May 9th, in this city of Braga, in the lodgings of Reverend Doctor Jao d'Abreda Rocha, ecclesiastic judge and general vicar of this court, the archbishop of Braga: there the Reverend Judge was alerted by D. Gastao Coutinho, general of the province of Entre Douro e Minho, of the fact that many people going towards the area of Porto acting upon a warning given in this city that the enemies were approaching in that area, saw some signs on the moon, in which there appeared a Wafer and two human figures that looked like Angels; and that the said judge know the truth of it: about which the Reverend Doctor ordered this writ to be made, that I sign; and he asked the following witnesses. Father Simao Alvares, apostolic notary, wrote it."

Source: Francisco Lopez Liureyro, *Favores do Ceo. Do braço do Christo que se despregou da cruz, & de outras marauilias dignas de notar. Dedicados ao III.me Senhor D. Rodrigo da Cunha...* (Lisbon: Antonio Alvarez, 1642); reprinted as *Favores do ceo a Portugal na acclamação do rei D. João IV e acabamento da oppressão dos reis Filippes..., por Francisco Lopes, livreiro lisbonense. Precedidos d'una noticia bibliographica do auctor, escripta pelo Professor Pereira-Caldas* (Porto: Ernesto Chadron and Braga: Eugenio Chadron, 1642, 1871).

237.

13 April 1641, Aragón, Spain: Apparitions

Domingo Sánchez, gardener at the Monastery of María de Aragón, who was sworn to chastity, requested permission to marry. On the night of April 13th, in bed with these thoughts, he saw a demon that pulled him out and dragged him for a long time around his lodgings, hitting him. The man sought help from the Virgin, who appeared, surrounded by lights and "aided him until dawn."

Source: José Pellicer de Salas y Tobar, in his *Avisos históricos*, a compilation of historical reports spanning the period May 24th 1639 to November 29th 1644.

238. 

4 May 1641, Madrid, Spain: Unexplained black cloud

At 9:00 P.M., "the sky being very calm, without there being a single cloud in it, (Jose Pellicer reported) an extremely black and dark cloud, that approached from somewhere between the east and the north, dilated and narrow, crossing between the west and Midday, that was stationary for some time" – giving the impression that Pellicer *may* have seen something resembling the 'cloud cigars' dealt with in modern reports.

Pellicer goes on to mention a burst of sound of unknown origin in the sky over Molina de Aragón. The people there heard "loud noises, bugles, drums, as if an invisible ferocious battle were happening but without anything to be seen."

Source: José Pellicer de Salas y Tobar, *Avisos históricos*, op. cit.

239. 

14 September 1641, Akhaltsike, Georgia
Blue wheel descending

Armenian chronicler Zacharia Sarcofag saw a strange phenomenon at sunset. The sky was not yet dark when suddenly "the ether on the eastern side was torn up and a big dark-blue light began to descend. Being wide and long, it came down approaching the Earth and it illuminated everything around, more brightly than the sun."

The forward part of the light "revolved like a wheel, moving to the north, calmly and slowly emitting red and white light, and in front of the light, at a distance of an open hand, there was a star the size of Venus. The light was still visible until my father had sung, weeping, six *sharakans*, after which it moved away. Later we heard that people saw this miraculous light up to Akhaltsike."

A *sharakan* is a brief prayer sung over two to three minutes, so the phenomenon would have lasted at least 15 minutes, according to researcher Mikhail Gershtein.

Source: Zacharia Sarcofag, *On the Fall of Light from the Sky*. Cited by M. B. Gershtein, "A Thousand Years of Russian UFOs," *RIAP Bulletin* (Ukraine) 7, 4, October-December 2001.

240.

3 July 1642, Olesa de Montserrat, Catalonia, Spain
Globe, changing its appearance

Joseph Aguilera and others saw a globe changing to "three moons," later an enormous light seen for one hour.

Source: Guijarro, Josep, *Guía de la Cataluña Mágica* (Barcelona: Ediciones Martínez Roca, 1999), 48.

241.

18 January 1644, Boston, Mass.: Luminous figures

Three men coming to Boston saw two "man-shaped lights" come out of the sea. About 8 P.M. several inhabitants of an area of Boston located near the sea saw a light the size of the full moon rise in the northeast. Shortly thereafter, another light appeared in the east. The witnesses observed a curious game of hide-and-seek between the two objects.

During this celestial ballet, several persons known to be sober and pious, who were aboard a boat between Dorchester and Boston, claimed they heard a voice in the sky uttering the following words in a most terrible voice: "Boy, boy, come away…"

These calls were repeated about 20 times, coming from various directions.

Source: John Winthrop, *Winthrop's Journal*, "History of New England" (1630-1649) (New York: Barnes and Noble, 1959).

242.

25 January 1644, Boston, Massachusetts
A Voice from the light

Luminous objects were seen sparkling, emitting flames. Again, the aerial ballet of the previous week was observed and a voice calling out: "Boy, boy, come away."

Source: John Winthrop, *Winthrop's Journal*, "History of New England" (1630-1649) (New York: Barnes and Noble, 1959).

243.

April 1645, Caudan, Brittany, France
A Procession of sky beings

A sixty-year old man named Jean Coachon, who lived in Calan, near Lanvodan and Vannes, stated that he had witnessed a procession of sky beings he called "angels," circling above the church, with the Virgin among them. This was related by the Lord of Lestour, who collected such stories.

Source: Eric Lebec, ed., *Miracles et Sabbats. Journal du Père Maunoir, missions en Bretagne (1631-1650)* (Paris: Les éditions de Paris, 1997), 85-86.

244.

11 November 1645, Location unknown ⓘ
Unidentified planetoid near Venus

A body large enough to be a satellite was seen near Venus from Naples by the astronomer Francesco Fontana. He made further observations of the "satellite" on December 25th 1645 and January 22nd, 1646. Jean-Charles Houzeau, director of the Royal Observatory of Brussels, baptized the satellite with the name "Neith" in the 1880s.

Source: Francesco Fontana, *Novae coelestium terrestriumq[ue] rerum observationes* ... (Naples, 1646).

245.

April 1646, St.Teath, Cornwall
Abductee becomes a healer

A woman named Anne Jefferies fell ill and claimed to have acquired healing powers after being abducted by six 'small people'. Anne Jefferies was the daughter of a poor laborer who lived in the parish of St. Teath. She was born in 1626, and is said to have died in 1698.

When she was nineteen years old, Anne went to live as a servant in the family of Mr. Moses Pitt, where she suffered a sudden loss of consciousness. A letter from Moses Pitt to the Right Reverend Dr. Edward Fowler, the Bishop of Gloucester, dated May 1st, 1696, explains how one day Jefferies had been knitting in an arbour in the garden when something so shocking happened to her "that she fell into a kind of Convulsion-fit." Soon afterwards members of the family found her writhing on the ground and carried her indoors, where she was taken to her bedroom and allowed to rest. When she regained consciousness she startled everyone gathered at her bedside by crying out, "They are all just gone out of the Window; do you not see them?" This and similar outbursts were immediately "attributed to her Distemper," her employers supposing she was suffering a bout of feverish 'light-headedness.'

Anne Jefferies remained in an unstable condition for some time, unable even to "so much as stand on her Feet." Gradually, however, she managed to recover from her sickness and by the following year was able to reassume her duties as a maid but she had not exactly become her old self again. Pitt writes that the first indication that Jefferies had acquired new skills came "one Afternoon, in the Harvest-time," when his mother slipped and broke her leg on the way back from the mill. A servant was told to saddle a horse and fetch Mr. Hob, the surgeon, from a nearby town. "Anne Jefferies came into the room and saw Mrs. Pitt with her leg outstretched. She asked her to show her the wound, which the woman did after some persuasion, and to rest the leg on her lap. Stroking it with her hand, Anne asked whether the woman was feeling any better. My Mother confess'd to her she did. Upon this she desired my Mother to forbear sending for the Chyrurgeon, for she would, by the Blessing of God, cure her leg."

What surprised Mrs. Pitt the most was not the maid's newfound healing powers but the fact that she seemed to know exactly when and where her fall had happened. Yet how could she? Moses writes that his mother demanded an explanation.

Anne said "You know that this my Sickness and Fits came very suddenly upon me, which brought me very low and weak, and have made me very simple. Now the Cause of my Sickness was this. I was one day knitting of Stockings in the Arbour in the Gardens, and *there came over the Garden-hedge of a sudden six small People, all in green Clothes, which put me into such a Fright and Consternation that was the Cause of this my great Sickness; and they continue their Appearance to me, never less that 2 at a time, nor never more than 8: they always appear in even Numbers, 2, 4, 6, 8.* When I said often in my Sickness, They were just gone out of the Window, it was really so; altho you thought me light-headed (…) And thereupon in that Place, and at that time, in a fair Path you fell, and hurt your Leg. I would not have you send for a Chyrurgeon, nor trouble your self, for I will cure your Leg."

From that time on, Anne Jefferies became famous throughout England as a faith-healer and fairy contactee. Moses Pitt writes that "People of all Distempers, Sicknesses, Sores, and Ages" travelled from far and wide to Cornwall to see the girl and receive her magical treatment. She charged no fee for her work."

Unfortunately, so many strange goings on and her growing reputation as a seer worried the local authorities. They sent "both the Neighboro-Magistrates and Ministers" to question the maid on the nature of her supernatural contacts. Despite hearing Anne Jefferies' "very rational Answers to all the Questions they then ask'd her," her interrogators concluded that the spirits she spoke to were "the Delusion of the Devil," and they "advised her not to go to them when they call'd her." Not long after this, the Justice of the Peace in Cornwall, John Tregagle Esq., issued a warrant for her arrest.

Jefferies spent three months in Bodmin Gaol. When she was finally freed it was decided that she could not return to the house of the Pitts, so she went to stay with Moses Pitt's aunt, Mrs. Francis Tom, near Padstow. There "she liv'd a considerable time, and did many great Cures," but later moved into her own brother's house and eventually married.

Source: Letter from Moses Pitt to the Bishop of Gloucester in Robert Hunt, *Popular Romances of West England* (1871).

246.

May 1646, The Hague, Netherlands
Fleet of airships, occupants

Unknown people and animals were seen in the sky of The Hague. A fleet of airships came from the southeast, carrying many occupants. It came close to the aerial spectacle. A huge fight ensued.

When the phenomenon vanished, people saw "something like a huge cloud that appeared at a place where nothing was visible before."

Source: *Signes from Heaven; or severall Apparitions seene and hearde in the Ayre...* (London: T. Forcey, 1646).

247.

21 May 1646, Newmarket & Thetford, England
Vertical pillar of light

"Betwixt Newmarket and Thetford in the foresaid county of Suffolk, there was observed a pillar or a Cloud to ascend from the earth, with the bright hilts of a sword towards the bottom of it, which piller (sic) did ascend in a pyramidal form, and fashioned it self into the forme of a spire or broach Steeple, and there descended also out of the skye, the forme of a Pike or Lance, with a very sharp head or point (…) This continued for an hour and a half."

Source: *Signes from Heaven: or severall Apparitions seene and heard in the Ayre...* (London: T. Forcey, 1646).

248.

1648, Edinburgh, Scotland ⓘ
Flight aboard a fiery coach

In the spring of 1670, Captain of the Town Guard and highly respected preacher Major Thomas Weir (ca.1596-1670) and his sister Jane Weir confessed to a series of terrible offenses. Thomas' confession began with a detailed summary of his sex crimes which was horrible enough in the eyes of the city officials in Edinburgh. But it was when he admitted to being a witch and a sorcerer that the authorities became truly anxious. Weir said that he and his sister had had dealings with demons and fairies, to whom they had duly sold their immortal souls.

The Devil appeared to Jane in the guise of a midget-like woman. Both she and her brother had been carried off by strange entities on several occasions. They said that in 1648 they were transported between Edinburgh to Musselburgh in a fiery "coach," and they had also been taken for a ride in a similarly fiery "chariot" from their house in the West Bow (a z-shaped street near Edinburgh Castle) to Dalkeith.

It is interesting that Thomas Weir was driven to coming clean about his private life because of the guilt he felt from having consorted with devils. Major Weir was an active member of a strict Protestant sect. Betraying God was, for him, his least forgivable crime. However, he was old and sick and he had been an important figure in society for as long as people could remember, so at the beginning he had trouble persuading the courts to arrest him. When at last he and his sister were remanded in custody she alone was convicted of witchcraft, while he was "only" found guilty of fornication, incest and bestiality (!).

Jane Weir was hanged and burnt at the stake at Grass Market on April 12th, 1670, and her brother the day before. Tradition holds that both refused to repent on the scaffold, crying out that they wished to die as shamefully as they deserved. When requested to pray on the eve of his execution, Major Weir answered, screaming, “Torment me no more–I am tormented enough already!” This gives the impression that he was convinced of the physical nature of his acts and of his contact with malign spirits, as does his reply on the scaffold when asked to beg God for mercy: “Let me alone – I will not – I have lived as a beast, and I must die as a beast!” Jane Weir’s final words were along the same lines.

Source: Charles McKay, *Memoirs of Extraordinary Popular Delusions and the Madness of Crowds* (1841).

249.

1650, Limerick, Ireland: Flying globe with light beam

A luminous globe brighter than the Moon shed a vertical light on the city, and then it faded as it passed over the enemy camp.

Source: Dominic O’Daly, *History of the Geraldines* (1665).

250.

Circa 1650, Fisherton Anger, Wiltshire, England ⓘ Contact with Spirits

A woman named Anne Bodenham, formerly a servant to Dr. Lamb of London, was accused of witchcraft and commerce with devils. She was eventually put to death in 1653 at the age of 80.

The cleric who recorded her trial noted that:

“Arrived at the place of execution, she attempted to go at once up the ladder, but was restrained. Mr. Bower pressing her to confess, she steadfastly refused, and cursed those who detained her.”

Fig. 21: The spirit creatures in Anne Bodenham's magic, emerging from a circle of fire.

A maid testified at the trial that she had seen Anne Bodenham invoke the Devil. After Anne had made a circle with a stick, and filled it with burning coals,

> "*Then appeared two spirits in the likeness of great boys with long shagged hair, and stood by her looking over her shoulder, and the Witch took the maid's forefinger of her right hand, in her hand, and pricked it with a pin and squeezed out the blood, and put it into a pen, and put the pen into the maid's hand, and held her hand to write in a great book, and one of the spirits laid his hand or claw over the witches, whilst the maid wrote, and when she had done writing whilst their hands were together, the Witch said Amen, and made the Maid say Amen, and the Spirits said Amen, Amen.*

> *"And the Spirits hand did feel cold to the maid as it touched her hand, when the witches hand and hers were together writing."*

Source: James Bower, *The Tryal, Examination and Confession of mistris Bodenham, before the Lord chief Baron Wild, & the Sentence of Death pronounc'd against her, etc.* (London: printed for G. Horton, 1653). See also *Doctor Lamb revived, or, Witchcraft condemn'd in Anne Bodenham a Servant of his, who was Arraigned and Executed the Lent Assizes last at Salisbury... by Edmond Bower an eye and ear Witness of her Examination and Confession* (London: printed by T.W. for Richard Best, and John Place, 1653).

251.

22 December 1651, Almerdor, Holland
Flying dutchmen

Dutch sailors saw a fleet of ships in the air, with many people and soldiers.

Source: *A report made before the harbormaster* (Seville: Juan Gómez de Blas, 1652).

252.

May 1652, Near Rome, Italy
Huge object drops strange matter

A single luminous object, 80 meters in size, was seen in the air. A mass of "gelatinous matter" fell to the ground.

Source: *Edinburgh Philosophical Journal* 1 (October 1819): 234.

253.

1656, Cardiganshire, Wales: Bedroom visitation

In a letter written in 1656, John Lewis of Cardiganshire (Wales) described the experience of an acquaintance of his:

"A man lay in bed at night while his family were all fast asleep. Just after midnight "he could perceive a light entering [his] little room." Suddenly a dozen or so little beings "in the shape of men, and two or three women, with small children in their arms" walked in.

"The room seemed different somehow. It was illuminated, and appeared to be wider than before. The beings began to dance around and tuck into a special feast, inviting the witness to try the meat. This went on for four hours, and in the meantime, "he could perceive no voice" except for the occasional whisper in Welsh "bidding him hold his peace." He found it impossible to wake up his wife. Finally, the party of little spirits moved their dancing on to another room, and then departed. Until the man cried out at last and woke up his family, for some unexplained reason "he could not find the door, nor the way into bed."

John Lewis described the man as "an honest poor husbandsman, and of good report: and I made him believe I would put him to his oath for the truth of this relation, who was very ready to take it."

Source: William E. A. Axon, *Welsh Folk-lore of the Seventeenth Century*. Y Cymmrodor Vol. XXI (1908), 116.

254.

1659, Leicester and Nottinghamshire, England
Flying coffin

Starting at 1 P.M. people observed an object "in the perfect figure and form of a black coffin, with a fiery dart and a flaming sword flying to and again, backwards and forwards the head of the said coffin, which was with great wonder and admiration beheld by many hundreds of people." This was seen until 3:15 P.M., when it broke up with great brilliance.

WONDERS
In the *North* and *Weſt* of
ENGLAND:
AS
They were communicated to divers
Honourable Members of *Parliament*, from ſeveral Countrey Gentlemen and Miniſters; concerning the ſtrange and prodigious flying in the Air of a Black *Coffin* betwixt *Leiceſter* and *Nottingham*, on *Sabbath* day laſt was a fortnight, with a flaming Arrow, and a Bloody Sword, caſting forth ſtreams of Fire, to the great wonder and aſtoniſhment of many Hundreds of People that beheld the ſparkling and glittering Rays, as far as *Newark*, *Beaver*, *Loughborough*, *Melton* and divers other places: With a *Conjectu-ration* thereupon, what theſe dreadful Signs from Heaven, may denote and ſignifie to the People on Earth this preſent Summer.

Fig. 22: Wonders in England

Source: *The five strange wonders, in the north and west of England as they were communicated to divers honourable members of Parliament, from several countrey gentlemen and ministers, concerning the strange and prodigious flying in the air of a black coffin betwixt Leicester and Nottingham, on Sabbath day last a fortnight, with a flaming arrow, and a bloody sword, casting forth firearms of fire ...* (London: W. Thomas, 1659).

255.

1660, New England, American Colonies
Aerial phenomena to the rescue of Puritanism

Some remarkable phenomena having appeared in the air, one of which is described as "resembling the form of a spear, of which the point was directed towards the setting sun, and which, with slow majestic motion, descended through the upper regions of the air, and gradually disappeared beneath the horizon," (note: possibly the zodiacal light) the magistrates and clergy availed themselves of the deep impression which these signs created, to promote a general reformation of manners among the people.

For this purpose, they published a catalogue of the principal vices of the times, in which were enumerated "a neglect of the education of children, pride displayed in the manner of cutting and curling the hair, excess of finery, immodesty of apparel, negligent carriage at church, failure in due respect to parents, profane swearing, idleness, and frequenting of taverns, and a sordid eagerness of shopkeepers to obtain high prices."

Source: James Silk Buckingham, *America, historical, statistic, and descriptive* (London: Fisher, Son & co., 1841), 259.

256.

14 March 1660, London (Westminster) England
Peculiar cloud

Boat passengers saw a dark, then bright cloud dropping fire over Westminster. About 8 P.M. they observed "a white bright cloud which gave such a light that they could plainly see the windows of the Parliament House, and people walking to and fro upon Westminster Bridge".

The cloud was seen to "drop down fire several times upon Westminster Hall and then it removed and (flew) over the Parliament Hall and did drop down fire upon that also several times".

Source: *Eniaytos terastios Mirabilis annus, or, The year of prodigies and wonders being a faithful and impartial collection of severall signs that have been seen in the heavens, in the earth, and in the waters; together with many remarkable accidents, and judgements befalling divers persons, according as they have been testified by very credible hands: all which have happened within the space of one year last past, and are now made publick for a seasonable warning to the people of these three kingdoms speedily to repent and turn to the Lord, whose hand is lifted up amongst us* (London, 1661).

257.

August 1660, Statford Row, near London, England
A Great ship in the air

The likeness of a "great ship" was seen in the air. It decreased in size and eventually disappeared.

The worthy chronicler does not fail to inform us that "this is testified by an able Minister living not far from the place, who received the information from the spectators themselves".

Source: *Mirabilis Annus* (1661)

258.

September 1660, London, England
Multiple unknown lights

"A gentleman of good quality and an Officer of Eminency in the late King's army and now a Justice of the Peace in the Country" reported seeing a bright light in the Southwest, along with six smaller ones. "Whilst he with several others, were with some admiration beholding them, they all fell down perpendicularly and vanished."

Source: *Mirabilis Annus* (1661).

259.

3 October 1660, Hull, England
Large tapered flying object

The soldiers on guard at the South Blockhouse saw a large fiery object tapering off at one end and leaving a narrow stream behind. It was so brilliant that they could read fine print or take up a pin from the ground by its light. This object was in sight for half an hour.

Someone who was approaching Hull that same night, coming from Lincolnshire, confirmed the first report: "He saw a very great light in the sky, whereby he could perfectly discern his way, though it was exceedingly dark."

The whole relation–continues our chronicler–"is signified by letters from several eminent men in Hull who spoke with the eyewitnesses, as also by some inhabitants of London, who upon occasion have been at Hull since that time, and there from very good hands have received credible information concerning the premises."

Source: *Mirabilis Annus* (1661).

260.

11 October 1660, Hertford, England
Flying circle with appendages

A person of very good note and credit awoke at 4 A.M. to see "a flashing like fire against his window, and fearing some house near him had been on fire, he immediately arose and went to the window."

He saw a large object with a circle around it, and two appendages above and below it, from which great flashes were indeed emitted. This object remained in view for several hours, and was observed by others in the town.

Source: *Mirabilis Annus* (1661).

261.

12 October 1660, London, England: Two unknowns

Around 4 P.M. people saw an object going through the air from west to east with a great noise.

It was shaped "like a beesome" according to witnesses. Immediately afterwards, another object of the same shape, but smaller, flew overhead on the same trajectory.

Source: *Mirabilis Annus* (1661).

262.

30 October 1660, Austy, England: Multiple objects

Several persons who were going to Ware Market suddenly saw a terrible flash of lightning, after which the night became brighter and brighter, until a great fiery thing appeared in the East and ascended. Three star-like objects fell from it vertically. The large object changed to a crescent shape and remained in view until dawn.

The same thing was seen by five people going from Hertford to London: they saw the flash of lightning, which was as bright as the noonday sun and made it seem that their horses were on fire. One suspects a fireball or very bright meteor, but the story disproves this hypothesis: "Within a little space, this Body of Fire rose up again into the Air, with a tayl (sic) about a Pole long, and went Eastward, where at last it fixed itself in the sky like a star".

Source: *Mirabilis Annus* (1661).

263.

30 October 1660, Yelden, England: Bright object

A bright object seen in the sky for two hours by a credible witness. The record reads:

"Very early was seen a great Star which…gave so great a light, that some inhabitants here…could see to do business in the house by the light of it; one credible person here beheld it two hours together, and at last saw it turn into the perfect form of a Roman S, and then presently it divided in the middle, and one half went to the north-east, and the other to the south-west, and so by degrees disappeared."

Source: *Mirabilis Annus* (1661).

264.

10 November 1660, Oxford, England: Humanoid

A scholar named Allen, of Magdalen College in Oxford, who was in bed, heard a noise like the sound of geese. He got up and looked out of a window on the side of a bridge, but saw nothing. As he went back to bed he saw a strange man at the door, apparently dressed as a bishop!

"At first he was not much affrighted, but called to it and abjured it to speak. The Bishop immediately rose up and approached towards his bed, at which the young man was exceedingly terrified, and crying out murder, murder, it vanished. He since says that he saw and heard something which he will discover to no one."

Source: *Mirabilis Annus* (1661).

265.

30 November 1660, Ilford, Essex, England
Dogfight in the sky

Very early in the morning two men saw a fiery cloud in the southwest. From under it appeared two bright objects as large as the moon, which began a dogfight in the atmosphere. One of them eventually grew dimmer while the other increased in size and remained in view for two hours, "a great part of which time they saw streaming from it… streams of fire and streams of blood." It then diminished until it was no larger than an ordinary star.

Source: *Mirabilis Annus* (1661).

266.

1 Dec. 1660, Hounsditch, England: Unknown moon

At 5 A.M. an inhabitant of Hounsditch saw an unexplained, bright object the size of the moon in the eastern sky.

Source: *Mirabilis Annus* (1661).

267.

1661, Goult, Vaucluse, France
A luminous figure heals a sick man

Antoine de Nantes, a messenger from Goult, who was gravely ill, caught sight of a marvelously beautiful child who hovered above a fiery halo. When this figure vanished, the man found himself fully healed. A chapel was consecrated two years later and "miracles became commonplace after that date."

Source: Louis Leroy, *Histoire des pélerinages de la Sainte Vierge en France* (Paris, 1873), 30.

268.

February 1661, Darken, Surrey, England
Flying cathedrals

A "discreet sober gentleman" saw a strange cloud in the evening sky, and two objects he compares to cathedrals or churches, "having upon it diverse goodly Pinnacles, and each of them a long streamer flying upwards upon it, and as he beheld it, he thought it grew up to a greater splendor and glory." The other object was darker.

After a while, the large one emitted puffs of vapor and disappeared, while the smaller one grew and became brighter. The witness was called into his house and could not observe the end of the phenomenon.

Source: *Mirabilis Annus* (1661).

269.

20 March 1661, Canterbury, England
A Star with an opening

A very large "star" with an "opening" underneath, from which issued streams of fire was seen for thirty minutes.

Source: *MIRABILIS ANNUS SECUNDUS; or, a second year of prodigies. Being a true and impartial collection of many strange signes and apparitions, which have this last year been seen in the heavens, and in the earth, and in the waters. Together with many remarkable accidents and judgements befalling divers persons, according to the most exact information that could be procured from the best hands; and now published as a warning to all men speedily to repent, and to prepare to meet the Lord, who gives us these signs of his coming...* (London, 1662).

270.

April 1661, Chard, Somersetshire, England
Multiple Objects

Several witnesses saw a narrow, long dusty cloud from which three very bright spots descended and joined.

Source: *Mirabilis Annus Secundus* (1662).

271.

April 1661, Between Ilford and Romford, England
Maneuvering light

About 10 P.M. Captain Chelmford, of Ipswich, and another man riding to London saw a fiery light with a green-white glow that changed direction. It approached at great speed, emitting light beams. When it was exactly overhead it suddenly changed direction again and disappeared at the horizon. Upon arriving in London, the two travelers had a notarial deed drawn up, recording their experience.

Source: *Mirabilis Annus Secundus* (1662).

272.

23 April 1661, Bednall-Green, England
Pillar containing lights

People saw a great pillar of fire with smaller objects (compared to "burning coals") within it, and at 10 o'clock that night "several persons near Pickadilly saw strange fiery clouds and other objects very terrible to the spectators, from some of whose mouths we received the information".

Source: *Mirabilis Annus Secundus* (1662).

273.

29 June 1661, Eastberry, Berkshire, England
Dark objects

The Sun was obscured by a great number of dark balls passing in front of it. Other objects looked like crosses.

Source: *Mirabilis Annus Secundus* (1662).

274.

October 1661, Bristol, England: Figures in a "cloud"

A cloud was seen rising out of the river. It opened up three times, revealing various figures inside.

Source: *Mirabilis Annus Secundus* (1662).

275.

April 1662, Tedworth, Wiltshire, England
The humanoid threat?

Numerous incidents of bedroom visitations, knocks and reports of witness paralysis by lights and humanoid entities.

Source: Rev. Joseph Glanvill, *Saducismus Triumphatus* (1681).

276.

11 May 1662, near Salt Ash, Cornwall, England
A great star and a black square object

At St. Stephens near Salt Ash, a "very great star" was reported, with the likeness of two red "legs" and a black square object. The star moved to and fro.

Source: *Mirabilis Annus Secundus* (1662).

277.

Late July 1663, Saint Martin, Brittany, France
Flying red cross

In the parish of Saint-Martin, near Quimper, a man named François Carré, from Bodeau, saw a red cross in the sky. It seemed to fly away from Saint Martin church and headed towards the chapel of Saint Michel.

Jérome de Lestour, a priest in Caudan in the diocese of Vannes, reports that "fearing to be the victim of an illusion, Carré called his wife without saying anything else than to look in the same direction. 'Do you see anything?' he asked. 'Yes, a red cross heading towards the chapel of Saint-Michel,' she answered."

Source: *The diary of Jesuit Father Julien Maunoir*, written in 1672, kept in the library of the Society of Jesus in Rome. Translation of this passage is by Yannis Deliyannis.

278.

15 August 1663, Roboziero near Bieloziero, Russia
Fiery object with burning beams

Farmer Levko Federov and others saw a fiery object estimated at 40 meters in diameter, with burning beams. It returned one hour later. A formal report from Saint Cyrille monastery reads:

"To His Highness the Archimandrite Nikita, to His Eminence the Starets Paul, to their Highnesses the Starets of St. Cyril Monastery, Most Venerable lords, salutations from your humble servant Ivachko Rievskoi.

"The farmer Lievko Fiedorov, from the village of Mys of Antusheva of your monastery estate Losy, has related to me the following facts: On this Saturday, the 15th day of August of the year 7171 (that is 7171 year of old style or 1663 of modern style), the faithful from the district of Bieloziero, Robozierskaya volost, had assembled in great number in the church of the village of Roboziero, in the present holyday of Assumption of the Virgin Mary.

"While they were inside, a great sound arose in the heavens and numerous people came out of God's house to watch it from the square. There, Lievko Fiedorov, the farmer in question, was among them and witnessed what follows which for him was a sign from God. At noon, a large ball of fire came down over Roboziero, arriving from the clearest part of the cloudless heavens. It came from the direction whence winter comes, and it moved toward midday (south) along the lake passing over water surface. The ball of fire measured some 140 ft. from one edge to the other and over the same distance, ahead of it, two ardent rays extended. The people seeing the terror of God gathered in the church and prayed to God and the Blessed Virgin, with tears and crying, and the big fire and two smaller ones disappeared.

"Less than an hour later, the people again came out to the square and the same fire suddenly reappeared over the same lake, from the same place where it first disappeared. It darted from the south to the west and must have been 1500 ft. away when it disappeared. But it appeared, in a short while, back again, from that another place, moving this time to the west; the third time the same fire ball appeared more terrific in width, and disappeared having moved to the west and it had been remaining over

Roboziero, over water, for an hour and a half. And the length of the lake is about 7000 ft., and the width is 3500 ft.

"As the fire ball was coming over water, peasants who were in their boat on the lake, followed it, and the fire burned them by the heat not allowing them to get closer. The waters of the lake were illuminated to their greatest depth of 30ft and the fish swam away to the shore, they all saw that. And where the fire ball came the water seemed to be covered with rust under the reddish light; it was then scattered by the wind and the water became clean again.

"And I, your humble servant, sent a message to the priests in Robozierskya district, exactly for this reason, and they responded to me with a letter confirming that they had such sign in that day. And you, most Venerable lords, would know about this. And this Robozierskaya district is about 6 miles away from Antusheva village of your monastery estate Losa."

Source: *Arkheografischeskaia Kommissiia,* Vol. 4 (covering years 1645-1676) (Saint Petersburg, 1842), 331-332. Courtesy of Thomas Bullard.

279. 

February 1665, Vienna, Austria: Flying box-like object

An object resembling a coffin was seen in the air, causing much anxiety.

Source: Walter G. Bell, *The Great Plague in London in 1665* (New York: Dodd, Mead & Co., 1924).

280. 

8 April 1665, Stralsund, Germany: Domed flying saucer

Aerial ships and a saucer-shaped object with dome were reported flying over the church of this town located near the Baltic Sea, and hovered there till evening. Witnesses were left trembling, with pain in their head and limbs. The

Fig. 23: The Stralsund phenomenon

case is described in researcher Illobrand Von Ludwiger's book *Best UFO Cases – Europe* published by the National Institute for Discovery Science in 1998.

Several fishermen first reported seeing a big swarm of starling birds flying in the sky about 2 P.M., coming from the north over the sea. They changed to battle ships fighting one another. A lot of smoke developed. New ships kept appearing, small and big ones, and the battle lasted for a few hours. Once the initial vision had vanished, the scenario changed. Writer Erasmus Francisci (whose real name was Erasmus Finx) describes the scene:

"After a while out of the sky came a flat round form, like a plate, looking like a big man's hat...Its color was that of the rising moon, and it hovered right over the church of St. Nicolai. There it remained stationary till evening. The fishermen, worried to death, didn't want to look further at the spectacle and buried their faces in their hands. On the following days they fell sick with trembling all over and pain in head and limbs. Many scholarly people thought a lot about that."

The Berliner *Ordinari und Postzeitungen* also wrote about the vision: "One of these fishermen had been sick on his feet. All of the citizens who have observed this are reliable. Yesterday, Herr Colonel von der Wegck and Docter Gessman interrogated two of the 6 fishermen. May God change this miracle for the best."

German researcher Von Ludwiger adds: "What the fishermen saw was a plate with a dome (man's hat) orange in color (like the rising moon) which hovered motionless for a long time and acted on the witnesses as if they became sick from strong radiation. (…) Erasmus Francisci hesitated to believe this account, because he could not find a suitable explanation: 'I read it at that time in the usual printed newspaper. But, to tell the truth, I didn't believe in that story, and I thought the fishermen had fished it out of the air or from a deceived imagination....'"

Francisci reported this account because between 1665 and 1680 several battles took place between the Swedes and the Prussians, and the spectacle could be given the meaning of a sign for an imminent war. Francisci states (p. 625): "After the sea was colored with so much blood after that time, the affair now seems to me believable. What the disk-like thing means to the good city shouldn't be hard to guess, if

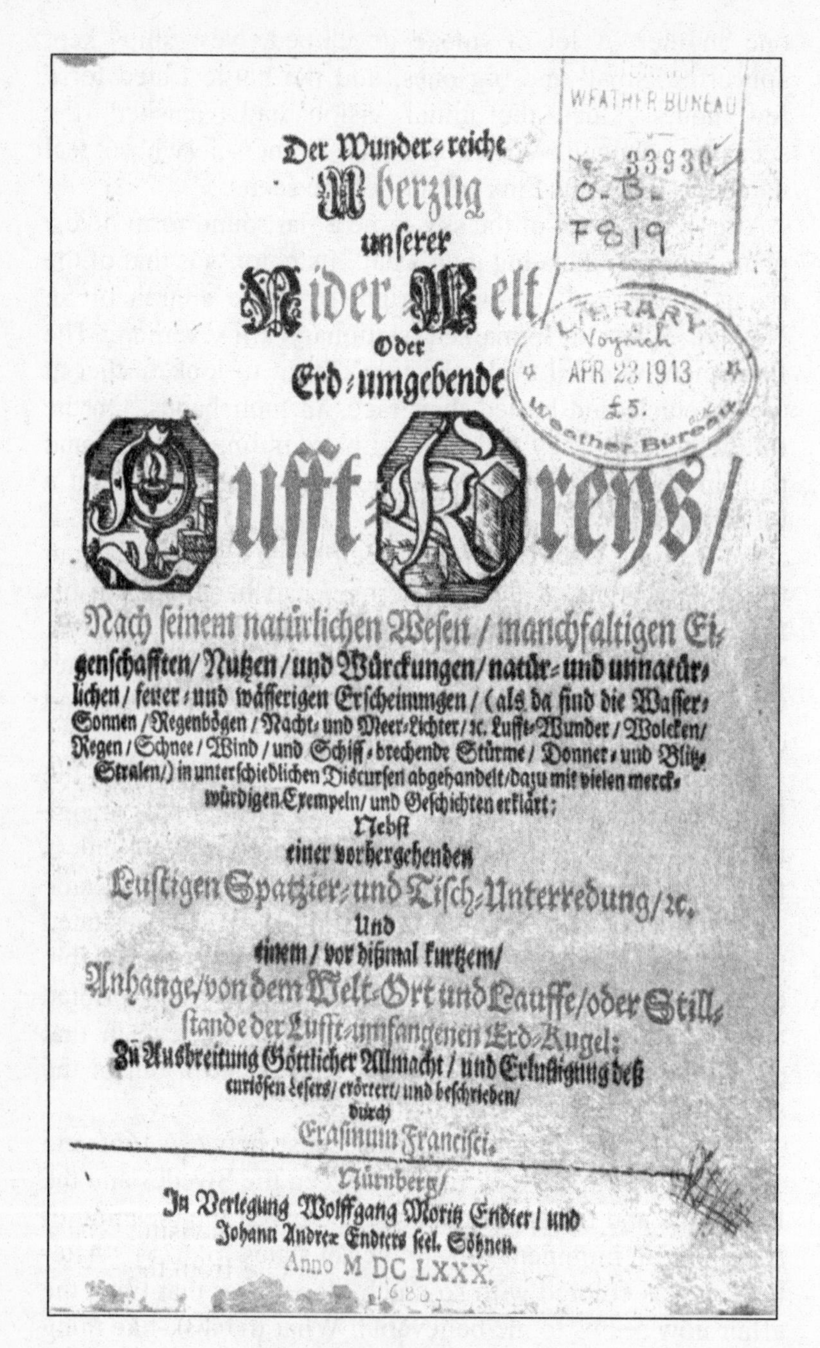

Der Wunder-reiche
Uberzug
unserer
Nider-Welt/
Oder
Erd-umgebende
Lufft-Kreys/

Nach seinem natürlichen Wesen / manchfaltigen Eigenschafften/ Nutzen / und Würckungen/ natür- und unnatürlichen / feuer- und wässerigen Erscheinungen/ (als da sind die Wasser-Sonnen / Regenbögen / Nacht- und Meer-Liechter/ ꝛc. Lufft-Wunder / Wolcken/ Regen / Schnee / Wind / und Schiff-brechende Stürme / Donner- und Blitz-Stralen/) in unterschiedlichen Discursen abgehandelt/ dazu mit vielen merckwürdigen Exempeln/ und Geschichten erklärt;

Nebst
einer vorhergehenden
Lustigen Spatzier- und Tisch-Unterredung/ ꝛc.
Und
einem / vor dißmal kurtzem/
Anhange/ von dem Welt-Ort und Lauffe/ oder Stillstande der Lufft-umfangenen Erd-Kugel;
Zu Ausbreitung Göttlicher Allmacht / und Erlustigung deß curiösen Lesers/ erörtert/ und beschrieben/
durch
Erasmum Francisci.

Nürnberg/
In Verlegung Wolffgang Moritz Endter / und Johann Andreæ Endters seel. Söhnen.
Anno M DC LXXX.

WEATHER BUREAU
33930
O. B.
F819
LIBRARY
Voynich
APR 23 1913
£5.
Weather Bureau
1680

Fig. 24: "Der Wunder-Reiche"

one remembers how the tower of St. Nicolai Church was destroyed in 1670 during wartime....'"

There is no question that the account is authentic, although the date may be incorrect by a few days. The title page of Francisi's book is given on the next page, as preserved in Berlin's Staatsbibliothek. The authors are indebted to researchers Yannis Deliyannis and Isaac Koi for tracking down important details of the case. After publication of von Ludwiger's book, some skeptics argued the observation could be accounted for by a mirage.

Source: Illobrand von Ludwiger, *Best UFO Cases – Europe* (Las Vegas, Nevada: National Institute for Discovery Science, 1998), quoting Erasmus Francisci (1680), and *Ordinari und Postzeitungen* (No.65) of April 10th, 1665. He also quotes from E. Buchner, *Medien, Hexen, Geisterseher* (16 Bis 18 JH), 42-43 (Munich: Albert Langen, 1926); Francisci, *Erasmus. Der wunder-reiche Ueberzug unserer Nider-Welt/Order Erd-umgebende* (Nürnberg, 1680).

281.

26 May 1666, Tokyo (Edo), Japan: Flying figure ⓘ

A mysterious light "20 feet long" shaped like a man flew towards the East. This is another example of a tantalizing report from Asia, about which we need more detailed information.

Source: Morihiro Saito, *The Messenger from Space*.

282.

20 January 1667, Gjov, Faeroe Islands
Luminous visitor

Jacob Olsen, 24, was awakened by a luminous visitor who healed him. He saw him again later, coming from the sky.

Source: Jacobsen Debes, *Færoæ & Færoa Reserata* (1673).

283.

15 November 1667, Mittelfischach, Germany
Sign of wonder

An engraving preserves the sighting of a "terrible sign of wonder" that took place during sunrise, and was seen for several hours in the sky over the town of Mittelfischach.

The image shows the sun shining through a break in the clouds while a group of people watch a formation of round lights. There is a scene of battle in the sky, and three crosses among dark nebulosities. The village is shown in detail to the left, with its church and a few houses.

Source: *Abriss des Erschrecklichen wunderzeichens, so sich den 15. Novembr. 1667 beim dorff Mittelfischach am Firmament des Himmels bey auff gehender Sonn etlich stunden lang sehen lassen.* [s.l.] (1667). [Goethe Universitätsbibliothek Frankfurt-am-Main, Einblattdr.G.Fr. 11]

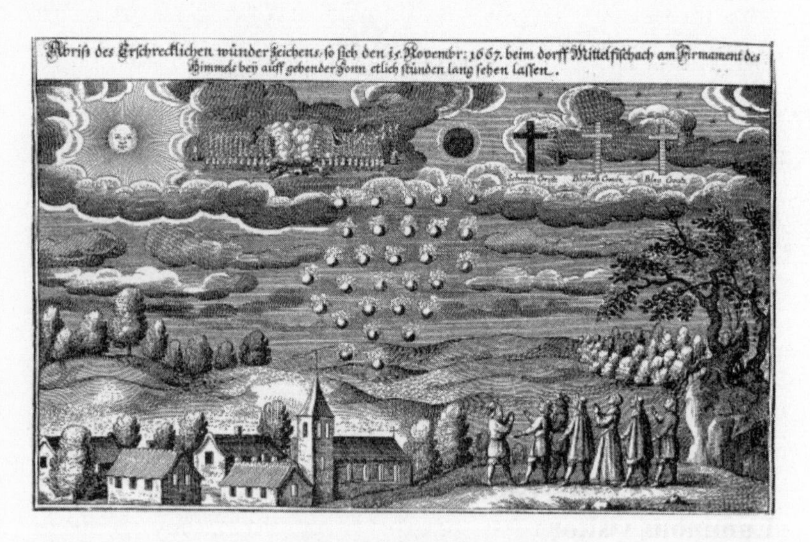

Fig. 25: Mittelfischach phenomenon

284.

Late December 1667, Bayárcal, Spain
Procession of lights

Bayárcal was a focus of attention for the inquisition. Among many testimonies relating to strange lights was that of Juan Muñoz, a tailor from Santander. In his sworn statement he said that in 1667, around Christmas time, at midnight, he saw a cross, behind which there was a banner, followed by four lights like wicks that flashed on and off. He supposed it to be the priest carrying the Viaticum (the Christian Eucharist given to a person in danger of death), though it seemed to be too bright.

As the lights moved, and he was also walking home, he reached a point some fifteen steps from them. He arrived home, quite anxious about what he had seen, but before closing the door he turned around to take another look. He then saw the lights pass in front on the church. Too frightened to investigate what the luminous thing was, he shut the door and swore he wouldn't even tell anyone about what he had seen. However, when he heard other neighbors relate they had seen the phenomenon, he told them what he had witnessed. They all agreed that such things had been seen many times before, and that the only possible explanation was that God sent them as signs to commemorate the lives of martyrs from the area.

The fact that Muñoz was new to the village and had never heard of the phenomenon was taken as proof that such things were not a mere figment of the imagination.

Processions of lights were seen in many Spanish villages, and in other European countries. Sometimes they would be seen over the rooftops, but usually at ground level, and could range in size from tiny to several meters across.

Source: Francisco A. Hitos, *Mártires de la Alpujarra en la Rebelión de los Moriscos* (1568). Republished by Apostolado de la Prensa, Madrid (1935).

285.

April 1670, countryside near London, England
Jane Lead's contact

An English woman named Jane Lead has her first contact with a bright cloud with a brilliant woman inside. Jane Lead was a British Christian mystic who lived from 1623 to 1704. Information about her early life is sparse but her family is known to have hailed from Norfolk. Born as Jane Ward in 1623, she married William Lead (or Leade) at the age of 21, and had four daughters by him.

According to her own writings, during a dance at a Christmas party, when she was 15 years old, Lead heard a miraculous disembodied voice. It said "Cease from this, I have another dance to lead thee in, for this is vanity." She interpreted this as a sign that she should devote her life to a spiritual cause, and in later life this decision led her into the study of theology, philosophy and alchemy.

In April 1670, as Lead reflected about the nature of Wisdom,

> There came upon me an overshadowing bright Cloud, and in the midst of it the Figure of a Woman, most richly adorned with transparent Gold, her hair hanging down and her Face as the terrible Crystal for brightness, but her Countenance was sweet and mild. At which sight I was somewhat amazed…

From this moment on, Jane Lead's life would be full of visions. Years later she would write about actual sightings that she had and several abductions by a group of beings who, by her own admission, were neither angels nor demons. (As noted by Jesse Glass in an article on Jane

Lead's mystic experiences, she often uses the term 'Magia' when referring to them). She calls her abduction experiences *Transports* throughout her diaries.

Source: Works of Jane Lead, especially her book *A Fountain of Gardens*. "Printed and Sold by J. Bradford, near Crowder's Wall," London 1696. Four Volumes. The original edition is very rare but most of Jane Lead's works can be found faithfully reproduced in on-line archives.

286.

18 August 1671, Regensburg, Germany
Signs in the clouds

Signs in the sky: An engraving shows an amazed crowd staring at ships in the sky, various mythical animals and

Wunderzeichen, zu Regenspurg gesehen am 18. Augusti. 1671.
Titulo XIII.

Fig. 26: Regensburg phenomenon

armies arrayed for battle. This engraving is cut from a book, *The Relationis historicae semestralis vernalis continuatio* (1672 edition) by Jacobus Francus and Sigismundus Latomus.

Source: *Wunderzeichen, zu Regenspurg gesehen am 18. Augusti 1671.* [Goethe Universitätsbibliothek Frankfurt-am-Main, Einblattdr.G.Fr. 12], engraving cut from Francus, Jacobus & Latomus, Sigismundus. *Relationis historicae semestralis vernalis continuatio* (1672). University of Frankfurt, Collection of Gustav Freytag (Einblattdr.G.Fr.12).

287.

25 January 1672, Paris Observatory, France
Unknown planetoid orbiting Venus

The great astronomer and planetary observer Giovanni Domenico Cassini, who was director of Paris Observatory at the time, recorded the presence of an object that seemed to be a satellite of Venus. He would not announce this discovery until he saw the object again, in 1686.

This supposed satellite was later named "Neith."

Source: "The Problematical Satellite of Venus," in *The Observatory* 7 (1884): 222-226.

288.

8 February 1672, off Cherbourg, France
Triple sky ships

Captain Isaac Guiton reports that a "star" came down; it split into two "ships", while a third one appeared later. The original reads: "An hour past midday, by the calmest weather in the world, appeared to us a star over our heads, about fifteen feet long. From there it went and fell to the north, leaving some smoke that formed into two ships, each with two lights and the mizzen and their large sails folded, both sailing into the south. The one on the north side was larger than the southernmost one. And as they sailed thus, they separated by about four feet, and another ship formed in the middle, seemingly bigger than the others, all black, and turning its bow to the north without any sails, yet equipped with its masts and ropes, as if resting at anchor. This seemed to us to take over half an hour. After which, they vanished to the south without leaving any trace…"

Source: Cited by Michel Bougard in *La chronique des OVNI* (1977), 96.

289.

16 November 1672, Tokyo, Japan ⓘ
Flying lantern

An object resembling a lantern flew away to the east.

Source: Takao Ikeda, *UFOs over Japan.*

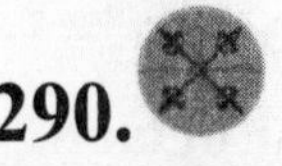

290.

1674, Japan, exact location unknown ⓘ
Fast-flying "dark cloud"

A dark, elongated cloud flies "like an arrow," on a N to SE trajectory. By definition, meteors are luminous, often described as "fiery." The description given here, of a dark object, seems to exclude the meteoritic explanation.

Source: *Brothers Magazine* I, 1, no original source quoted.

291.

23 May 1676, London, England: Bedroom visitation

In a diary entry, spiritual writer Jane Lead describes a visit from three mysterious figures.

"I saw one as in a Figure of a sprightly Youth, presenting himself near my Bedside, which amazed me, and I was afraid to take knowledge of him, who made out to me, as if he would draw my Aspect to him, but I could not find any Power for Speech with him. This disappearing, another in taller Stature, and more Manly Countenance, drew upon me, seeming to desire Familiarity with me; and then I looked when this Appearance would have spoke, but it was passive, and silent, only pleasant in its Countenance, who on a sudden withdrew. Then again was a presentation of a Person in a middle Stature, comely, sweet, and amiable for attraction; yet I being bounded in my Spirit, was hindered: otherwise I could have run with my Spirit into him."

While this apparition could be interpreted as an effect of schizophrenia this report would be seen as classic "bedroom visitation" by aliens in the context of today's abduction literature.

Source: *The Works of Jane Lead*, op. cit.

292.

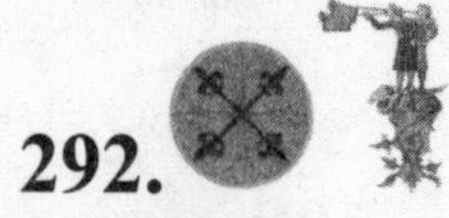

16 July 1676, London, England
Fiery object, a globe of light

Spiritual writer Jane Lead witnessed an object in the sky that she took to be "the eye of God." She wrote "This Morning…there appeared to me an Azure blue Firmament, so Oriental as nothing of this, in this Visible Orb could parallel with it. Out of the midst hereof was a most wonderful Eye, which I saw Sparkling, as with Flaming Streams from it. Which I am not able to Figure out, after that manner, in which it did present it self unto me. But according to this Form it was, as much as I am able to give an account of it, it was thus, or after this manner. There was a Flaming *Eye* in the midst of a Circle, and round about it a Rainbow with all variety of Colours, and beyond the Rainbow in the Firmament, innumerable Stars all attending this Flaming Eye."

Jane's style is as difficult and archaic as that of any 17th century mystic, but what she seems to be describing is a luminous or fiery ("sparkling…flaming") lenticular ("eye-shaped") object that flew over London on a sunny day ("the blue firmament"). The object emitted rays or jets ("flaming streams") and seemed to move within a wider rainbow-colored circle, perhaps accompanied by smaller objects

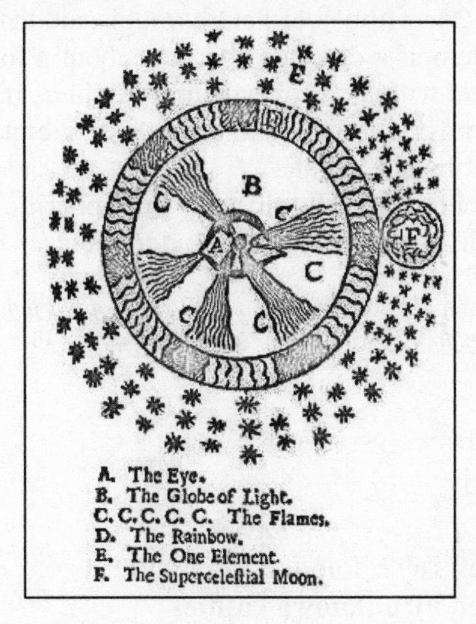

Fig. 27: Jane Lead's vision

which looked like stars. Her drawing shows an oval or eye-like thing emitting five shafts of flame or light, enclosed by a thick circle surrounded by stars. The circle's interior is labeled "The Globe of Light," suggesting something more substantial than a mere 'ring' or 'rainbow.'

Source: *The Works of Jane Lead*, op. cit.

293.

20 Sept. 1676, Uffington Fields, England: Wavy dart

A fiery 'meteor' in the shape of a dart moved with a wavy vibrating motion. At 7 P.M., according to Morton, an "unusual meteor" was seen by residents of Northamptonshire. Mr. Gibbon of Peterborough said that

"The stem at a distance appear'd about a foot and half in length and with a narrow stream of light as if were a String of Cord affix'd to it. It had a wav'd or vibrated motion. Its duration about a minute."

Mr. Gibbon allegedly first saw this apparition at the zenith as it made its way toward Uffington Fields.

Source: John Morton, *The Natural History of Northamptonshire; with some account of the antiquities* (London, 1712), 348.

294.

22 March 1677, London, England
Assaulted by unknown entities

Spiritual writer Jane Lead wrote that during the night she had been "cast as into a magical Sleep, where I saw my self carried into a Wilderness." There she found herself in a peaceful, natural environment. Before she could enjoy these circumstances, however, a being that she had seen before and two other female entities "did make a kind of Assault upon me; but one of the Females was more fierce, and did give my outward Skin a prick, as with a sharp Needle. Upon which I called for Angelical aid to succour me, or else too hard they would be. Whereupon I was parted from them, and saw them in that place no more…"

Lead writes that after this vision, before waking up in bed, she was told that she needn't worry, that it would not happen to her again.

Source: *The Works of Jane Lead*, op. cit.

295.

30 December 1677, at sea West of Granada, Spain
Unknown "star"

Pierre Boutard, an officer aboard the ship *La Maligne* notes in the logbook that "on Thursday the thirtieth day of December 1677 in the morning about 4 hours, we have seen a star in the direction of northwest ¼ west, ending southeast ¼ east, but carrying (such) a great light that all on board thought there was widespread fire, but it was accompanied by over 200 rays carrying such a light, that we believed we were all lost. We dropped anchor about 9 or 10 in the morning in the small bay of Grenada."

Source: Michel Bougard, *La chronique des OVNI* (1977), 97.

296.

9 February 1678, London, England
Landing of a large ship

In a diary entry entitled *A Transport*, Jane Lead writes:

"In the Morning after I was awaked from Sleep, upon a sudden I was insensible of any sensibility as relating to a corporeal Being, and found my self as without the clog of an Earthly Body, being very sprightly and airy in a silent place, where some were beside my self, but I did not know them by their Figures, except one, who went out, and came in again: and there was no speaking one to another, but all did set in great silence."

Lead's ordeals could not be closer to the situation of a modern abductee: She is woken up in a disoriented trancelike state, possibly confusing reality with a dream or a recent half-forgotten memory; around her are 'figures' she does not know, except for one; there is an eerie silence; next, Lead recalled seeing a gold-colored craft "come down" to "a pretty distance" from where she was.

"It was in the form of a large Ship" with four golden wings. The ship "came down with the greatest swiftness as is imaginable." She asked some of the figures beside her if they could see what she could, and mysteriously they said they couldn't! No doubt puzzled by their answer, Jane looked again and *saw herself* in front of the others, "leaping and dancing and greatly rejoicing to meet it."

Bar the detail about a third-person view of herself when the ship landed, this is the kind of account given by people whose cases fill countless UFO books today, and whose stories are often taken at face value.

It is not sufficient to accuse abductees of confabulation and of sharing science-fiction fantasies *because the same 'fantasies' have been reported and believed for hundreds of years*, since long before the popularisation of the genre. Was Lead's vision a muddled memory of an earlier experience?

Lead's diary entry of February 9th concludes: "But when I came up to it [the Ship], then it did as suddenly go up again, withdrawing out of sight, unto the high Orb from whence it came. After which I found my self in my Body of sense, as knowing I had been ranging in my Spirit from it for a while, that I might behold this great thing."

Source: *The Works of Jane Lead*, op.cit.

297.

17 September 1680, Lisbon, Portugal
Landing of a hairy occupant

A large black cloud-like mass landed in a field, releasing a huge hair-covered being that appeared – and disappeared.

Source: A photocopy of the pamphlet, written by Julio Alberto de la Hinojosa, was reproduced in its entirety in *Fenómenos Celestes en el Pasado: Siglos VIII al XIX*, published by the Centro de Estudios Interplanetarios, Barcelona, 1995.

298.

17 November 1684, Saint Aubin, Brittany, France
Tear-shaped object

About 10 A.M. a priest from Lannion saw "a flame in the shape of a teardrop, as big as one's hand, coming down from the sky. Its motion was extremely slow, for it took no less than seven to eight minutes to reach the horizon. It seemed a bit bluer. Its tail threw off sparks, and it was on the opposite side from the sun."

Source: *Histoire de l'Académie Royale des Sciences* (1684), 419.

299.

9 July 1686, Leipzig, Germany
Unknown astronomical object

About 1:20 A.M. a brilliant object, half the apparent size of the Moon, was observed hovering for a full 15 minutes. The observer was "the late Mr. Gottfried Kirch, for many years a diligent observer of the heavens, perfectly well instructed in astronomical matters," according to Rev. Edward Polehamton, who notes:

"A fire ball with a tail was observed, in 8 ½ quarter degrees of Aquarius and 4 degrees north, which continued immoveable for half a quarter of an hour, having a diameter equal to half the moon's diameter. At first, the light was so great that we could see to read by it; after which, it gradually vanished in its place. This phenomenon was observed at the same time in several other places; especially at Schmitza, a town distant from Dantzig eleven German miles, towards the south, its altitude being about 6 degrees above the southern horizon.…Whence, by easy calculus, it will be found, that the same was not less than sixteen German miles distant in a right line from Leipsic, and above 6 ½ such miles perpendicular above the horizon, that is at least thirty English miles high in the air. And though the observer says of it, *immotus perstitit per semi-quandrantem horae*, it is not to be understood that it keeps its place like a fixed star, all the time of its appearance; but that it had no very remarkable progressive motion. For he himself has, at the end of the said *Ephemerides*, given a figure of it, whence it appears that it darted obliquely to the right-hand, and where it ended, left two globules or nodes, not visible but by an optic tube."

Source: Gotfried Kirch, *Ephemerides* (contained as an appendix to the ephemeride for the year 1688). Quoted by E. Polehamton in *The Gallery of Nature and Art, or a Tour through Creation and Science* (1815).

300.

28 August 1686, Paris Observatory, France
Mystery planetoid near Venus

A second observation by Cassini of the supposed satellite of Venus, which would later be named "Neith." Venus was a morning "star" at the time, with heliocentric longitude 59° and elongation 38°. The object was estimated to be ¼ the diameter of Venus and it showed the same phase as Venus. Cassini then revealed his two sightings.

Source: "The Problematical Satellite of Venus," in *The Observatory* 7 (1884): 222-226.

301.

Circa June 1688, Yunan Province, China
Flying umbrella

A large yellow "umbrella-like" object rose from the ridge and came down again, with many lights:

"In the year 27 under the reign of emperor Kangxi of the Qing dynasty, my brother-in-law Bixilin went to his home in the mountains, 20 kilometers from the city of Kunmin. While staying there, he saw every day at noon, when the weather was clear, a large yellow cover like an umbrella that rose slowly above a ridge. This object threw such brilliant lights that he dared not look at it directly. It rose and got lost into the clouds. A little while later it would come down, always slowly, going up and down in the same way. At nightfall, the flying object lost its yellow color and turned paler and blurry. It disappeared completely when the sky was dark."

Source: Shi Bo, *La Chine et les Extraterrestres*, op.cit., 36.

302.

20 December 1689, England, exact location unknown
Strange object

About 4:45 A.M., a fiery object shaped like a half-moon changed into a bright sword and "ran westward."

Source: "Diary of Jacob Bee of Durham," reprinted in *Six North Country Diaries*, vol. CXVIII, J.C. Hodgson, ed. (Durham: Surtees Society, 1910).

303.

6 May 1692, Edo (Tokyo), Japan: Three unknowns ⓘ

In broad daylight, three luminous objects like the sun, moon and a star appeared, sparkling "in an unearthly way."

Source: *Inforespace* 25.

304.

1693, Hamburg, Germany: Round machine ⓘ

A very luminous, round "machine" with a sphere at its center, crossing the sky.

Source: Researcher Winkler (in a catalogue published by the Fund for UFO Research) cites Peter Kolosimo. Unfortunately we have no specific reference for this case, which could refer to an ordinary meteor. There could also be some confusion with case 310 below.

305.

Sept. 1693, Bowden Parva, Northampton, England
Unknown, complex object

"The top of it was in Form of the letter W: And had a Lift or String of Light appendant to the lower Part of the W, about a Yard and Half in Length. It continued some time, and was seen by several round the country."

Source: John Morton, *Natural History*, op. cit.

306.

December 1693, Egryn, Merionethshire, Wales
Unexplained fiery phenomenon

A "fiery exhalation" came from the sea and set fire to the hay with "a blue weak flame." The fire, though easily extinguished, "did not the least harm to any of the men who interposed their endeavour to save the hay, though they ventured (perceiving it different from common fire) not only close to it, but sometimes into it."

Source: The oldest report comes from a letter dated January 20th, 1694 by a certain Maurice Jones to the author of the additions to *Cambden's Brittania.* The letter was published in *The Philosophical Transactions*, vol. XVIII (1694), pp. 49-50, along with a chronology and names of witnesses of the events.

307.

23 July 1694, London, England
Figures moving within a fiery circle

Jane Lead describes something she calls an "enclosed Principle" surrounded by a fiery circle:

> Within it there appeared bright Beryl Bodies walking up and down, and with them did appear as in a Looking glass, like as a round Globe, the Personal Glory of our Mighty King, who moved as they moved. The Ground they went upon, was paved as with Sparkling Stones, with Veins of Gold, which cast forth a mighty Lustre.

Source: *The Works of Jane Lead*, op. cit.

308.

25 November 1696, Tobolsk, Russia
Human form in the sky

An object split into four luminous parts, the center being dark, with a human form inside. The report reads: "On Sunday, at 2 P.M., there appeared in Tobolsk a sign in the sun: it split into four parts, as it were, and the rays from the sun were light, but in the middle it was dark. And among these parts one could see in the darkness something like a man with extended arms."

Source: Cherepanov, *Aerial Fears of Tobolsk in Olden Days–from the Siberian Olden Days.* (Tobolsk, 1882). Quoted in *Zvesdochtets* (Moscow, 1990), 214-215.

309.

28 November 1696, Tobolsk, Russia
Double object descending

At 5 A.M. "there was a sign in the east: from a dark cloud there was suspended something like an iron-clamp with a great fire that shone brightly and descended down to the very earth."

Source: Cherepanov, *Aerial Fears of Tobolsk in Olden Days – from the Siberian Olden Days* (Tobolsk, 1882), op. cit.

310.

4 November 1697, Mecklenburg & Hamburg, Germany
Two wheels in the sky

Two enormous, glowing wheels are shown in a picture with crowds watching the sky. This was primarily seen between 6 and 7 P.M.: "A great fiery ball in the shape of a cannon ball was seen floating... which phenomenon or air sign also above the horizon of Hamburg at the same time but below was seen floating a cross shape..." After a quarter of an hour the ball emitted a bang. It then "disappeared from the sight of many thousands of spectators." After the same delay the cross-shape, described as having "a sulfur-gleaming, bright shining terrible lightning" also departed.

Fig. 28: Mecklenburg phenomenon

Source: *Nachdencklich-dreyfaches Wunder-Zeichen: 1. Eines groß-erschröcklich-feurigen Cometen, 2. Eines entsetzlichen Feuer-Kugel Lufft-Zeichens, 3. Einer sehr ungestalten Fontange-Mißgeburt ...,* [s.l.], 1697 [Goethe Universitätsbibliothek Frankfurt-am-Main, Einblattdrucke G. Freytag].

311.

1699, St. Didier, Vaucluse, near Avignon, France
Merging globes in the sky

A priest saw a large light and three globes coming from the sky and merging together: "As I arrived near the oratory I saw the sky open, a great light appeared and soon I observed three globes of fire. The middle one was higher than the other two. I thought, 'here are the lights I have been told about.' Immediately I fell to my knees and thanked God for such a great marvel. At the same time, two more lights appeared, but a bit higher than the place where the chapel is located (…) The two globes merged with the middle one and vanished."

Source: *L'histoire du diocèse d'Avignon* by Abbot Granget, cited by Michel Bougard, *La chronique des OVNI* (1977), 99-100.

Epilogue to Part I-D

From the scientific observations of several astronomers to the visions of Jane Lead, the 17th century is especially interesting to a researcher of unusual aerial phenomena because it gives us a complete template by which to parse the claims and sightings of today's witnesses.

This section of our Chronology records stories of abductions by little people (interpreted as "fairies" in Celtic countries, but similar in stature and behavior to today's Aliens) as well as reports of partial paralysis and occasional healing powers among humans exposed to these phenomena. Such enigmas continued to be seen in the light of theology, to the grave detriment of poor witnesses accused of commerce with demons, but a new philosophical movement would soon remove the old religious backdrop in favor of a revolutionary, "experimental" mode of thinking.

The observations recorded in *Mirabilis Annus*, an important document we have quoted on several occasions, provide a good illustration of the context of the prodigies and their interpretation for political or religious purposes. In his thoughtful analysis (*An Age of Wonders, Prodigies, Politics and Providence in England 1657-1727*, Manchester University Press, Manchester and New York, 2002, 27-30) William E. Burns notes:

> *Mirabilis Annus* made clear its political point very early on. Rather than merely adducing specific prodigies to demonstrate divine displeasure, it adopted an apocalyptic tactic of delegitimizing the regime through the sheer quantity of prodigies alleged to have taken place in the preceding 'Year of Wonders'

Whatever the physical nature of the unexplained objects that triggered the sightings, their interpretation allowed critics of the régime to vent their opposition:

> The prodigies that *Mirabilis Annus* actually recounted, which divided into the four categories of prodigies of air, fire, earth and water, and judgments on particular individuals, continued to undermine the regime's legitimacy by depicting it as sinful and weak. One prominent technique for this was the use of historical parallels.

Aerial phenomena were also used as symbols of desired events, historical changes that the compiler of the changes wished to see happen. Again, in the words of Burns:

> *Mirabilis Annus* claimed that a Surrey gentleman had a vision of a glorious cathedral in the sky beside a small church with a star inside it. The cathedral vanished, while the small church, whose star suggests the glory of God, was exalted. This symbolically represented the hope of dissenters that the Church of England would be overthrown and that the small gathered churches of the dissenters would triumph over it. Even less subtle was the appearance of a black cloud dropping fire over Westminster Palace and the Parliament House.

We are left with the fact that the interpretation of the reported events is generally biased by the writer who recounts the cases, but that may be the price we have to pay for obtaining any knowledge of the underlying phenomena in the first place. As to the actual explanation for the sightings, it is left for us to discover.

Early in the 17th century Descartes and Pascal in France, and Francis Bacon in England, had already introduced new methods of inquiry into the order of nature. By the end of the period people were beginning to think in new ways, inspired by the progress of science based on observation.

The Age of Reason was imminent: in 1703 Isaac Newton would be elected President of the Royal Society; in 1705

Edmund Halley would predict that the comet last seen in 1682 would return in 1758 (it did) and in 1707 French inventor Denis Papin would invent the high-pressure boiler that would lead to the first steam-powered ship and would pave the way to James Watt's steam engine and the industrial revolution.

The new impetus in science in the closing years of the seventeenth century parallels a worldwide evolution in classical literature, in education and in the arts.

When the year 1700 comes around, there are literate people everywhere who are eager to read intelligent reports of new ideas and discoveries. Magazines circulate throughout Europe and America; new journals are born. Naturally, reports of unusual aerial phenomena continue to thrive in this new enlightened culture. They are now reported in the pages of well-edited periodicals like the *Gentleman's Magazine*, the *Annual Register* or the *Philosophical Transactions* of the Royal society of London, "*giving some account of the present undertakings, studies, and labours of the ingenious in many considerable parts of the world.*"

As we shall see, these "undertakings and labours" often had to do with an attempt by "the ingenious" to understand phenomena that were beyond the physics of the time – and still present us with a most interesting challenge today.

PART I-E

Eighteenth-Century Chronology

Often called "The Century of Enlightenment," the eighteenth century is characterized by intense interest for the rational study of nature, systematic investigation of "meteors," the rise of an international community of scientists and "natural philosophers," experiments with electricity, Benjamin Franklin's demonstrations of the nature of lightning, the wide development of navigation, the worldwide recognition and imitation of the Royal Society, and early attempts to fly culminating in the first manned balloons.

The search for new planets gave rise to numerous observations of unknown bodies by competent astronomers, both professionals and amateurs, eagerly reported in considerable detail in the pages of the new scientific journals and publications dedicated to an enlightened elite.

The eighteenth century belonged to Newton and Lavoisier, to the triumph of Reason. Unlike modern "rationalists," however, intellectuals who considered themselves enlightened were dedicated to careful observation of nature and did not recoil before its more mysterious aspects. On the contrary, unusual aerial phenomena were carefully documented, published and commented upon with an openness of mind that is sorely lacking in our "modern" era of institutionalized science.

312. 

August 1700, Sahalahti, Eastern Finland ⓘ
Abducted by a disk

An old man, a smith named Tiittu, is said by a local story to have disappeared shortly after a flying disk hovered over the village. His son went to search for him, and met a being he perceived as a "bear" who said he had flown off.

"After Tiittu had gone to the forest, the same day villagers saw a huge disc hovering above the village. It stayed without moving for a moment, then started to fly out to the direction where Tiittu had gone to. Villagers believed that it was a mark of the end of the world. They were horrified.

"For two days they stayed inside praying, singing religious songs and confessing their sins. Only in the third day they were calm enough to go back to their normal work. When Tiittu didn´t return, the villagers started to look for him. In the forest Tiittu´s son suddenly met a big being looking like a bear. The being started to speak in Finnish: 'Don´t be afraid. I can tell you that you are looking for your father in vain. You saw that 'sky ship' like a rainbow—it took your father up to the heights, to another, better world, where lives a race much higher than your people. Your father feels good there and doesn´t miss his home.' The bear disappeared, and they stopped looking for Tiittu.

"All the people of Sahalahti were talking about the mysterious case. Then they got a new priest, who announced in the church: "This story speaks of sinful witchcraft, and it represents the imagination of drunken and mad people, so you´d better forget it."

Source: Finnish researcher, Tapani Kuningas, published this story in the Finnish magazine *Vimana* (No. 3-4, 1967) and later in his book *Ufoja Suomen taivaalla* (Kirjayhtymä, Helsinki, 1970). He claimed the story was a local tradition in Sahalahti, in East Finland. However, no confirmation for this exists except for a single letter, which is now lost.

313.

1701, Cape Passaro, Sicily, Italy: Hovering light

Witness C. De Corbin reports observing a very bright light in the sky, hovering for two hours in spite of a strong wind.

Source: Abbé J. Richard, *Histoire Naturelle de l'Air et des Météores* (1771).

314.

September 1702, Japan, exact location unknown ⓘ
Red residue from a sun-like object

An object like a red sun was seen in the sky, dropping cotton-like filaments.

Source: *Brothers Magazine* I,1.

315.

1704, Hamburg, Germany: Sparkling flying boats

People saw the sky "crisscrossed with sparkling boat-like objects" chasing one another, blending and separating, multiplying in plain view. We have too little information to conclude they saw an aurora borealis.

Source: Yves Naud, *UFOs and Extraterrestrials in History* (Geneva: Ferni, 1978), vol. II, 176.

316.

28 October 1707, Hidaka County, Wakayama, Japan ⓘ
White light

During a tsunami that struck the coast, a luminous object like a white ball appeared in the waves.

Source: Takao Ikeda, *Nihon nu ufo* (Tokyo: Tairiku shobo, 1974).

317.

18 December 1707, Southern coast of England
Huge cylinder

A huge cylinder and an odd cloud moved along with nocturnal lights, low on the horizon. The phenomenon was described by "the Worshipful Charles Kirkham, Esq." as "a long dark Cloud of a Cylindrical Figure which lay horizontally, and seemed to divide the Brightness into two almost Equal Parts. It had little or no motion, tho' the Wind blow'd brisk. But on a sudden there appear'd a swelling Brightness in that Cylindric Cloud, which broke out into Flames of a pale-coloured Fire."

The flames lasted less than half a minute, with "the Cloud from whence they proceeded still keeping its first Position, and not diminish'd. It was wonderfully frightful and amazing."

Source: Rev. John Morton, *Natural History of Northamptonshire* (1712), 349-350.

318.

11 May 1710, London, England: Man in a flying object

At 2:00 A.M. multiple witnesses saw "a strange comet" which seemed to be carried along with two black clouds. "After which," according to the report, "follow'd the likeness of a Man in a Cloud of Fire, with a Sword in his Hand, which mov'd with the Clouds as the other did, but they saw it for near a quarter of an Hour together, to their very great surprise..." The scene was depicted in a woodcut.

Fig. 29: Sighting by nightwatchmen in London

Source: *The Age of Wonders: or farther and particular Discriptton[sic] of the remarkable, and Fiery Appartion[sic] that was seen in the Air, on Thursday in the Morning, being May the 11th 1710. also the Figure of a Man in the Clouds with a drawn Sword; which pass'd from the North West over toward France, with reasonable Signification thereon; and the Names of several Inhabitants in and about the City of London, that saw the same, and are ready to Attest it. Also an Account of several Comets that have appear'd formerly in England, and what has happen'd in those Years* (London: J. Read, 1710?). [British Library, 1104.a.24]

319.

2 April 1716, Tallin, Baltic Sea: Clouds fighting

Two large dark clouds engaged in combat, and many smaller fast clouds.

The phenomenon was observed over the Baltic Sea, near Revel (modern Tallinn). The reports come from various official documents and ship logbooks. It was the second day of Easter, at around 9:00 P.M., when a dense or black cloud appeared in the sky. Its base was wide but its top was pointed, and it seemed to travel upwards quickly, "so that in less than three minutes its angle of elevation reached half of a right angle." As the cloud appeared "there manifested in the WNW direction an enormous shining comet that ascended up to about 12 degrees above the horizon." At this moment, a second dark cloud rose from the north, approaching the first one: "There formed between these two clouds, from the north-eastern side, a bright light in the shape of a column that for a few minutes did not change its position..."

One version states that this column of light remained still for around ten minutes. Then the second cloud moved very quickly through the column, "and hit the other cloud that was moving from the east." The collision produced "great fire and smoke" for about fifteen minutes, "after which it began to gradually fade and ended with the appearance of a multitude of bright arrows reaching an [angular] altitude of 80 degrees above the horizon."

Source: M. B. Gershtein, "A Thousand Years of Russian UFOs," *RIAP Bulletin* (Ukraine) 7, 4, October-December 2001. The two accounts provided here were made by Baron de Bie, the ambassador of the Netherlands, and Russian Commander N. A. Senyavin.

320.

6 March 1717, at sea southwest of Martinique
Hovering object

A solid object like a mast hovers two feet above the water. In his log Chevalier de Ricouart, captain of the frigate *La Valeur*, noted: "At two in the morning we were making some progress in a southeast direction. We saw something like the mast of a ship pass alongside, standing up about two feet above the water."

Source: Michel Bougard, *La chronique des OVNI* (Paris: Delarge, 1977), 104.

321.

19 March 1719, Oxford, England: A physicist puzzled

Very bright, whitish and blue object moving from the west in a straight line at 8:15 P.M., much slower than a meteor. Multiple witnesses all over England, including the Vice-President of the Royal Society, physicist Sir H. Sloane, who saw it travel over 20 degrees in "less than half a minute." Although listed as a meteor, the slow speed is most curious.

Source: Sir Edmund Halley, "An Account of the Extraordinary Meteor Seen All Over England," *Philosophical transactions of the Royal society of London* 30 (1720): 978-990.

322.

16 January 1721, Bern, Switzerland
Three globes emerge from a pillar of fire

At night, there "was perceived a great Pillar of Fire standing over the Mountains, near that City, to the Westward of its Fortifications, which advancing by little and little toward the City, burst at length, without making any great Noise, and then three Globes of Fire was seen to Issue out of it, which took each of them a different Way, and at length disappeared."

Source: Anon., *An account of terrible apparitions and prodigies which hath been seen both upon Earth and Sea, in the end of Last, and beginning of this present Year, 1721* (Glasgow: Thomas Crawford, 1721), 5-7.

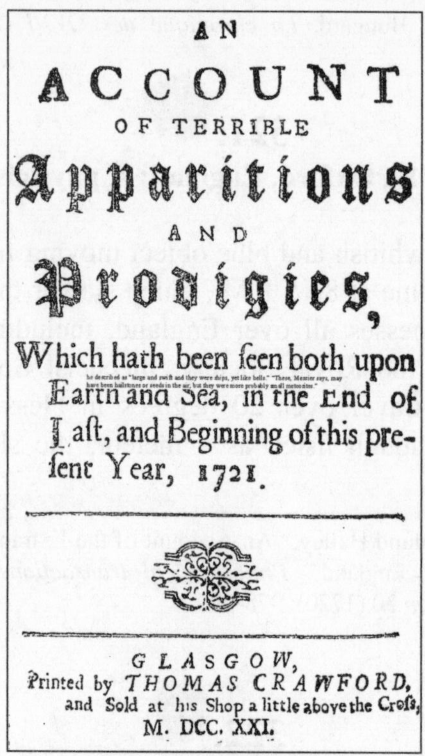

AN

ACCOUNT

OF TERRIBLE

Apparitions

AND

Prodigies,

Which hath been ſeen both upon

he described as "large and swift and they were ships, yet like bells." "These, Messier says, may have been hailstones or seeds in the air, but they were more probably small meteorites."

Earth and Sea, in the End of Laſt, and Beginning of this preſent Year, 1721.

GLASGOW,
Printed by THOMAS CRAWFORD,
and Sold at his Shop a little above the Croſs,
M. DCC. XXI.

Fig. 30: Prodigies in Bern

323.

29 October 1726, Vilvoorde, Brabant, Belgium ⓘ
Terrifying objects

About nine o'clock at night and for two consecutive hours were seen in the sky "horrible and strange meteors" that came among the clouds like lightning and disappeared in the same way. Their aspect was most terrifying.

In the absence of a more complete description, we cannot exclude the notion that witnesses may have observed an aurora borealis.

Source: J. Nauwelaers, *Histoire de la Ville de Vilvorde*, vol. 2 (Paris, 1941).

324.

1729, Finis Terrae Cape, Galicia, Spain
Strangers from the sky

A local story claims that three men came out of a cloud, had a meal at the market, took off and flew south.

Source: Benito Jeronimo Feijoo, *Teatro crítico universal (1726-1740),* Volume Three (1729). Text from the Madrid edition of 1777, 86-87.

325.

1 October 1729, Noes, Uppland, Sweden: Fiery globe

Two hours prior to sunrise, M. Suen-Hof saw red vapors in the sky, which stretched in wide bands from north to south, then proceeded to gather together into a fiery globe about two feet in diameter. The globe kept moving in the same direction where the reddish vapors had appeared. It emitted sparks and was as bright as the sun. After moving through a quarter of the sky it disappeared abruptly, leaving thick black smoke and a burst of sound similar to cannon shot.

Source: Sestier, *La Foudre et ses formes*, T.I., 222. Cited by Camille Flammarion, *Bolides Inexpliqués par leur aspect bizarre et la lenteur de leur parcours–Bradytes,* in *Etudes et Lectures sur l'Astronomie* (Paris: Gauthier-Villars, 1874), T.5, 143.

326.

2 November 1730, Salamanca, Spain ⓘ
Globe of fire with beams

Torres himself was a witness to the incident. He wrote that at 11:30 P.M. he saw, from Salamanca, "an amazing Globe of fire," as large as a building. On each side of the globe were two luminous beams or columns which seemed to rise and fall, "becoming brighter as they moved." The columns changed from green to red and the light from the phenomenon illuminated the surrounding area. At two o'clock in the morning the columns joined together but the spectacle did not disappear until 4:30 A.M.

Though often cited as a UFO, many consider the event to have been an aurora borealis.

Source: Diego de Torres Villaroel, *Juicio, i Prognostico del Globo, i Tres Columnas de Fuego* (Madrid: Manuel Caballero, 1730).

327.

9 December 1731, Florence, Italy
Unexplained luminous "cloud"

"A luminous cloud was seen, driven with some violence from east to west, where it disappeared below the horizon." A contemporary author named Bianchini speaks of several strange luminous spheres making a whirring sound.

Source: Robert Mallet, "Catalogue of recorded earthquakes from 1606 BC to AD 1850," *Annual Report of the British Association for the Advancement of Science* (1852), and "Third Report on the facts of earthquake phenomena," id., 1853 and 1854.

328.

17 March 1735, London, England: Unexplained light

Dr. John Bevis observed an unknown light in the sky. It remained stationary for one hour.

Source: John Bevis, MD. "An Account of a Luminous Appearance in the Sky, seen at London…" *Philosophical Transactions* (1739-1741), 41: 347-349.

329.

5 December 1737, Sheffield, England
Beams of hot light from a luminous body

At about 5 P.M. a peculiar phenomenon was seen. The witness (astronomer Thomas Short) described it as "a dark red cloud that made its appearance, with a luminous body underneath that sent out very brilliant beams of light."

It did not look anything like aurora borealis, because the light beams were moving slowly for some time, then stopped. Suddenly the air was so hot that he had to take off his shirt, although he was outside.

Source: Thomas Short, "An Account of Several Meteors, Communicated in a Letter from Thomas Short, MD to the President," *Philosophical Transactions (1683-1775),* 41 (1739-1741): 625-630.

330.

6 December 1737, Bucharest, Romania: Red intruder

In the afternoon, an object only described as a "Symbolic form," blood-red in color, appeared from the west. After remaining in the sky for two hours it split into two parts that shortly joined again and went back towards the west.

Source: Ion Hobana and Julien Weverbergh, *Les Ovni en URSS et dans les Pays de l'Est* (Paris: Robert Laffont, 1972), 287-288, citing Biblioteca Academiei Române, BAR ms. rom. 2342, fol. 3-4.

331.

23 February 1740, Toulon, France ⓘ
Rising purple globe plunges, releases balls of fire

During the night of 23 to 24 February people saw a purple "globe of fire" that rose gradually, and then appeared to plunge into the sea, where it rebounded. Reaching a certain height, it blew up and spread several balls of fire over the sea and the mountains. It made a sound like that of a violent thunderclap or a bomb as it burst. The witnesses reported the event to the Marquis de Caumont.

Source: *Histoire de l'Académie des Sciences*, 1740.

332.

23 October 1740, England
Unknown planetoid orbiting Venus

Astronomer and mathematician James Short, one of the most prolific telescope makers of the 18th century, reported his observation of what he thought was a satellite of Venus (later called "Neith" by Hozeau). The heliocentric longitude of Venus was 68° and its elongation 46°.

Source: "The Problematical Satellite of Venus," in *The Observatory* 7 (1884): 222-226.

333.

16 December 1743, London, England
Slow, waving 'rocket'

A correspondent of the Royal Society reports on an unusual sighting in these terms:

"As I was returning home from the Royal Society to Westminster, (at) 8h 40m, being about the Middle of the Parade in St. James Park, I saw a Light arise from behind the Trees and Houses in the S. by W. point, which I took at first for a large Sky-Rocket; but when it had risen to the Height of about 20 Degrees, it took a motion nearly parallel to the Horizon, but waved in this manner, and went on to the N. by E. Point over the Houses.

"It seemed to be so very near, that I thought it passed over Queen's Square, the Island in the Park, cross the Canal, and I lost Sight of it over the Haymarket. *Its Motion was so very slow, that I had it above half a Minute in View*; and therefore had Time enough to contemplate its Appearance fully, which was what is seen in the annexed Figure."

"A seemed to be a light Flame, turning backwards from the Resistance the Air made to it. BB a bright Fire like burning Charcoal, enclosed as it were in a open Case, of which the Frame CCC was quite opaque, like Bands of Iron. At D issued forth a Train or Tail of light Flame, more bright at D, and growing gradually fainter at E, so as to be transparent more than half its Length. The Head seemed about half a Degree in Diameter, the Tail near 3 Degrees in Length, and about one Eighth of a Degree in Thickness."

Note: Given such a precise observer, it is difficult to call this phenomenon an ordinary meteor.

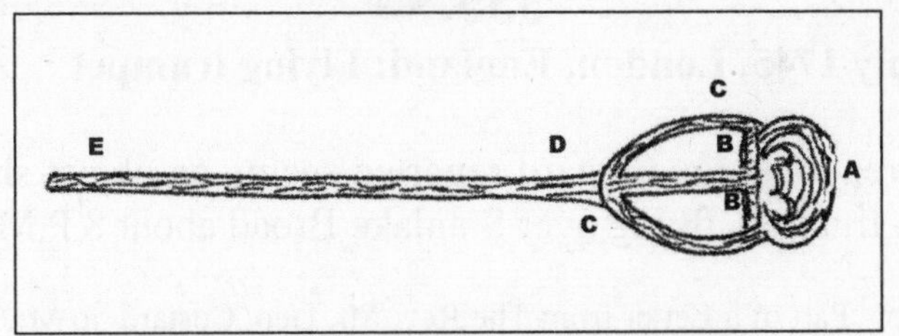

Fig. 31: "Waving rocket" in London

'Slow' meteors are known to exist but they are poorly explained: Camille Flammarion called them *bradytes* but he acknowledged they were extremely rare. If this happened today we would suspect a satellite re-entry, but there was no such thing in 1743.

Source: *Philosophical Transactions of the Royal Society of London* 43 (1745): 524.

334.

23 June 1744, Knott, Scotland: Armed men in the sky

Twenty-six witnesses, including a judge, observed a troop of armed men in the sky above a hill: "A man named D. Stricket, then servant to Mr. Lancaster, of Blakehills, saw, one evening about 7 o'clock, a troop of horses riding leisurely along Souter Fell in Cumberland."

After he called his master, "Mr. Lancaster discovered the aerial troopers," who became visible near a place called Knott. They were in sight for two hours and "this phenomenon was seen by every person (twenty-six in number) in every cottage within the distance of a mile."

Source: Statement attested before a magistrate by Lancaster and Stricket on the 21st of July, 1745. See "Phantom Armies" in Milwaukee (Wisconsin) *Sentinel*, November 1, 1871 citing *A Folio of Apparitions and Wonders*, preserved in the British Museum; also, *Arminian Magazine, consisting chiefly of extracts and original treatises on universal redemption* 18 (May, 1795): 244-245.

335.

14 July 1745, London, England: Flying trumpet

Reverend George Costard reported seeing an object shaped like a trumpet, flying over Stanlake Broad about 8 P.M.

Source: "Part of a Letter from The Rev. Mr. Geo. Costard to Mr. John Catlin, concerning a Fiery Meteor seen in the Air…" *Philosophical Transactions* 43 (1744-1745): 522-524.

336.

5 August 1748, Aberdeen, Scotland
Three globes of light

Eleven witnesses swore before the city council of Aberdeen they had observed three globes of light, men and armies in the sky, at 2 P.M. in a valley located five miles west of the city. They first assumed the three globes of light were meteorological in nature, but their intensity increased, and twelve tall men then appeared, dressed in bright clothes. They walked across the valley, followed by two armies that appeared to re-enact the battle of Culloden, near Inverness, which had taken place on 16 April 1746.

In cases of "armies in the sky" and heavenly battles we generally suspect an aurora, but the timing of this sighting (early afternoon) excludes this interpretation.

Source: *Flying Saucer Review* 32, vol. 17, no. 6 (1971), citing a letter by Roger Sandell, in *Culloden* by John Prebble (chapter 7). Another letter dated September 5, 1748, relates this story and mentions the "three globes of light". This source, signed R. F. (Robert Forbes), cites an extract from an older letter dated August 20, 1748, in which "a gentleman of Aberdeen" writes to his correspondent in Edinburgh about the visionary battle that took place on August 5, 1748. See *The Lyon in mourning, a collection of speeches, letters, journals, etc. relative to the affairs of Prince Charles Edward Stuart, by the Rev. Robert Forbes, A.M., Bishop of Ross and Caithness, 1746-1775.* Edited from his Manuscript by Henry Paton, vol. II (Edinburgh: Scottish History Society, 1895), 181-182.

337. 

1752, Kazan, Russia: Abducted in a flying cauldron

A man named "Yashka" reportedly met a stranger dressed in white who took him to a flying cauldron. He believed he visited another world, and then returned to the Earth.

Source: Vadim Chernobrov, *Synodic Archives* (Kazan University, 1909), No. 635:135, V.XXV.

338. 

15 April 1752, Stavanger, Norway: Flying octagon

"An octagonal luminosity in the sky emitted fireballs from its angles."

Source: Alexis Perey, *Sur les tremblements de terre de la péninsule scandinave* (Paris, 1842), 17. Perey draws from *La Gazette*, June 10th, 1752.

339. 

1 June 1752, Angermannland, Sweden
Bright streak emits balls of light

Between 4 and 5 A.M., luminous "balls of fire" emerged from a bright streak in the sky extending from the northeast to the southwest, for 12 to 13 miles along the coast.

Source: Robert Mallet and John William Mallet, *The Earthquake Catalogue of the British Association* (British Association for the Advancement of Science, London: Taylor & Francis, 1858).

340.

15 August 1754, Amsterdam and Chiswick, England
Sphere at ground level

After sunset a strange sphere, with an apparent diameter equal to that of the full moon, was observed shooting blinding bright beams, and descending close to ground level.

Source: *The Gentleman's Magazine* 25 (1755): 461-462.

341.

29 December 1758, Colchester, Essex, England ⓘ
Wandering oval object

At 8.00 P.M., an object described by contemporary eyewitnesses as looking like a huge football seemed to descend from the sky. It then "vanished like a squib without a report."

Source: *London Magazine* 27 (1758): 685.

342.

20 May 1759, unknown location ⓘ
Unexplained satellite of Venus

Astronomer Andreas Mayer reported an observation of a planetoid object seemingly orbiting Venus.

Source: Mayer's observation first appeared as a very brief footnote in his book, *Observationes veneris gryphiswaldenses* (1762), 16-17. The full report was first published by Johann-Heinrich Lambert in 1776 in *Astronomisches Jahrbuch oder Ephemeriden für das Jahr 1778* (Berlin, 1776), 186.

343.

16 September 1759, Lönmora, Sweden
Abducted for four days

The following handwritten text is recorded in the parish book of Ramsberg, Sweden:

"In the evening of 16 September 1759, the crofter Jacob Jacobsson's eldest son Jacob, 22 years old, had crossed the lake, Vastra Kiolsjon, to the crofter Anders Nilson at Lönmora, to deliver the food packet for him and his father for the following day's work in the Woods. Coming back across the lake, as he pulled the boat upon the shore, something strange happened to him.

"A large and broad road appeared before him. He followed it and soon reached a large red mansion, in his own words, 'with grander buildings than Gamlebo.' Soon he found himself seated on a bench by the door in a big chamber. He saw a chubby little man with a red cap on his head, sitting at the end of a table, and crowds of little people running back and forth. They were in every way like ordinary men, but of short stature. A bit taller than the rest was a fine-looking maiden, who offered him food and drink. He said, 'No, thank you.' The Little people asked him whether he wanted to stay with them, and he answered, 'God, help me back home to my father

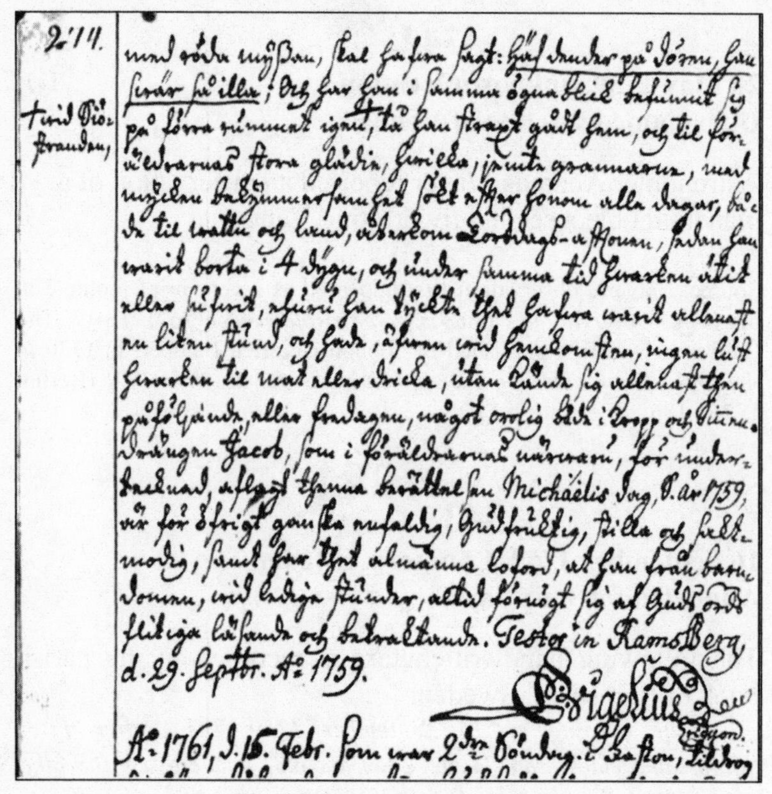

Fig. 32: Lönmora manuscript

and mother!' Then the man with the red cap said, 'Throw him out, he has such an ugly mouth!'

"In the next instant he was back by the lake shore, and from there he returned home. His parents greeted him with pleasure. They had been very worried; together with the neighbors they had searched the woods and the lake for him. Four days and nights had passed without a trace of him. When he finally came back on Thursday evening he had not eaten or slept for four days, yet he had no desire for food or drink. He thought he had been away only for a little while. The following day everything was normal except for an uneasy feeling in his body and mind.

"Jacob made this statement to me in the presence of his parents on St. Michael's Day 1759. This boy has quite a simple, pious, meek and gentle character. He is praised by everyone; all his life he has

been known to take pleasure in reading and contemplating God's words whenever he has some spare time."

Source: *Ramsberg sockens kyrkobok*, E1:1, 1786-1774, handwritten entry by Reverend Vigelius. The book is kept at Landsarkivet, Uppsala, Sweden. Translation by Clas Svahn.

344.

7 May 1761, France: Planetoid orbiting Venus

Prominent French astronomer J. L. Lagrange observed an object that seemed to be in orbit around Venus. He announced that its orbital plane was perpendicular to the ecliptic. Venus was then an evening "star" at 207° heliocentric longitude and 34° elongation.

Source: "The Problematical Satellite of Venus," *The Observatory* 7 (1884): 222-226.

345.

6 June 1761, unknown location
Planetoid orbiting Venus

Astronomer Scheuten reported an object that he observed while tracking Venus in transit across the disk of the Sun. The planet was accompanied by a smaller dark spot on one side, which followed Venus in its transit.

Source: "The Problematical Satellite of Venus," *The Observatory* 7 (1884): 222-226.

346.

26 December 1761, Weyloe, Denmark
A pale object emits a beam

"The following letter was received from Weyloe, in the diocese of Copenhagen:

"On the 26th of December last, about ten at night, there arose a great storm. I did not go to bed, and about four minutes past two in the morning, I observed a sudden light across my windows, which I took for lightning: the storm at this time increased not a little. I kept my eye fixed at my window; and at four o'clock I perceived a ray of light which seemed to come in a horizontal direction from the moon, to appearance about a toise and half (nine feet) in length, and about the thickness of a man's arm. Rays darted from it on each side.

"Running into my garden, I saw a ball of fire, about the size of a common ball, running gently from south to north. At first the ball was of a pale colour, like the sun covered with clouds, and threw out many rays. It grew more and more red, and smaller, and in two minutes disappeared without noise or smoke. My astonishment was the greater, as the tempest ceased soon after, though it had been accompanied with such violent blasts of wind, that many imagined they felt the shock of an earthquake. I have spoken to a dozen of people, who also saw it. Of all the phaenomena I have seen in Norway, I remember none equal to this, nor attended with like circumstances."

Given the weather environment, one could hypothesize globular lightning, but the description of multiple beams is highly unusual.

Source: *The annual register, or a view of the history, politics, and literature, for the year 1761,* 5th Ed (London: J. Dodsley, 1786), 67.

347. 

February 1762, Nuremberg, Germany
Unknown astronomical object

Single object, "a black round spot" passing in front of the Sun, as observed by Mr. Staudacher. He missed it the next day, and commented, "Perhaps this is a new planet."

Source: "Observations of the transits of intra-mercurial planets or other bodies across the Sun's disk." *The Observatory* (1879): 135.

348. 

9 August 1762, Basel and Solothurn, Switzerland
Slow-flying spindle in the Sky

Two witnesses at separate observatories (Rostan in Basel and Croste in Solothurn) reported a vast spindle-shaped cigar in slow flight in front of the Sun.

Monsieur de Rostan, an astronomer and member of the Medicophysical Society of Basel, Switzerland, observed the object with the aid of a telescope as it eclipsed the sun. This object could be observed daily for almost a month from Lausanne and also by a second astronomer in Sole, near Basel. Monsieur de Rostan traced its outline with a *camera obscura* and sent the image to the Royal Academy of Sciences in Paris. The drawing has not been preserved, unfortunately, but there is no doubt that it once existed and was regarded with some amazement. As this is historically an important incident the original report is produced in full below:

An account of a very singular phenomenon seen in the disk of the sun, in different parts of Europe, and not in others.

"The 9th of August, 1762, M. de Rostan, of the economic society at Berne, of the medico-physical society at Basle, while he was taking the sun's altitudes with a quadrant, at Lausanne, to verify a meridian, observed that the sun gave but a faint pale light, which he attributed to the vapours of the Leman lake; however, happening to direct a fourteen foot telescope, armed with a micrometer, to the sun, he was surprised to see the eastern side of the sun, as it were, eclipsed about three digits, taking in a kind of nebulosity, which environed the opaque body, by which the sun was eclipsed.

"In the space of about two hours and a half, the fourth side of the said body, whatever it was, appeared detached from the limb of the sun; but the limb, or, more properly, the northern extremity of this body, which had the shape of a spindle, in breadth about three of the sun's digits, and nine in length, did not quit the sun's northern limb. This spindle kept continually advancing on the sun's body, from east towards west, with no more than about half the velocity with which the ordinary solar spots move; for it did not disappear till the 7th of September, after having reached the sun's western limb.

"M. Rostan, during that time, observed it almost every day; that is to say, for near a month; and, by means of a camera obscura, he delineated the figure of it, which he sent to the royal academy of sciences at Paris.

"The same phenomenon was observed at Sole, in the bishopric of Basle, situated about five and forty German leagues northward of Lausanne. M. Coste, a friend of M. de Rostan, observed it there, with a telescope of eleven feet, and found it of the same spindle-like form, as M. de Rostan, only it was not quite so broad; which, probably, might be owing to this, that growing near the end of its apparition, the body began to turn about, and present its edge.

"A more remarkable circumstance is, that at Sole it did not answer to the same point of the sun as it did at Lausanne: it therefore had a considerable parallax: but what so very extraordinary a body, placed between the sun and us, should be, is not easy to divine. It was no spot, since its motion was greatly too slow; nor was it a planet or comet, its figure seemingly proving the contrary. In a word, we know of nothing to have recourse to in the heavens, whereby to explain this phenomenon; and, what adds to the oddness of it, M. Messier, who, constantly observed the sun at Paris during the same time, saw nothing of such an appearance."

Source: "Natural History: An Account of a Very Singular Phenomena Seen in the Disk of the Sun, in Different Parts of Europe, and Not in Others," *Annual Register* 9 (1766): 120-121.

349. 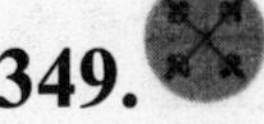

19 November 1762, Location unknown: Planetoid ⓘ

Planet-like body passing in front of the sun: "Lichtenberg saw, with the naked eye, a great round spot of about one twelfth the diameter of the Sun, traverse a chord of 70° in 3 hours."

Source: R. C. Carrington, *Monthly Notices of the Royal Astronomical Society*, 20, January 1860.

350. 

4 March 1764, unknown location ⓘ
Mystery satellite of planet Venus

The supposed satellite of Venus was observed again as the planet was an evening "star." Its heliocentric longitude was 59° and its elongation was 30°. There were no less than eight observations of this object during 1764.

Source: "The Problematical Satellite of Venus," in *The Observatory* 7 (1884): 222-226.

351.

28 March 1764, unknown location ⓘ
Mystery satellite of planet Venus

Another reliable observation of a planetoid object apparently orbiting Venus. The planet's heliocentric longitude was 98° and its elongation was 35°.

Source: "The Problematical Satellite of Venus," *The Observatory* 7 (1884): 222-226.

352.

May 1764, near Gotha, Germany: Unknown object

Single object, seen passing in front of the sun by Mr. Hoffmann. It was a large round spot of about one fifteenth the diameter of the Sun, crossing it slowly north to south.

Source: "Observations of the transits of intra-mercurial planets or other bodies across the Sun's disk," *The Observatory* 29 (1879): 135.

353.

13 June 1765, Mount Prospect, Inishannon, Ireland
Sky throne

"Last Monday Evening, between eight and nine o'Clock, an extraordinary Phaenomenon was seen from Mount-Prospect, near Inishannon, by several Gentlemen and Ladies. A most superb Throne appeared in the Northern Sphere, enclosed by a broad Circle of a Gold Colour, with a Lion in the front Protecting the Throne, which appearance lasted about half an Hour, and went off by slow Degrees. The Evening was very Serene, and the Sky all around appeared quite black. We are assured of the Truth of this Relation by People of Veracity."

Source: *The Public Register*, or *Freemans Journal* (Dublin, Ireland) 15 June, 1765.

354.

8 September 1767, Perthshire, Scotland
Large luminous pyramid leaves damage in its wake

"We hear from Perthshire, that an uncommon phaenomenon was observed on the water of Isla, near Cupor Angus, preceded by a thick dark smoke, which soon dispelled, and discovered a large luminous body, like a house on fire, but presently after took a form something pyramidal, and rolled forwards with impetuosity till it came to the water of Erick, up which river it took its direction, with great rapidity, and disappeared a little above Blairgowrie. The effects were as extraordinary as the appearance.

"In its passage, it carried a large cart many yards over a field of grass; a man riding along the high road was carried from his horse, and so stunned with the fall, as to remain senseless a considerable time. It destroyed one half of a house, and left the other behind, undermined and destroyed an arch of the new bridge building at Blairgowrie, immediately after which it disappeared."

Source: Letter from Edinburgh dated 8 Sept. 1767, in *The Annual Register*, 1767.

355.

4 January 1768, Copenhagen, Denmark
Unidentified planetoid orbiting Venus

Astronomer Christian Horrebow reported an observation of "a small light, that was not a star" which appeared to be in orbit around Venus. This object, named "Neith" by M. Hozeau of Brussels observatory, was never identified with certainty and was certainly not a natural satellite.

Source: H. C. F. C. Schjellerup, "On some hitherto unknown observations of a supposed satellite of Venus," *Copernicus* 2 (Dublin, 1882): 164-168.

356.

24 October 1769, Oxford, England ⓘ
Hovering intruder

An object like a "house on fire" seen in the sky for an hour. It moved up and down with jets of gas, rumbled.

Source: John Swinton, "An account of a very remarkable Meteor seen at Oxford", *Philosophical Transactions*, 60 (London, 1771): 532-535.

357.

8 May 1775, Waltham Abbey, Hertfordshire, England
Light ball

"At 8:30 P.M. a remarkable phenomenon was observed by a gentleman at Waltham Abbey.

"A meteor, resembling a nebulous star, appeared just above the moon, passed eastward, with a slow motion, parallel to the ecliptic, through an arch of about 5 or 6 degrees, and then disappeared. It subtended an angle of 6 or 7 minutes, and was of the same brightness and colour with the moon."

Source: *The Annual Register* (London, 1776): 116.

358.

17 June 1777, France, location unknown
An unidentified Messier object

During a lunar eclipse, astronomer Charles Messier observed dark objects moving in parallel directions, which he described as "large and swift and they were ships, yet like bells."

"These, Messier says, may have been hailstones or seeds in the air; but they were more probably small meteorites."

Source: "Observations of the transits of intra-mercurial planets or other bodies across the Sun's disk," *The Observatory* 29 (1879): 136.

Fig. 33: French astronomer Charles Messier

359.

5 February 1780, Bussières, France: Flaming dragon

About 6 P.M. a flaming “dragon” was seen in the sky for 15 minutes, illuminating everyone below.

Source: French UFO magazine, *Lumières dans la Nuit* 338.

360.

March 1783, Japan, location unknown: Low-flyers ⓘ

For several days, people reported luminous objects flying north to south “just over the rooftops.”

Source: *Brothers* I, 1. No original source provided.

361.

24 June 1784, China, exact location unknown Oscillating ‘star’

A big star appeared suddenly in the southeast, scintillating. It rose and came down three times. Another star repeated the same motion and was said to have fallen on a village.

Source: Shi Bo, *La Chine et les Extraterrestres*, op.cit., 44.

362.

11 September 1787, Edinburgh, Scotland
Wandering globe

About 8:30 P.M. people saw a fiery globe larger than the sun in a northerly direction. It proceeded horizontally to the east, about 15 to 20 degrees in elevation. Then it descended to the horizon, rose again higher than before with short waves in its trajectory and finally moved west and was lost to view behind a cloud, where it seemed to explode.

Source: John Winthrop, "An Account of a Meteor Seen in New England, and of a Whirlwind Felt in That Country: In a Letter to the Rev. Tho. Birch, D. D., Secretary to the Royal Society, from Mr. John Winthrop, Professor of Philosophy at Cambridge in New England," *Philosophical Transactions* 52 (1761-1762): 6-16.

363.

12 June 1788, Zamora, Spain: Two large flying globes

A letter from an Irish clergyman at the University of Zamora mentions that between 4 and 5 A.M., "a most alarming and singular phenomenon appeared in the southeast quarter of the Heavens. Two large globes of fire, seemingly about the bulk of a Bristol barrel, were seen to move horizontally for a few minutes at the height of seven or eight degrees from the surface of the earth.

They approached and dashed violently against each other, till some kind of centrifugal force separated them, after which they steered different courses; one moving East South East and the other West by North. As many persons were then up in the town, numbers repaired to an adjacent hill for the advantage of prospect."

The 'meteors' proceeded slowly in their course for about twenty minutes. The one on the southeast quarter burst with a crack that might be heard at ten miles distant. The other continued gradually descending till it was lost to sight.

Source: *London Times*, Thursday, July 10, 1788.

364.

12 November 1791, Göttingen, Germany
Object in front of the sun

Single witness (astronomer Lichtenberg): An object is observed passing in front of the Sun.

Source: *Philosophy Magazine* 3 (1899).

365.

19 January 1793, England, exact location unknown
Opaque body

A long opaque body was seen stationary over the center of the sun by many witnesses.

Source: *Gentleman's Magazine & Historical Chronicle* 63 (1793).

366.

28 December 1793, Bucharest, Romania
Unexplained "moon"

In the evening, a man who was dining in Bucharest about 7:30 P.M. reported that "the moon has accomplished a miracle," making a journey along the sky in half an hour.

Source: Ion Hobana and Julien Weverbergh, *Les Ovni en URSS et dans les Pays de l'Est* (Paris: Robert Laffont, 1972), 222. The authors cite *Biblioteca Academiei Române,* BAR ms. rom. 2150, fol. 1110.

367.

20 August 1794, Balasore, India
Phenomenon in the heavens

At 7:45 P.M. a number of witnesses saw an oversized meteor, brighter than any of the planets. As it descended "it made short and frequent pauses, at which times it appeared far more brilliant than while it was in motion."

The object was lost to sight behind the hills, but not for long: "We expected to have seen no more of it: but in about two minutes after we observed it again, ascending above the hills, where it balanced and waved several times, in a horizontal direction, North and South: it then sank again, illuminating the hills in its declination as before. It rose and fell a second and a third time, with little variation in its movements, after which we saw it no more and all around was darkness."

Source: *The Star* (London, England), Saturday, March 14, 1795, issue 2049.

368.

27 January 1795, Quangxi prov., Linggui area, China
Crash of a large maneuvering light

A large "star" in the Southeast rose and fell three times, followed by another one that "crashed in a village."

Source: Shi Bo, *La Chine et les Extraterrestres*, op.cit., 44.

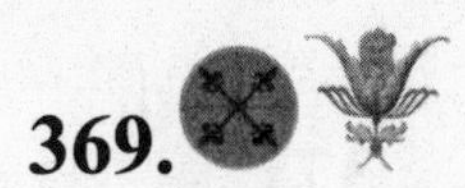

369.

12 Oct. 1796, New Minas, Bay of Fundy, Nova Scotia
Airships

Fifteen "ships" seen in the air moving east, with ports on the side. A man aboard one ship extended his hand.

The incident is mentioned in the five-volume diary of loyalist merchant and Judge Simeon Perkins (1734-1812):

"A strange story is going that a fleet of ships have been seen in the air in some part of the Bay of Fundy. Mr. Darrow is lately from there by land. I enquired of him. He says that they were said to be seen at New Minas, at one Mr. Ratchford's by a girl about sunrise and that the girl being frightened called out and that two men that were in the house went out and saw the same sight, being fifteen ships and a man forward of them with his hand stretched out. The ships made to the eastward. They were so near that the people saw their sides and ports. The story did not obtain universal credit, but some people believed it. My own opinion is that it was only in imagination as the clouds at sunrise…"

Source: C. B. Fergusson, ed., *The Diary of Simeon Perkins 1790-1796* (Toronto: The Champlain Society, 1961), 430.

370.

18 January 1798, Tarbes, France
Astronomical anomaly

Astronomer D'Angos observed an object passing in front of the Sun. It was "a slightly elliptical, sharply defined spot, about halfway between the centre and edge of the Sun, which passed off about 25 minutes afterwards."

Source: "Observations of the transits of intra-mercurial planets or other bodies across the Sun's disk," *The Observatory* 29 (1879): 136.

371.

10 Sept. 1798, Alnwick, Northumberland, England
Cylinder

A particularly graphic case of a shape-changing UFO was reported by a Northumberland schoolteacher, Alexander Campbell, and a friend. According to *The Annual Register* for 1798 when the object first appeared high up in the south-western sector of the sky it seemed to be no bigger than a star, but as it came closer it "expanded into the form and size of an apothecary's pestle."

"It was then obscured by a cloud, which was still illuminated behind; when the cloud was dispelled, it re-appeared with a direction south and north, with a small long streamer, cutting the pestle a little below the centre, and issuing away to the eastward. It was again obscured, and, on its re-appearance, the streamer and the pestle had formed the appearance of a hammer or a cross; presently after the streamer, which made the shaft to the hammer, or stalk to the cross, assumed two horns to the extreme point, towards the east, resembling a fork. It was then a third time obscured, but when the cloud passed over, it was changed into the shape of two half moons, back to back, having a short thick luminous stream between the two backs; it then vanished totally from their sight. It is observable that every new appearance became brighter and brighter, till it became an exceedingly brilliant object, all the other stars, in comparison, appearing to be only dim specks."

The sighting lasted some five minutes in all.

Source: *Inforespace* 28, quoting from *The Annual Register* 83 (London, 1798).

372.

July 1799, Bruges, Belgium
Unexplained maneuvering "meteor"

An unusual "meteor" crosses the sky towards the south and returns north, then makes a 45-degree turn to the northwest, proving it was no natural object, and certainly not a meteor.

Source: T. Forster, "A Memoir on Meteors of Various Sorts," *Philosophical Magazine and Journal of Science,* July-Dec. 1847.

373.

13 November 1799, Gerona, Catalonia, Spain
Maneuvering stars

Numerous brilliant "shooting stars" are seen to cluster together, and then separate.

Source: A. Quintana, "Assaig sobre el clima d'Olot.," *Servei Meteorològic de Catalunya*, Generalitat de Catalunya, Notes d'Estudi 69: 3-88 (Barcelona, 1938).

Epilogue to Part I-E
The Era of Human Flight Begins

By the end of the 18th century political events reflected a nearly-universal thirst for knowledge and liberty, and the rejection of authoritarian principles. The American Declaration of Independence in 1776, and the French Revolution of 1789, had ushered in a new, often troubled series of intellectual movements inspired by science.

In 1797, German astronomer Olbers had published a method for calculating the orbits of comets, thus removing them from the realm of cosmic enigmas to treat them rationally as solar system objects. Lithography was invented the following year by Aloys Senefelder, another German. In 1800 William Herschel discovered infrared solar rays and Volta produced electricity from batteries of zinc and copper.

Nothing could now stop the rise of science and technology: in 1801 Bichat published his *Anatomie Générale* while Lalande released a catalogue of 47,390 stars. An American engineer, Robert Fulton, built the first submarine, the Nautilus. London became a city of 864,000 and Paris claimed 547,000 inhabitants. It was the time of Beethoven, Paganini, and Haydn. The world had made a momentous transition from the Age of Reason to the Enlightenment.

As accurately noted by the people's chronicle of our own times, Wikipedia, "The Enlightenment was a time when the solar system was truly discovered: with the accurate calculation of orbits, the discovery of the first planet since antiquity, Uranus by William Herschel, and the calculation of the mass of the sun using Newton's theory of universal gravitation. These series of discoveries had a momentous effect on both pragmatic commerce and philosophy. The excitement engendered by creating a new and orderly vision of the world, as well as the need for a philosophy of science that could encompass the new discoveries, greatly influenced both religious and secular ideas. If Newton could order the cosmos with natural philosophy, so, many argued, could political philosophy order the body politic."

This was also the period when humanity began to challenge gravity as well, and made the first attempts to fly. In the late 1780s, as enthusiasm for the "Industrial Revolution" was felt, the Montgolfier brothers invented the *montgolfière*, or hot air balloon. They were the sons of a paper manufacturer at Annonay, near Lyon. When playing with inverted paper bags over open fire they found that the bags rose to the ceiling. This led them to experiment further with larger bags made of other materials. During 1782 they tested indoors with silk and linen balloons.

On December 14, 1782 they succeeded in an outdoor launch of an 18 m^3 silk bag, which reached an altitude of 250 m. On June 5, 1783, as a first public demonstration, they sent up at Annonay a 900 m^3 linen bag inflated with hot air. Its flight covered 2 km, lasted 10 minutes, and had an estimated altitude of over 1600 m. The subsequent test sent up the first living beings in a basket: a sheep, a duck and a cockerel, to ascertain the effects of higher altitude. This was performed at Versailles, before Louis XVI of France, to gain his permission for a trial human flight.

On November 21, 1783, the first free flight by humans was made by Pilâtre de Rozier and the Marquis d'Arlandes, who flew aloft for 25 minutes about 100 m above Paris for nine kilometers. (Karl Friedrich Meerwein with his flapping "ornithopter" probably preceded this event in 1781, but it never became a viable means of flight.)

Any study of unexplained aerial phenomena after the year 1800 must take into account not only the possible explanations we have already discussed (atmospheric effects, optical illusions, aurorae, meteors, comets, globular lightning, mystical visions, and hallucinations) but other causes, from simple balloon observations to over-excited press reports and hoaxes inspired by the passion of the early days of human flight.

PART I-F
Nineteenth-Century Chronology

The first half of the nineteenth century, which culminated in the worldwide extension of the Industrial Revolution that had begun around 1770 in England, was marked by a vast increase in scientific education. Curiosity towards all the phenomena of nature was encouraged; observatories and laboratories sprang up in every nation, and it became fashionable to report original contributions to the knowledge of science or, as it was called, "natural philosophy."

In the course of their observations scientists, both amateur and professional, noted unknown phenomena and reported them without fear of censorship or ridicule. In contrast with the rigid adherence to conformism in the name of rationalism that plagues the modern academic community, there is a pleasant sense of freedom and curiosity when one reads the reports of that era. Astronomers were eager to attach their name to discoveries of comets, new planets or unusual phenomena, leading to open, unbiased examination of any novel report.

The search for a planet (tentatively named "Vulcan") whose orbit would place it between Mercury and the Sun is a case in point. It was motivated by the irregularities in the motion of Mercury. Celestial mechanics had become sophisticated enough for astronomers of the time to record such minute differences–hence the need for actual observations of new planetoid bodies that could account for an effect on Mercury.

When Le Verrier, the celebrated director of Paris observatory whose brilliant calculations led to the discovery of Neptune in 1846, tried to prove his hypothesis about the existence of an intra-mercurial planet, he actually encouraged serious observers the world over to come forward with any sighting of unknown objects in the vicinity of the sun. As Le Verrier told the French Academy of Sciences on 2 July 1849: "I felt profound surprise, as I worked on the theory of Mercury and saw that the mean motion of that planet, as determined by observations of the last 40 years, was notably weaker than indicated by the comparison of older data with modern ones. My attempts to reach a theory that would resolve this have been unsatisfactory so far." (quoted in *L'Evolution de l'Astronomie au XIXe Siècle*, by Pierre Busco. Paris: Larousse, 1912).

Ten years later, in a celebrated letter to Faye, Le Verrier stated he had reached a solution, calling for the existence of one or more intra-mercurial planets. He went on to call for careful observation of any unusual object passing in front of the sun:

"The present discussion should confirm astronomers in their zeal to scrutinize the surface of the sun every day. It is most important that any spot of regular shape, as small as it is, which would happen to be seen on the disk of the Sun, be tracked for some time with the greatest care, in order to ascertain its nature through knowledge of its motion."

This invitation to observe the sky for anomalous objects sent hundreds of professionals and well-equipped amateurs to rummage through records of past observations and to spend more time at the telescope. As a result, many of the references we have accumulated in this section no longer come from obscure local papers but from the mainstream scientific literature, from the *Comptes Rendus* of the French Academy of Sciences to *Philosophy Magazine* or the *Quarterly Journal of the Royal Institute*.

The many calculations never led to the discovery of the intra-mercurial planet, much to Le Verrier's chagrin.

Early in the 20th century, Einstein's relativity theory accounted nicely for the perturbations of Mercury, and astronomy no longer needed the elusive planetoid! Even the good reverend Webb, whose classic books on *Celestial Objects for Common Telescopes* are still used as a reference by astronomers the world over, suffered the indignity of having his own observations of an unknown planet censored from recent printings, such as the popular paperback edition of 1962. All such data published in the nineteenth century were swept under the weighty rug of scientific oblivion, leaving only a few people like us to recompile these "damned" facts and ask: What was it?

374.

20 March 1800, Quedlinburg, Germany
Fast-moving object

A preacher and amateur astronomer named Fritsch reported an unknown object rapidly crossing the disk of the sun.

Source: E. Ledger, "Observations or supposed observations of the transits of intra-mercurial planets or other bodies across the Sun's disk," *The Observatory* 3 (1879-80): 136.

375.

30 September 1801, London, England
Unusual formation

Between five and six in the morning, a very bright object was seen in the East. It was "shaped something like a cross; this was accompanied by two smaller ones like stars, one towards the left, which was also bright; and one just perceptible a little below it, the whole moving fast towards the South."

Source: *London Times*, 8 October 1801, 3.

376.

7 February 1802, Quedlinburg, Germany
Dark celestial body

An unknown dark body is again observed crossing the Sun, "having rapid motion of its own" by amateur astronomer Fritsch. The tiny spot crossed the Sun in a northwesterly direction and accelerated.

Source: E. Ledger, "Observations or supposed observations of the transits of intra-mercurial planets or other bodies across the Sun's disk," *The Observatory* 3 (1879-80): 136.

377.

10 October 1802, Magdeburg, Germany
Astronomer's report of a dark moving object

Le Verrier reports an unknown dark object seen by Fritsch, rapidly crossing the sun's disk.

He writes: "Fritsch, at Magdeburg, saw a spot moving 2 minutes of arc in 3 minutes of time, and not seen after a cloudy interval of 4 hours."

Source: Le Verrier, "Examen des observations qu'on a présentées, à diverses époques, comme pouvant appartenir aux passages d'une planète intra-mercurielle devant le disque du Soleil," *Comptes Rendus of the French Academy of Sciences* 83 (1876): 583-9, to 587-8.

378.

15 August 1803, Ormans near Evilliers, France
Luminous globes and a religious discovery

Two luminous globes emit sun-like rays and hover over a tree. Witnesses: Pierre Mille, from Malcôte, with his three daughters and a local craftsman.

The whole family was on its way to church for the Feast of the Assumption when they saw two small lights in front of an oak tree, inside which a small statue of the Virgin Mary was found upon investigation.

Some time before (at Easter) the youngest daughter of Pierre Mille had seen the Virgin accompanied by two small floating lights at the same spot, "on the path between Maizières and Ornans."

Source: Abbé Louis Leroy, *Histoire des Pélerinages de la Sainte Vierge en France*, Tome II (Paris, 1874), 265.

379.

27 June 1806, Geneva, New York, USA
Dark object crossing the lunar disk

At 1:00 A.M. astronomer W. R. Brooks, director of Smith Observatory, recorded the passage of a long, dark object that crossed the disk of the moon in 3 to 4 seconds, moving west to east. It did not appear to be a bird.

Brooks was observing with a two-inch telescope at magnification 44X. The object was about one third the apparent size of the moon.

Source: *Science Magazine*, 31 July 1896; *Scientific American* 75: 251.

380.

July 1806, Maine: White globe and glowing specter

In 1806 the Rev. Abraham Cummings set out to investigate the apparition of a ghost. A serious scholar with a Master's degree from Brown University, he was sure the tales would turn out to be fraudulent. The philosopher C. J. Ducasse reproduces Cummings' testimony as follows:

"Some time in July 1806, in the evening, I was informed by two persons that they had just seen the Spectre in the field. About ten minutes after, I went out, not to see a miracle for I believed they had been mistaken. Looking toward an eminence twelve rods distance from the house, I saw there as I supposed one of the white rocks. This confirmed my opinion of their spectre, and I paid no attention to it.

"Three minutes after, I accidentally looked in the same direction, and the white rock was in the air; its form a complete globe, with a tincture of red and its diameter about two feet. Fully satisfied that this was nothing ordinary I went toward it for more accurate examination.

While my eye was constantly upon it, I went on for four or five steps, when it came to me from the distance of eleven rods, as quick as lightning, and instantly assumed a personal form with a female dress, but did not appear taller than a girl seven years old. While I looked upon her, I said in my mind "you are not tall enough for the woman who has so frequently appeared among us!" Immediately she grew up as large and tall as I considered that woman to be.

"Now she appeared glorious. On her head was the representation of the sun diffusing the luminous, rectilinear rays every way to the ground. Through the rays I saw the personal form and the woman's dress."

Cummings wrote that the entity was encountered on scores of occasions, and in his report he included thirty affidavits from witnesses to prove it. In all cases a small luminous cloud appeared first and then grew until it took the form of the deceased woman. Afterwards it would take its exit in much the same way.

Source: C. J. Ducasse, *Paranormal Phenomena, Science and Life After Death* (New York: Parapsychology Foundation, 1969); Abraham Cummings, *Immortality Proved by the Testimony of Sense* (Bath, Maine, 1826).

381.

7 August 1806, Rutherford, North Carolina
Flying things, white beings

"Patsey Reaves, a widow woman, who lives near the Appalachian Mountain, declared, that about 6 o'clock P.M., her daughter Elizabeth, about 8 years old, was in the cotton field, about ten poles from the dwelling house, which stands by computation, six furlongs from the Chimney Mountain, and that Elizabeth told her brother Morgan, aged 11 years, that there was a man on the mountain. Morgan was incredulous at first, but the little girl affirmed it, and said she saw him, rolling rocks or picking up sticks, adding that she saw a heap of people.

"Morgan then went to the place where she was, and called out, [saying] that he saw a thousand or ten thousand things flying in the air. On which Polly, daughter of Mrs. Reaves, a good four years, and a Negro woman, ran out to the children and called Mrs. Reaves to see what a sight yonder was.

"Mrs. Reaves says she went about 8 poles towards them, and, without any sensible alarm or fright, she turned towards the Chimney Mountain, and discovered a very numerous crowd of beings resembling the human species but could not discern any particular members of the human body, nor distinction of sexes; that they were of every size, from the tallest men down to the least infants; that there were more of the small than of the full grown, that they were all clad with brilliant white raiment; but could not describe any form of their garment; that they appeared to rise off the mountain south of said rock, and about as high; that a considerable part of the mountain's top was visible about this shining host, that they moved in a northern direction, and collected about the top of Chimney Rock.

"When all but a few had reached said rock, two seemed to rise together and behind them about two feet, a third rose. These three moved with great agility towards the crowd, and had the nearest resemblance of two men of any before seen. While beholding those three her eyes were attracted by three more rising nearly from the same place, and moving swiftly in the same order and direction. After these, several others rose and went toward the rock."

The sighting went on for about an hour, during which time Mrs. Reaves sent for Mr. Robert Siercy. The latter was reluctant to come as "he expected to see nothing extraordinary," but after a second messenger was sent to him he finally arrived. At first he glanced at the mountain without seeing anything strange, but when he took a second look "he said he saw more glittering white appearances of human kind that ever had he seen of men at any general view." He noticed there were entities of different sizes, "that they moved in throngs," and that they "moved in a semicircular course between him and the rock." Two of the larger beings seemed to go before the others at a distance of about 20 yards, where "they vanished out of sight, leaving a solemn and pleasing impression on the mind, accompanied with a diminution of body strength."

Source: *Statesville* (North Carolina) *Landmark* June 15, 1883. The report was formally made on 7 August 1806 and presented to Mr. J. Gates, Editor of the *Raleigh Register and State Gazette*, where it was first published the following September.

382.

22 July 1808, Maine, USA
An old diary describes a maneuvering light

The diary of school teacher Cynthia Everett, who taught in Maine during the early 1800s, contains the following entry:

"About 10 o'clock I saw a very strange appearance. It was a light which proceeded from the East. At first sight, I thought it was a Meteor, but from its motion I soon perceived it was not. It seemed to dart at first as quickly as light, and appeared to be in the atmosphere, but lowered toward the ground and kept on at an equal distance sometimes ascending and sometimes descending. It moved round in the then visible Horizon. (it was not very light) and then returned back again."

Dr. Ranlett, a historian at the State University College at Potsdam, finds it significant that Cynthia Everett did not explain what she witnessed as a natural phenomenon, although she was well educated and had firsthand knowledge about the night sky. "She was the kind of person who would have explained it as natural phenomenon, if she could have."

Source: *New York* (Ogdensburg) *Journal*, March 29, 1978. The article describes the work of Dr. Judith Becker Ranlett: while studying the diary of her husband's great-great grandmother, she found this unusual sighting.

383.

1 September 1808, Moscow, Russia
Radiant "plate" flying over the Kremlin

Alexander Afanasyev, of the manuscript department of the Russian State History Museum, found a document in the personal archive of a Moscow senator Peter Poludensky. "On September 1, 1808 at 8 o'clock and 7 minutes after noon, in the sky, clear and sown with stars, a phenomenon appeared, incomparable in its beauty and rigor, as well as in radiance and enormous size, to anything seen before. As we noticed it, attracted by the loud cracking sound, it was rising in an arch over the horizon, from 55' to almost 90'. Having passed this distance in an instant, it stopped among the clouds as if over the Kremlin and looked like a long straight plate some nine arshin (6.35 meter) long and half arshin (0.35 meter) thick.

"Then on its front edge, turned to the South-West, an oval flame flared, some two arshin (1.4 meter) long and one and a half arshin (about one meter) thick, with a flame that can only be compared to the radiance of burning phosphor.

"Floating in a circle without open fire or sparkle, it nonetheless lighted everything around as broad daylight; then the flame went out, the light disappeared, but the bright plate remained and quite smoothly went perpendicularly upwards, reached the stars and still could be seen for some two minutes and then, without disappearing, it became invisible due to the extraordinary height."

A sketch was attached, depicting the flying object. Afanasyev ruled out the possibility of a hoax based on the age of the paper and the writing style.

Source: Moscow daily *Komsomolskaya Pravda*, 2 July 2006, and V. A. Bronshten, "There was in a clear sky... (Явилось на небе ясном…)", *Vokrug Sveta,* no. 9 (2528), September 1984, 55-56.

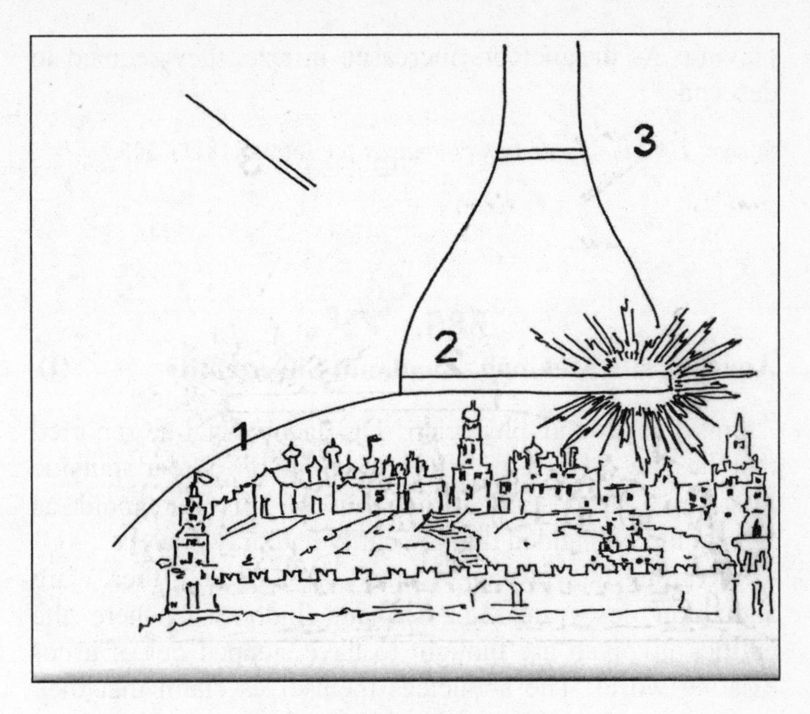

Fig. 34: Moscow phenomenon

384.

10 August 1809, London, Hatton Garden, England ⓘ
Aerial ballet of dazzling lights

John Staveley reports he saw many objects moving around a black cloud: "They were like dazzling specks of light, dancing and traipsing thro' the clouds. One increased in size until it became of the brilliance and magnitude of Venus on a clear evening. But I could see no body in the light. It moved with great rapidity, and coasted the edge of the cloud. Then it became stationary, dimmed its splendor, and vanished. I saw these strange lights for minutes, not seconds. For at least an hour, these lights, so strange, and in innumerable points, played in and out of this black cloud. No lightning came from the clouds where these lights were playing. As the meteors increased in size, they seemed to descend."

Source: *The Edinburgh Annual Register for 1809* 2 (1811): 508.

385.

August 1810, Meklong, Thailand: Silver entity ⓘ

A missionary and physician, Dr. Jacob Hazlitt, reported that he saw a man in silver clothing on a road outside Meklong. He described the skin of this humanoid as 'gleaming' and added that the entity had only one eye.

According to writer Ahmad Jamaludin, there are abduction cases in Malaysia and Indonesia where the entities involved are thought to have stepped out of a co-existing world. The abductees themselves claim that they were taken to a different world "on the other side of our reality." They were not subjected to medical examinations. No messages were given and the motive behind the abduction was never known.

When they returned home, the witnesses are said to have suffered the same side-effects as a typical UFO abductee, including temporary amnesia, extreme thirst, tiredness and emotional upset. The locals called these beings the *Bunian People*, but where the word '*Bunian*' comes from is unclear. They were said to dress in a similar way to the local people.

Source: Ahmad Jamaludin, *Alien Encounters*, No. 17, October 1997. Unfortunately the story is given without a precise original reference.

386.

September 1810, Thailand ⓘ
Abduction by one-eyed humanoid

A woman claimed she was awakened by an unknown force one night and was surprised to hear that the surrounding area was devoid of animal sounds. Something was not right: Looking out the window, the lady beheld a strange humanoid in her back yard. She claimed that the being only had one eye and was dressed in a suit that seemed to be made out of metal. The episode ended with the woman claiming to have been abducted to a 'palace of lights'.

Source: Ahmad Jamaludin, *Alien Encounters*, No. 17, October 1997.

387.

19 September 1810, Brezeau, Holland
Strange globe absorbs water

A remarkable incident occurred in the Dutch village of Brezeau. The 36th volume of the *Philosophical Magazine* reported that between 5:00 P.M. and 6:00 P.M., "a luminous meteor appeared to the south, and about the distance of a quarter of a league from the small commune of Brezeau: persons who attentively examined it assert that it was nearly a quarter of an hour in collecting, floating over the place where it was first seen; and that when all its parts had united, it appeared all at once as a very considerable globe of fire, taking a northerly direction."

The phenomenon "spread terror among the inhabitants of the village, who believed their houses would be burnt, and they themselves perish." It was followed by thick fog. Curiously, "in crossing a river it absorbed water, of which some afterwards fell as rain."

It is difficult to imagine a natural phenomenon with these characteristics. Its duration is also an anomaly, as it lasted forty-five minutes before turning into a column of fire and rising towards the sky.

Source: "Meteor seen in Holland", *The Philosophical Magazine* 36 (1810): 395-396.

388.

20 March 1812, near Manosque, France
Four entities inside lights

On the road from Villeneuve to Manosque, seven travelers in the coach to Digne were scared when they observed a luminous ball that hovered over the path close to the coach. The object split into four lights. Four human figures were seen, enclosed inside the objects, which looked like lanterns. Terrified, the travelers chose to stop at the inn of *Quatre Tours* rather than going all the way to Manosque.

Source: Louis-Joseph-Marie Robert, *Notice historique sur le tremblement de terre du village de Beaumont, département de Vaucluse, et examen des causes qui ont pu déterminer dans un pays non volcanique, 128 secousses successives dans 75 jours...*, (Aix: Augustin Pontier, 1812), 10.

389.

8 April 1813, Atlantic Ocean
Shell-shaped floating monster

As the ship "Niagara" was about latitude 43 north, longitude 65 west, a large hump was seen on the Southern horizon, "bearing N.W. distance 6 or 8 miles ahead, *which we supposed the hull of a large ship bottom up*. When within a gun shot of it, we discovered that it had motion."

On nearer approach the sailors thought the object must be a giant fish, "apparently 200 feet in length, about 30 feet broad, and from 17 to 18 feet high in the centre." Whatever it was, the floating object was covered with a shell. "Near the head on the right side was a large hole or archway, covered occasionally with a fin which was at times 8 or 10 feet out of the water." The crew intended to send a boat to make further discoveries, "but was deterred from the dreadful appearance of the monster."

Source: Log Book of the ship *Niagara*, captain Merry, traveling from Lisbon to New York, cited in the Plattsburgh [New York] *Republican*, 14 May, 1813.

390.

25 July 1813, Portsmouth, Virginia, USA
A letter to Thomas Jefferson

Carpenter Edward Hansford wrote to Jefferson on July 31, 1813 to describe the object which he and a Baltimore citizen named John L. Clark had witnessed:

"*A Ball of fire as full as large as the sun at Maridian (sic) which was frequently obscured within the space of ten minutes by a smoke emitted from its own body, but ultimately retained its brilliancy, and form during that period, but with apparent agitation. It then assumed the form of a turtle which also appeared much agitated and as frequently obscured by a similar smoke. It descended obliquely to the West, and raised again perpendicular to its original hite (sic) which was on or about 75 degrees.*"

Edward Hansford lived in Norfolk County during the Revolutionary War, working on forts erected by the Commonwealth. In 1802, he was appointed harbormaster for the District of Norfolk and Portsmouth.

Source: "Edward Hansford to Thomas Jefferson, July 31, 1813" in *The Thomas Jefferson Papers* Series 1. General Correspondence, 1651-1827. Library of Congress, Digital Archive "American Memory" (image 1031)
http://memory.loc.gov/ammem/collections/jefferson_papers/index.html

391.

5 September 1814, near Agen, France
Slow-moving, round object

At 11 A.M., by clear sky and as a stiff breeze was blowing, a slow moving, perfectly round white object with a grayish center appeared at great height northwest of the town. Although first described as a "cloud," it became completely stationary in spite of the wind and remained in that position until about noon, when it suddenly sped off to the south, revolving on its own axis, and emitting rumbling noises that culminated in an explosion.

A shower of stones was released, after which the "cloud" stopped again, and slowly faded away. The explosion was heard throughout the region and terrified the inhabitants.

Source: *Annales de Chimie et de Physique* 25, tome XCII, Oct. 1814; *Philosophical Magazine & Journal* 44 (1814): 316 and 45 (1815): 23-26.

392.

25 September 1817, Lanuéjols, France: Large disk ⓘ

At 6:25 P.M. three travelers who had just visited the roman monument near Lanuéjols and were going through the "Causse de Mende" to reach that town saw in the south-southeast (towards Villefort) a luminous object, "reddish in color, in the shape of a disk three or four times larger than the full Moon." It disappeared 2 or 3 minutes later. The sky was hazy and a fine rain was falling.

Source: *Le Midi Libre*, 25 September 1954. We have not traced an earlier source.

393.

6 January 1818, Ipswich, England
Unknown object near the sun

Mr. Capel Lofft, an English writer (1751-1824) and Mr. Acton report a strange object near the sun, visible three and a half hours. They saw "a small, uniformly opaque, subelliptical spot, moving more rapidly over the Sun than Venus in transit. Before sunset it disappeared and certainly seemed of a cometary or planetary character. This is a well-attested instance."

Source: Capel Lofft. "On the appearance of an opaque body traversing the sun's disc," *Monthly Magazine* 45, March 1 1818, 102-3.

394.

3 August 1818, Worthing, England
Motionless light in the sky

Mr. Thomas Young saw "a very bright meteor" near Cassiopeia at 11:15 P.M. This object started from a point 19 degrees from the pole at 65 degrees in right ascension. It moved to 17 degrees from the pole and 80 degrees in right ascension and remained motionless for a full minute.

Source: Thomas Young, "Observations d'un météore lumineux." *Annales de Chimie et de Physique* 9, (Paris, 1818): 88-90.

395.

26 June 1819, Buchholz, Germany: Multiple planetoids

Astronomer J.W. Pastorff (1767-1838) observed what he thought was a "comet" close to the Sun, but Olbers pointed out it could not have been a comet. The same day, Gruithuisen (observing from Holland) reported three unknown bodies crossing the disk of the Sun, "…viz, one near the middle of the Sun, and two small ones without nebulosity near the western limb."

It is notable that this observation, initially published in Reverend Webb's well-known astronomy handbooks, has been deleted from recent editions!

Source: "New planets," *Annual of Scientific Discovery* (1860): 410-11, at 411.

396.

9 October 1819, Augsburg, Germany
Enormous planetary intruder

An enormous mass passed in front of the Sun. Mr. Stark, canon of Augsburg, reported this observation, which is quoted by Le Verrier in *Monthly Notices of the R.A.S.*, February 1877. It was "a well-defined round spot, about the size of Mercury, not to be seen the same evening."

Source: "Observations of the transits of intra-mercurial planets or other bodies across the Sun's disk," *The Observatory* 29 (1879): 136.

397.

Ca. 1820, Hopkinton, New Hampshire, USA
Close encounter

At dusk a young man was followed home in a patch of forest called "The Lookout" near Putney Hill (also known as Gould Hill) for almost two miles by several glowing balls. They would stop whenever he stopped to look back at them, and then resume their movement as he started walking again, but never came closer than within 50 feet.

Such glowing balls had been seen in the area since 1750, moving slowly in mid-air, so one may suspect a natural phenomenon. The date given here is a rough estimate.

Source: Charles Chase Lord, *Life and Times in Hopkinton, N.H.* (Concord, NH: Republican Press Association, 1970).

398.

12 February 1820, Augsburg, Germany
Trans-solar traveler

A circular well-defined spot, with an orange-gold tint, not seen again the same evening, is recorded to have been seen by two independent observers, viz. Stark and Steinhübel. It crossed the sun in about 5 hours.

Source: R. C. Carrington, "On some previous observations of supposed planetary bodies in transit over the Sun," *Monthly Notices of the Royal Astronomical Society* 20 (March 1860): 192-4.

399.

27 May 1820, Spello (Perugia), Umbria, Italy ⓘ
Unknown stars in a moving triangle

Half an hour before sunset, while Saint Caspar (San Gaspare Del Bufalo) was preaching in the public square, people saw a cross of three stars in the sky to the east. It came over of the head of the Servant of God and formed a triangle, "one (light) was up and the two side ones at a lower level". Before this unusual phenomenon the people remained stunned; many fell into deep commotion and even the Bishop, Monsignor Lucchesi, who was present, "was astonished and amazed".

Some doubts can be expressed about the event. The correspondence of Gaspare has been published yet the event is mentioned nowhere in his letters, even though we have letters dated from multiple dates in May 1820. He sent two letters to his hierarchy on the 27th on the 28th of May, 1820, neither of which letters mention such an event.

Source: Biography of Saint Caspar (1786-1837). Despite featuring in UFO archives we have yet to trace a source, though some reference to the event appears here: *Gaspar Del Bufalo, A close-up acquaintanceship.* Depositions of V. Severini, G. Menicucci, B.

Panzini at the processes for the canonization of St. Gaspar Del Bufalo (Rome: Pia Unione Preziosissimo Sangue, 1992), 49.

400.

7 September 1820, Embrun, France
Arago's unknown formation: Military precision

Astronomer François Arago, director of Paris observatory, reports that "numerous people have seen, during an eclipse of the moon, strange objects moving in straight lines. They were equally spaced and remained in line when they made turns. Their movements showed military precision."

Source: François Arago, *Oeuvres Complètes de François Arago* (Paris, 1857), v.11, 575-8 ; "Etoiles filantes en plein jour", in *Annales de Chimie et de Physique*, vol. 30 (1825): 575-8.

Fig. 35: French astronomer François Arago (1786-1853)

401.

13 February 1821, Paris, France: Luminous globe

People leaving a theatre observed a luminous globe in the air, which did not vanish until daybreak. The translator notes sceptically that this comes from "a French publication, as an example of the superstitious opinions which even in our times prevail to a considerable degree, in an otherwise enlightened country."

Source: *Adams Sentinel* (Gettysburg, Pennsylvania), May 2nd 1821.

402.

23 October 1822, Buchholz, Germany: Two objects

Astronomer Pastorff sees two round spots passing in front of the Sun.

Source: "New Planets," *Annual of Scientific Discovery* (1860): 409-11, at 411.

403.

5 December 1822, Aberdeen, Scotland
Globular intruder

"Soon after six o'clock, a most extraordinary meteor was observed, almost due north from Aberdeen. When first seen, it had the appearance of a large ball of the moon's diameter; but descending towards the horizon, it formed the shape and appearance of a luminous fiery pillar; soon after which it ascended, and assumed its original globular form – again descended a little, and began to extend itself as before, when it suddenly vanished."

Source: *The Edinburgh Advertiser*, 10 December 1822.

404.

22 May 1823, Hereford, England
Bright unknown near Venus

Bright shining object observed near Venus, again reported by an experienced astronomer, Reverend T. W. Webb.

Fig. 36: Reverend Webb

Reverend T.W. Webb was the author of *Celestial Objects for Common Telescopes*, a very popular reference book for serious amateur astronomers.

Source: *Nature* 14: 195.

405.

12 August 1825, at sea near Hawaii, Pacific
Large red round object, wide illumination

English naturalist Andrew Bloxam and others saw a large red luminous object rise, illuminating everything. It fell out of sight, rose and fell again: "About half past 3 o'clock this morning the middle watch on deck was astonished to find everything around them suddenly illuminated.

"Turning their eyes eastward they beheld a large, round, luminous body rising up about 7 degrees apparently from the water to the clouds, and falling again out of sight, and a second time rising and falling: it was the color of a red-hot [cannon] shot and appeared about the size of the sun...It gave so great a light that a pin might be picked up on deck."

Source: *The Diary of Andrew Bloxam* (Honolulu, 1925). As reported in *UFO Investigator* (NICAP) 4, No. 5 (March 1968).

406.

1 April 1826, Saarbrücken, Germany: Gray object

A grayish object, whose size was evaluated at over 1 meter, rapidly approached the ground with a sound like thunder and "expanded like a sheet."

Source: *American Journal of Science and Arts* 26 (July 1834): 133; *The Quarterly Journal of Science, Literature, and Art, by the Royal Institution of Great Britain* 24 (July to Dec. 1827): 488; E.F.F. Chladni, "Ueber eine merkwürdige meteorische Erscheinung, am 1. April 1826, nicht weit von Saarbrücken", in *Annalen der Physik* 83, no. 7 (Leipzig, 1826): 372-377.

407.

1827, Tietjerk, Friesland, Holland Fiery man from the sky

A man named Lieuwe Klaasens and a pastor saw a fireball land nearby, taking form as a fiery man who flew up.

Source: Kornelius Ter Laan, *Folkloristisch woordenboek van Nederland en Vlaams België,* ('s-Gravenhage: G. B. Van Goor zonen, 1949).

408.

March 1828, Mount Wingen, Australia Cigar-shaped object lands

A mysterious flying object was said to have descended upon Mount Wingen at the Burning Mountain Nature Reserve. It was "cigar-shaped and had a funny silver colour" and made a loud banging noise. According to the report, "when it landed it set fire to all the vegetation and killed the cattle."

Allegedly, tall strangers appeared in the town at the same time. "They never said anything but always pointed to the things they wanted."

The event must have caused quite a stir as the folk of Wingen began linking it with strange disappearances among them: "Quite often people just disappeared and dogs and domesticated animals disappeared too," wrote the informant, referring to the tale his grandfather used to tell.

Source: *Australian Post*, June 17, 1989, and W. Chalker, *Project 1947: Australian Aboriginal Culture & Possible UFO Connections* (1996).

409.

17 July 1829, Kensington, Pennsylvania, USA
Bright red object crossing the Delaware

Between 11 P.M. and midnight "a meteor of rather singular character" arose from the neighbourhood of the Schuylkill, passed over Kensington and the river Delaware, and disappeared behind the woods of Jersey.

"A long trail of light, like that of a shooting star was seen to follow it in the beginning of its ascension; large sparkles that separated themselves from it and descended slowly, were distinctly visible until hidden from view by the tops of the houses. Its motions were rapid, irregular, and wavering, like the fluttering of a kite or the rocking of an air balloon.

"Its appearance was of a deep red colour, and remarkably brilliant, seemingly about half the size of the moon. It arose until it crossed the Delaware, when it appeared but an inconsiderable speck scarcely discernable, and then descended with astonishing velocity until within a short distance of the horizon, where it remained stationary for a few moments. "Suddenly it became exceedingly large and brilliant, sparkles again separated from the main body, and descended as before. It soon after became dim and disappeared behind the trees. Altogether, I should suppose it was visible about fifteen or twenty minutes."

Source: *Hazard's Register of Pennsylvania* 4, 3 (July 18, 1829): 48.

410.

23 July 1830, Whinny Park, near Cupar, Fife, Scotland
Two unknown flashing lights and a beam

As he was travelling from Auchtermuchty to Letham, reverend Alexander Espline noticed a peculiar light hanging in the air above Whinny Park, the property of a wealthy man named James Millie. As he came closer, Espline saw there were actually two lights of unequal brilliance. The smaller one emitted a beam, after which both lights started flashing. Scared by the display, Espline ran away. Two days later, the body of James Millie was found near the site.

An extract from the Edinburgh Observer, published as a broadside, related that Millie was a middle aged man living in a remote area in Whin (Whinny) Park near Cupar, and was murdered sometime in June 1830 by his servant, John Henderson. Henderson was arrested on Sunday 25 July, 1830. It is interesting that the light was not considered related to the murder at the time. Indeed, it tells that "the path was so beaten that, but for an accidental circumstance, the discovery would probably never have been made."

Flickering lights, often an ignis fatuus, emerge in many medieval stories to indicate a burial or significant site.

Source: Elliot O'Donnell, *Ghosts with a purpose* (Rider, 1951); "Horrid murder! A full, true and particular account of that most atrocious and horrid murder..." (1830), National Library of Scotland, NLS F.3.a.13 (108).

411.

14 November 1832, Tyrol, Germany ⓘ
Hovering object

A ball of bright light accompanied by "falling stars" hovered for 15 minutes. Our primary source is based on a letter received from Bruneck (Tyrol, Germany). While it is true the phenomenon occurred at the usual time of the Leonids meteor shower, the event does not naturally involve a ball of light suspended in the sky for a quarter of an hour, so the case merits our attention.

Source: *The Annual Register or a view of the History, Politics and Literature of the year 1832* (London, 1833): 444-445.

412.

16 March 1833, North Carolina
A very slow "meteor" changes course

At 6:35 P.M., during twilight, a man observed an object as bright as Venus, about the same elevation but a little to the right of it. It was running in a northerly direction until it changed course, running parallel to the horizon. It assumed a serpentine shape and became stationary extending over 12 to 15 degrees and retaining its brilliance for about two minutes. The witness reports: "It continued gradually to fade, appearing more and more like a thin whitish cloud; and at 6:40 the last vestige of it disappeared, being visible just 5 minutes."

Source: Boston (Mass.) *Investigator*, 17 May 1833.

413.

13 November 1833, Niagara Falls, N.Y., USA
Hovering square

A large luminous square object was observed in the sky for an hour. It remained stationary, and then went away slowly.

Source: *American Journal of Science* 25: 391.

414.

1834, Cologne (Köln), Germany
Bright object splits in two

Bright object flying NE-N parallel to the horizon, reappeared and split in two.

Source: François Arago, *Astronomie Populaire*, vol. IV (Paris, 1840): 266.

415.

11 May 1835, Sicily, Italy
Astronomer's sighting

Unknown luminous object reported by astronomer Cacciatore. It was observed on four consecutive days.

"Cacciatore noted what he first believed was an eighth magnitude star on May 11, 1835; but, with his next observation, on May 14, 1835, its position had changed relative to another star, and he thought the object either a comet or planet beyond Uranus. Clouded skies prevented further observations until June 2, 1835; but, by then, the object had been lost."

Source: "Supposed new planet," *American Journal of Science*, S. 1, 31 (1837): 158-9, and "Cacciatore's supposed planet of 1835," *Nature* 18 (July 4, 1878): 261.

416.

6 October 1835, Cosenza, Italy
Maneuvering pyramid

A pyramid-shaped meteor appears and heads off towards a mountain, leaving a "gloomy tail." It first appeared as a lighted object seen flying West of Cosenza. It rose into the air and changed shape, leaving a vaporous trail, moving slowly towards the south. It followed a parabolic curve and disappeared towards Fiumefreddo harbor.

Source: "Casistica dei Fenomeni Straordinari" in *Orizonti Sconosciuti* 5, "Periodo: 1819-1857" (1976); Nicola Leoni, *Della Magna Grecia e delle tre Calabrie* (Napoli, 1844), 325-326.

417.

12 January 1836, Cherbourg, France
Spinning disk whistles

A "luminous body, seemingly two-thirds the size of the moon" was witnessed at 6:30 P.M. "Central to it there seemed to be a dark cavity." The object was traveling at around half a mile per second at an altitude of 1000 feet or so and seemed to rotate on its axis. It cast shadows on the ground as it whistled past.

Source: *Rept. British Assoc. for the Advancement of Science* 77 (1860). The object was not 'doughnut-shaped' as many have written.

418.

1 July 1836, Szeged, Hungary ⓘ
Light globes and a lady in white

Globular lights, poltergeist effects and apparitions of an entity resembling "a lady in white" and a dwarf-sized Franciscan monk. The case mainly concerns a so-called haunted house and objects being thrown around by unseen hands. Such manifestations are not uncommon in ufological literature and have been related by some researchers to the abduction phenomenon.

Source: The story comes from a letter by a Dr. V. Stantsky which was sent to Justinius Kerner and published in the latter's periodical: *Magikon, Archive für beobachtungen aus dem Gebiete der Geisterkunde und des magnetischen und magischen Lebens*, vol. 3 (Stuttgart, 1846), 223-237; William Howitt, "Throwing of Stones and Other Substances by Spirits," *The Spiritual Magazine*, vol. VI, 3 (Feb. 1865): 55-56.

419.

8 July 1836, Saratov Province, Russia
Hovering globe

At 10 P.M. there appeared, almost on the horizon to the north, a globe-shaped whitish mass as large as the moon; for several minutes it hovered in the air, after which it slowly descended to the ground and disappeared, leaving a zigzag trail.

Source: Mikhail Gershtein, *Potu storonu NLO* (Beyond the other side of UFOs) (Moscow: Dilya ed., 2002), 159, citing *Utkin S.NLO 200 let nazad?* (UFO 200 years ago?) in the newspaper *Zarya Molodezhi*, Saratov, 3 Feb. 1990.

420.

1 November 1836, Buchholz, Germany
Unexplained objects

Two unexplained objects crossing the face of the sun, reported by astronomer Pastorff: they were of unequal size, changing position relative to each other.

Source: *Annual of Scientific Discovery* (1860): 410.

421.

1837, Scarborough, England
Frightening lights at ground level

By a clear starlight night Mr. White, chief officer of the preventive service of the Scarborough station ("a most respectable authority") was proceeding from his house to a cliff where one of his men, named Trotter, had the lookout.

According to a letter from his son to a science magazine, "He passed a plantation in his way, in which he heard a loud crash among the trees, as if it had been the fall of an aerolite (…) He saw before him what he thought were balls of fire, about the size of an orange, appearing and disappearing with an undulating motion, about five or six feet from the ground; not accompanied by any noise, nor did they move over the hedges; but he observed other luminous appearances shooting across the road and sky, emitting a hissing noise like a rocket, but not so loud.

"The same appearances (particularly the latter) had so frightened the man, that he had actually hid himself for fear of them."

Source: *The Magazine of Natural History* (Longman, Orme, Brown, Green and Longmans, 1837): 550-551.

422.

16 February 1837, Buchholz, Germany
Uncorrelated planetoid

Another unexplained object crossing the face of the sun, reported by astronomer Pastorff.

Source: *Annual of Scientific Discovery for 1860* (Boston, 1867): 410.

423.

29 August 1837, Tirgu-Neamt, Romania
Luminous sphere

During the night a luminous sphere was observed by local people. It came closer to the ground at dawn, illuminating the fields with an intense reddish glow.

Source: *Albina Românească*, 2 Sept. 1837.

424.

1838, India, location unknown: Disk with appendage

A flying disk, about the apparent size of the moon but brighter, from which projected a hook-shaped appendage, was reported by G. Pettitt. It was visible about 20 minutes.

Source: Baden Powell. "A catalogue of observations of luminous meteors," *Annual Report of the British Association for the Advancement of Science* (1849): 1-53, at 2, 44.

425.

2 October 1839, Rome, Italy: Unidentified planet

Astronomer De Cuppis of the Royal College: unknown body similar to a planet passes in front of the sun. This is one of the main observations selected by Le Verrier to compute his orbit of Vulcan. The object was "a perfectly round and defined spot, moving at such a rate that it would cross the sun in about 6 hours."

Source: E. Dunkin, "The suspected Intra-Mercurial planet." *Monthly Notices of the Royal Astronomical Society* 37 (February 1877): 229-30.

426.

9 April 1843, Greenville, Tennessee, USA
Lighthouses in the sky

According to the Greenville (Tennessee) *Miscellany*:

"About eight o'clock, there was seen in the south-western sky a luminous ball, to appearance two feet in circumference, constantly emitting small meteors from one or the other side of it. It appeared in brightness to outrival the great luminary of day.

"On its first appearance it was stationary one or two minutes, then, as quick as thought, it rose apparently thirty feet, and paused – then fell to the point from whence it had started, and continued to perform this motion for about fifteen times. Then it moved horizontally about the same distance, and for nearly the same space of time. At length it assumed its first position; then rose again perpendicularly about twelve feet, and remained somewhat stationary, continuing to grow less for an hour and a quarter, when it entirely disappeared."

Source: *Brother Jonathan*, Vol. 5:2 (May 13, 1843): 55.

427.

3 October 1843, Warwick, Ontario, Canada
Flying men

Charles Cooper, a farmer, saw something strange crossing the sky in the middle of the afternoon:

"On the third day of October, as I was labouring in the field, I saw a remarkable rainbow, after a slight shower of rain. Soon after, the bow passed away and the sky became clear, and I heard a distant rumbling sound resembling thunder. I laid by my work, and looked towards the west from whence the sound proceeded, but seeing nothing returned to my labour.

"The sound continued to increase until it became very heavy, and seemed to approach nearer. I again laid by my work, and looking towards the west once more, to ascertain its cause, I beheld a cloud of very remarkable appearance approaching, and underneath it, the appearance of three men, perfectly white, sailing through the air, one following the other, the foremost one appearing a little the latest. My surprise was great, and concluding that I was deceived, I watched them carefully.

"They still approached me underneath the cloud, and came directly over my head, a little higher up than the tops of the trees, so that I could view every feature as perfectly as of one standing directly before me. I could see nothing but a milky-white body, with extended arms, destitute of motion, while they continued to utter

doleful moans, which, I found as they approached, to be the distant roar that first attracted my attention. These moans sounded much like Wo-Wo-Wo! I watched them until they passed out of sight. The effect can be better imagined than described. Two men were labouring at a distance, to whom I called to see the men in the air; but they say they did not see them. I never believed in such an appearance until that time."

Source: Eli Curtis, *Wonderful Phenomena* (New York, 1850).

428.

4 October 1844, location unknown ⓘ
Unknown planetoid

Astronomer Glaisher: A luminous object as bright as Jupiter, "sending out quick flickering waves of light."

Source: "Astronomical puzzle," *John Timbs' Year-Book of Facts in Science and Art* (1843): 278.

429.

29 March 1845, London, England
Orange object, hovering

At 11 P.M. Mr. Goddart observed an unusual object and reported it to an English journal: "The sky was perfectly clear and the stars sparkled (…) My attention was suddenly diverted by a weak light in the constellation of Canes Venatici, resembling a speck of fog about 4 in magnitude but clearly of a yellow color. I immediately pointed my telescope toward it, which gives small but very clear and bright vision. The meteor appeared as a fog of four stars, with the center of an orange color. From Alpha Can.Ven. it moved slowly towards Coma Berenices, gaining ever more brilliance. It took two minutes before it went out."

Source: *L'institut, journal général des sociétés et travaux scientifiques de la France et de l'étranger. 1ère section, Sciences mathématiques, physiques et naturelles,* vol. 13, no. 590 (Paris, 1845): 148.There is a reference to this observation in Poggendorf's catalog in *Annalen der Physik und Chemie* (1854).

430.

11 May 1845, Capodimonte Observatory, Naples, Italy
Black objects

Astronomer Schumacher wrote to Gauss on 18 September 1845: "Erman sends me a report of Capocci about bodies which he has seen pass in front of the sun on 11 to 13 May." Ernesto Capocci (1798-1864) was an Italian astronomer. Heinrich Christian Schumacher (1780-1850) was the founder of *Astronomische Nachtrichten*.

Source: Correspondence between mathematician Carl Friedrich Gauss and astronomer Schumacher. *Briefwechsel zwischen C.F. Gauss und H.C. Schumacher*, edited by C.A.F. Peters, vol. 5 (Altona, 1863), 46-47; *Report of the nineteenth meeting of the British Association for the Advancement of Science (year 1849)* (London, 1850): 46.

431.

18 June 1845, between Malta and Turkey
Three luminous objects rise from the sea

"At 9:30 P.M. the brig *Victoria*, from Newcastle to Malta, in lat. 36° 40' 56", long. 13° 44' 36" was becalmed, with no appearance of bad weather; when her top-gallant and royal masts suddenly went over the side as if carried away by a squall. Two hours it blew very hard from the east; and whilst all hands were aloft reefing topsails, it suddenly fell calm again, and they felt an overpowering heat and stench of sulphur. *At this moment three luminous bodies issued from the sea, about half a mile from the vessel, and remained visible for ten minutes.* Soon after it began to blow hard again, and the vessel got into a current of cold fresh air."

The geographic coordinates given in the report would place the ship 900 miles away from Antalya in Turkey.

Source: "Malta Mail", cited by *London Times*, 18 August 1845, and James Glaisher, et al. "Report on observations of luminous meteors, 1860-1861." *Annual Report of the British Association for the Advancement of Science* (1861): 1-44, at 30.

432.

18 June 1845, Ainab Mountain, Lebanon
Two large unknowns

"At Ainab, on Mount Lebanon, at half an hour after sunset, the heavens presented an extraordinary and beautiful though awful spectacle." Witnesses described the phenomenon as "composed of two large bodies, each apparently at least 5 times larger than the moon, with streamers or appendages from each joining the two, and looking precisely like large flags blown out by a gentle breeze." They appeared in the west, remaining visible for an hour, taking an easterly course, and gradually disappeared.

"The appendages appeared to shine from the reflected light of main bodies, which it was painful to look at for any time. The moon has risen about half an hour before, and there was scarcely any wind."

This phenomenon may well have been related to the first sighting for this date, above.

Source: James Glaisher, et al. "Report on observations of luminous meteors, 1860-1861," *Annual Report of the British Association for the Advancement of Science*, (1861): 1-44.

433.

7 June 1846, Darmstadt, Germany ⓘ
Slag residue falling

An object falling from the sky is found to be "only slag." In several well-attested modern cases when an unknown flying object dropped some molten metal, the residue was also found to be "nothing but slag," So while not representing an important sighting, the report should not be summarily discounted and we include it for its possible physical relevance.

Source: James Glaisher et al. "Report on observations of luminous meteors, 1866-67." *Annual Report of the British Association for the Advancement of Science* (1867): 288-430, at 416.

434.

25 August 1846, Saint-Apre, France
Bright globe emits 'stars'

At 2:30 A.M. Dr. Moreau was returning from a visit to a patient's home by warm, calm weather when he found himself bathed in the light coming from a globe that seemed to open up, emitting hundreds of star-like objects. This was observed for three to four minutes, after which the display slowed down and the globe disappeared.

Source: "Sur un météore lumineux," *Comptes-Rendus de l'Académie des Sciences* 23 (Paris, 1846): 549-550.

435.

19 September 1846, La Salette, France
Brilliant light and apparition

Two cowherds, an eleven year old boy, Maximin, and a fourteen year old girl, Mélanie Calvat, saw a sudden flash of light. She testified:

"I could see our cows grazing peacefully and I was on my way down, with Maximin on his way up, when all at once I saw a beautiful light shining more brightly than the sun. 'Maximim, do you see what is over there? Oh! My God!' At the same moment, I dropped the stick I was holding. Something inconceivably fantastic passed through me in that moment, and I felt myself being drawn. I felt great respect, full of love, and my heart beat faster. I kept my

eyes firmly fixed on this light, which was static, and as if it had opened up, I caught sight of another, much more brilliant light which was moving, and in this light I saw a most beautiful lady."

The "lady" was dressed in white and gold, with a cap of roses on her head. She was surrounded by a brilliant light and was weeping. The lady complained that Sunday was being desecrated and the peasants were blaspheming the saints in swearing. (The Curé d'Ars and other clergy were complaining about these very sins in their sermons). If there was no amendment, there would be great disaster, the harvest would fail and people would starve

The parish priest declared the lady to be the Blessed Virgin; the apparitions were later approved by the Bishop of Grenoble, and pilgrimages began. Mélanie became a nun and continued to receive revelations. Maximin tried unsuccessfully to become a priest but was always in debt.

Source: Mélanie wrote down her testimony in 1878. On 15 Nov. 1879 it was published with the "Imprimatur" of the Bishop of Lecce. In 1904, a few weeks before her death, it was reprinted "ne varietur" at Lyon.

436. 

November 1846, Rangoon River, China
Light beam and extreme heat

By a dark night, at about 7:30 P.M., a bright light appeared, accompanied by extreme heat, and moved rapidly in front of the ship where the witness, (the wife of the ship's owner) was standing with the captain and a 4 year old child.

The light did not come as a bolt of lightning but rather as a compact flame that terrified the witnesses. Several people in the vicinity felt the sudden increase in heat, although they were not in a position to see the light.

Source: Collingwood, *Philosoph. Magazine*, and *L'Institut*, 29 April 1868, 144.

437. 

19 March 1847, Holloway, London, England
Unknown object ascending

"On the evening of Friday, March 19, "A" and I left Albion road, [Holloway], about half-past eight. Not any stars were then visible, but when we were in Highbury place, "A" called my attention to what we thought a fire-balloon ascending slowly. It was in the west, a little inclining to the south. As it passed on slowly to the west, its intense brilliance convinced me that it was not an earthly thing.

"When it appeared to be over Hampstead (but as high in the heavens as the sun is at six o'clock in the evening when the days are longest) it shot forth several fiery coruscations, and whilst we were gazing at it broke into an intensely radiant cloud: this cloud sailed on slowly, and we never took our eyes off it. At this time the stars were shining. When we were in the gravel path opposite to Highbury terrace, the cloud was higher in the heavens and more to the west. It cast a most brilliant light on the houses there, brighter than moonlight, and unlike any light I ever saw. It appeared of a blue tint on the bricks, but there was no blue light in the cloud itself.

"Suddenly, over the radiant cloud appeared another cloud still more brilliant, but I now felt so awe-struck, that I cannot say precisely how long they hung one over the other before the most wonderful sight happened. Perhaps they remained so for two or three minutes, when from the upper cloud a small fiery ball (about the size that the largest planets appear to the naked eye) dropped into the lower cloud, and was instantly absorbed. Soon after, another similar ball dropped from the upper to the lower cloud and then a ball apparently four or five times the size of the two preceding, fell from one cloud to the other.

"Shortly after this both clouds disappeared, apparently absorbed in the heavens, though I did see a few particles of the brilliant clouds floating about for a minute or so. Presently the moon appeared, considerably to the northward of the place where the clouds had hung. We then saw the bright light across the heavens, which you told me, was zodiacal light, which lasted for more than an hour."

Source: "Meteoric Stones," *The Living age,* vol. 56, No. 717 (Feb 20, 1858), 503. Also *Excelsior: Helps to Progress in Religion, Science and Literature*, vol. V (London: James Nisbet and Co., 1856).

438.

11 October 1847, Bonn, Germany
Fast-moving intruder

Single object, a small black spot rapidly crossing the disk of the sun, reported by astronomer Schmidt.

It was "neither a bird nor an insect crossing before the telescope."

Source: R. C. Carrington, *Monthly Notices of the Royal Astronomical Society* 20 (January 1860): 100-1.

439.

19 November 1847, Oxford, Wytham Park, England ⓘ
A large object makes two stops

A large object reported by Mr. Symonds was stationary at two points of its trajectory, which took seven minutes.

Source: *Report of the eighteenth meeting of the British Association for the Advancement of Science (year 1848)* (London, 1849): 9-10.

440.

1848, Arabian Sea, Arabia: Two wheels at sea ⓘ

A ship at sea was approached by two "rolling wheels" that exploded with a crashing sound:

"Sir W. S. Harris read a report from a ship towards which two fiery wheels described as rolling haystacks on fire had whirled. When they came close there was a horrible crack as two masts had ruptured under a violent shaking. A strong odor of sulphur was noted."

Source: Mentioned in *GEPA Bulletin*, new series no. 2 (Feb. 1965), 17, the original source was *The Athenaeum*, no. 1086 (19 August, 1848).

441.

9 March 1848, Oxford, Wytham Park, England ⓘ
Unidentified celestial body

Single unidentified object reported by Mr. Symonds.

Source: *Report of the eighteenth meeting of the British Association for the Advancement of Science (year 1848)* (London, 1849): 10.

442.

4 September 1848, Nottingham, England
Slow celestial object

At 8:59 P.M. Mr. Lowe observed an unidentified object from Highfield Observatory. The bright star-like source moved from Eta of Antinoüs to Pi of Sagittarius. It covered this distance in no less than 45 seconds, much too slow for a meteor. Its intensity was estimated at six times the brilliance of Jupiter.

Source: *Report of luminous meteors of the British astronomical association*, 1849, quoted by Flammarion in *Bradytes*, op. cit.

443.

18 September 1848, Inverness, Scotland
High velocity objects

Two large, bright lights that looked like stars were seen in the sky. Sometimes they were stationary, but occasionally they moved at high velocity.

Source: *The Times*, 19 Sept. 1848.

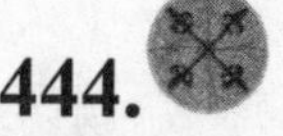

444.

5 February 1849, Deal, Kent, England
Two dark objects

Two dark objects seen crossing the disk of the sun by an observer named Brown.

Source: E. J. Lowe, "Meteors, or falling stars," *Recreative Science* 1 (1860): 130-8, at 138.

445.

13 February 1849, Reims, France
Star moving in the sky with sudden accelerations

Skilled amateur observer Coulvier-Gravier observed an unusual "star" at 7:30 P.M. It was a 3rd magnitude object first seen in the vicinity of Delta Cephei. It moved through a course of 20 degrees in the sky, with sudden accelerations and stops ("saccades").

Source: Rémi Armand Coulvier-Gravier, *Recherches sur les Météores* (Paris: Mallet-Bachelier, 1859), 292.

446.

12 March 1849, location unknown: Sighting of Vulcan

Astronomer Sidebotham reports an object crossing the disk of the sun. This is one of the observations judged reliable enough by Le Verrier to compute his orbit of Vulcan, the supposed intra-mercurial planet.

Source: E. Dunkin. "The suspected Intra-Mercurial planet," *Monthly Notices of the Royal Astronomical Society,* 37 (February 1877): 229-30.

447.

4 April 1849, Delhi, India: Very slow object

An extremely slow object was seen in the sky, dimming and brightening.

Source: *Report of the twentieth meeting of the British Association for the Advancement of Science (year 1850)* (London, 1851): 129.

448.

14 October 1849, Athens, Greece
Unknown celestial object

Astronomer Schmidt reports an unknown celestial object crossing the disk of the sun.

Source: E. Ledger, "Observations or supposed observations of the transits of intra-mercurial planets or other bodies across the Sun's disk," *Observatory* 3 (1879-80): 135-8, at 137.

449.

15 January 1850, Cherbourg, France
Light, swinging motion

About 7:45 P.M., by snowy weather, a bright light appeared above the trees. It was observed by Mr. Fleury, swinging about its base, which was in line with the horizon. It scintillated, seemed ready to disappear, then was re-ignited and finally disappeared. Small flashes continued to be seen, moving south.

Source: Camille Flammarion, *Bolides Inexpliqués par leur aspect bizarre et la lenteur de leur parcours – Bradytes,* citing Sestier, *La Foudre et ses formes*, T.I., 205.

450.

5 February 1850, Sandwich, Kent, England
Slow-moving "red-hot iron ball"

At 6:50 P.M. according to Mr. W. H. Weekes, a small luminous object appeared stationary near Orion and approached slowly on a straight line, growing to one third the apparent diameter of the moon. It went from a speck of light to a "red-hot iron ball," hovered for about three minutes and disappeared in a shower of fire. The object had remained stationary for 1 min 45 seconds, then moved horizontally for a full 45 seconds.

Source: *Report of the British Astronomical Association* (1851): 1-52 at 2-3, 38, and Baden Powell, *On Observations of luminous meteors*, op. cit.

451.

18 February 1850, Athens, Greece: Solar intruder

An unknown body was seen passing in front of the Sun. It was observed and reported by astronomer Schmidt.

Source: R. C. Carrington, "In the 10th number of Professor Wolf's Mittheilungen über die Sonnenflecken..." *Monthly Notices of the Royal Astronomical Society* 20 (January 1860): 100-1.

452.

13 March 1850, Paris, France
A flying object reverses course

Mr. Goulvier-Gravier, founder of a private observatory dedicated to meteoritics, observed a magnitude 3 "shooting star" that came from the southeast and reversed course at 4:45 A.M.

Source: Goulvier-Gravier, *Recherches sur les Météores* (Paris: Mallet-Bachelier, 1859), figure 88 on page 300.

453.

3 October 1850, Talcot Mt., near Hartford, CT, USA
Maneuvering object

According to a paper read by Professor Brocklesby before the American Association in 1851, an observer named Graylord Wells, who was on the eastern slope of Talcot Mountain saw an unidentified object:

"The evening was clear and the moon near the meridian, when Mr. Wells saw, a little south of west, and full 60 degrees above the horizon, a bright meteor apparently a foot in diameter. It shone with an orange hue, and below was a train which seemed to be 15 or 16 feet in length, fan-shaped, and possessing an apparent breath at its further extremity of two feet. The meteor rose from east to west with a slow and steady motion, and in its progress passed above or to the north of the moon. And when it had arrived on the eastern side, directly turned toward the southeast, and dropping below the moon, a part of its attendant train swept over the lunar disk."

The phenomenon gradually descended to the horizon in the southeast. The observer stated that this could not have lasted less than three minutes in moving the length of its train, and that the time of its visibility "could not possibly have been less than an hour, and was probably an hour and a half."

Source: The *Ohio Journal of Education* 2 (Dec. 1853): 411.

454.

10 October 1852. Reims, France
Wandering "star" makes a 90 degree turn

M.Coultier-Gravier, observing the sky from his private observatory to compile statistics on shooting stars, recorded an unusual "meteor" at 8 P.M. It was a third-degree object first seen near "nu" of Capricorn. It described a 30-degree trajectory in the sky, changing its direction from northwest to southwest.

Source: Coultier-Gravier, op. cit., 306 and fig. 97.

455.

1853, Paris, France: Red disk on slow trajectory

A witness named Amédé Guillemin, who lived on *Rue Amelot*, observed an object moving extremely slowly, horizontally above *Père-Lachaise* cemetery. The object was "a pale red disk."

Source: Flammarion, *Bradytes*, op. cit., 155.

456.

22 May 1854, unknown location: Multiple unknowns

A contributor named R. P. Greg reports that a friend of his saw an object equal in size to Mercury in the vicinity of that planet, and behind it an elongated object, and "behind that something else, smaller and round."

Source: Baden Powell "Report on observations of luminous meteors, 1854-55," *Annual Report of the British Association for the Advancement of Science* (1855): 79-100, at 94.

457.

21 January 1855, New Haven, Connecticut
Another red object

About 10 P.M. a man saw a brilliant red ball about two minutes in diameter, first visible about eight degrees below the guards in Ursa Minor. It seemed stationary at first, but in about fifteen seconds commenced moving slowly towards the east in an almost horizontal line, with what seemed like a slight undulatory motion. It passed below and about one degree from the star Benetuash in Ursa Major and disappeared in the distance, not far from Denebola in the constellation Leo. The observation had lasted ten minutes. The writer adds: "there was no explosion, nor was any scintillation thrown off at any time."

Source: *New Haven Palladium,* 23 January 1855.

458.

11 June 1855, Bonn, Germany: Dark unknown body

A dark body was seen crossing the disk of the sun. It was reported by astronomers Ritter and Schmidt.

Source: E. Ledger, "Observations or supposed observations of the transits of intra-Mercurial planets or other bodies across the Sun's disk," *Observatory* 3 (1879-80): 135-8, at 137.

459.

11 August 1855, Tillington, Sussex, England
Red wheel in the sky

At 11:30 P.M. a Mrs. Ayling and other witnesses watched in awe as a red wheel-like object with spokes emerged from behind some hills and remained visible in the sky for an hour and a half.

Source: Baden Powell, "Report on observations of luminous meteors, 1855-56," *Annual Report of the British Association for the Advancement of Science* (1856): 53-62, at 54-55.

460.

10 December 1855, Copenhagen, Denmark
Unexplained object

An object varying in size from the apparent diameter of the sun to that of a star was visible in the south-western atmosphere for 10 to 12 minutes. It "changed its configuration several times, having appeared now in one mass, then in two, then again in three, and so forth alternately, lighting up the heavens to a considerable distance."

Source: Copenhagen *Faedrelandet,* quoted by the *Manchester Guardian* of 5 January, 1856, under the heading "Prussia, from our own correspondent."

461.

1858, Jay, Ohio, USA: Silent vessel with passengers

Alerted by a sudden shadow over the place where they were standing, several witnesses including Mr. Henry Wallace are said to have looked up in time to see "a large and curiously constructed vessel, not over one hundred yards from the earth." A number of very tall people were seen aboard this craft, which the recorder of the event believes was "a vessel from Venus, Mercury, or the planet Mars, on a visit of pleasure or exploration, or some other cause."

Mr. Wallace reportedly added: "The vessel was evidently worked by wheels and other mechanical appendages, all of which worked with a precision and a degree of beauty never yet attained by any mechanical skill upon this planet (...) This was no phantom that disappeared in a twinkling...but this aerial ship was guided, propelled and steered through the atmosphere with the most scientific system and regularity, about six miles an hour, though, doubtless, from the appearance of her machinery, she was capable of going thousands of miles an hour."

Author Jesse Glass rediscovered the book containing this report and claimed he found evidence at the Ohio Historical Society regarding the existence of a man named Henry Wallace in Jay, Ohio at the time.

Research by C. Aubeck disclosed that there was indeed a Post Office at Jay from March 14th 1839 to March 23rd 1842, but its whereabouts had become unknown shortly afterwards. Indeed, the town did not figure in any gazetteer or Erie County history. Aubeck managed to pinpoint the location of Jay from comments made by the historian Henry Timman in his popular weekly column, *Just Like Old Times.* It was, he said, "on the township line between Milan and Huron" but nothing marks the spot today. According to census records a Henry C. Wallace lived in nearby Erie County in 1850, a fifteen year old lad from New York. By 1860 he must have either died or moved on, because he is not listed again in the state of Ohio. This Mr. Wallace was too young to have lived in Jay and in fact was registered as a resident of Florence, a different township.

We can only conclude that the claim rests on the veracity of names that cannot be verified today.

Source: Dr. William Earl, *The Illustrated Silent Friend, embracing subjects never before scientifically discussed* (New York, 1858).

462.

26 March 1859, Orgères, France: Sighting of Vulcan

Mr. Lescarbault, an amateur astronomer, has observed a body of planetary size crossing the disk of the sun. He wrote to Le Verrier, who came to Orgères to meet with him and to verify the records of the observation in view of

Fig. 37: French astronomer Urbain Le Verrier, discoverer of Neptune

computing an orbit for Vulcan, the intra-mercurial planet which he hypothesized. In his letter, Lescarbault wrote: *"The duration of the passage of the new planet was one hour seventeen minutes, and twenty seconds of sidereal time. I have the conviction that, some day, a black dot, perfectly circular, very small, will be seen again passing in front of the sun (…) This object must be the planet or one of the planets whose existence in the vicinity of the solar globe you have announced a few months ago, Mr. Director, using this same wonderful power of computation that made you recognize the existence of Neptune in 1846."*

Source: *L'Année Scientifique* (1878): 16.

463.

29 January 1860, London, England
Unknown planetoid

An unknown object of planetary size is reported by Mr. Russell and three other observers.

Source: F. A. R. Russell, "An Intra-Mercurial planet," *Nature* 14 (October 5, 1876): 505.

464.

1 March 1860, Moscow, Russia: Unexplained sky object

At 9:45 P.M. "a star to the southwest of the Great Bear suddenly commenced to wax larger, assuming at the same time the color of iron at a red heat, but without the appearance of any sparks or rays." It was observed in this condition until 11:30 P.M., growing to half the size of the moon. It then became dimmer, and by midnight it had disappeared. In its stead "a sort of black speck was to be noticed by the light of the other stars."

The writer adds: "It remains for the astronomers to describe, and poets to sing, the destruction of the luminary, which, for ought we know, may have been the abode of a race superior to our own."

Source: The Russian correspondent of the *London Telegraph*, quoted in *The Banner of Liberty* (Middletown, New York), 6 June 1860.

465.

17 July 1860, Dharamsala, India: Lights in the heavens

On the evening of the day when a remarkable meteor had fallen in the area, a man who was observing the sky about 7 P.M. saw a pattern of lights, each lasting for one minute or more, over places where there were no houses or roads:

"Some were high up in the air moving like fire balloons, but the greater part of them were in the distance in the direction of the lower hills in front of my house, others were closer to the house and between Sir Alexander Lawrence's and the Barracks. I am sure from some which I observed closely that they were neither fire baloons (sic), lanterns nor bonfires, nor any other thing of that sort, but *bona fide* lights in the heavens. Though I have made enquiries among the Natives the next day, I have never been able to find out what they were or the cause of their appearance."

Source: *The Canadian Journal of Industry, Science and Art*, Canadian Institute (1849-1914), vol 7 (1862): 197.

466.

24 September 1860, Nebraska City, USA
Three unexplained objects in apparent formation

After sunset Mr. Joel Draper and ferryman Mr. Beebout saw a bright object low in the West. "While gazing with amazement at that, which in size, color, brightness and shape resembled one-fourth of the sun taken from its edge, we soon discovered another spot further to the right and a little higher, which was about one-third the size of the first; then another directly above the first, one-third the size of the second. All of these we soon discovered to be moving towards the south, or to the left of their former position, with great rapidity (…) as they moved, they all retained the same relation to each other as when they first appeared."

The account goes on: "This took place after sunset, but, by means of the brightness of these bodies, it was as light as some ten or fifteen minutes before sunset. There were no clouds or vapors in the sky in that direction. They could not have been sun dogs or mock suns, for (such phenomena) remain, as long as they continue, in the same relative position to the sun."

Source: *Brooklyn Daily Eagle*, 12 October 1860.

467.

10 November 1860, Washington, D.C., USA
Three unknown objects flying over the Capital

Just about sunset several witnesses observed an object "the size of a balloon" moving with great rapidity in a south-westerly direction: "Notwithstanding the light of day was still strong and clear, the illumination of the object was brilliant and distinct (…) We heard, while gazing at this wonder, that two similar ones had passed previously. The one we saw, after moving south-westerly, at an angle with the path of the sun, took a course directly west, and straight from us; fading gradually and very rapidly until lost from sight.

Source: Washington D.C. *Herald* and *Brooklyn Eagle*, front page, Friday 16 November, 1860.

468.

1861, North Atlantic Ocean: Three luminous bodies

Three luminous bodies are reported to have come from the North Atlantic Ocean and stayed in view for no less than ten minutes during a squall.

Source: "A Catalogue of Observations of Luminous Meteors," *Report of the British Association*, 1861.

469.

4 October 1861, New York, USA
Mysterious object with occupants

About 6 P.M. a "mysterious balloon" passed over the city, with two men in it, from west to east at great height. Witnesses speculated that it was either "in the service of the traitors" or that it was "a device built by Professor Lowe, which had parted its fastenings," both impossible explanations.

Source: The *New York Times*, 5 October, 1861.

470.

12 August 1863, Madrid, Spain: Maneuvering object

"The night before last there was observed on the horizon a luminous body that appeared towards the east, and it was promptly thought to be a comet. Its colour was reddish, and on the top part there could be seen an appendix or crown, that was doubtlessly ablaze. It was stationary for a long time; but later it began to move quickly in different directions: horizontally, rising, and lowering."

The comet hypothesis is not tenable in this case.

Source: *Gaceta de Madrid*, 14 August 1863 (issue 226). Recorded by Franck Marie and Daniel Villain archives. Cited by Charles Garreau in *Alerte dans le Ciel*, 142.

471.

1 November 1864, Florence, Italy: Hovering white globe

"A white globe of fire many times larger than the full moon seemed hanging almost motionless in the air." Shades of orange and blue passed over its surface. After a full minute it suddenly disappeared, vanishing on the spot. The witness adds: "Only just before its disappearance a smaller ball was seen immediately below it, of a fiery orange colour, the first one appearing at that moment of the same hue."

Source: Madame Baldelli, "Large Fireball," *Astronomical Register* 3(1865): 53.

472.

4 December 1867, Chatham, England
A group of black disks

"On the afternoon of Monday the 4th, between the hours of 3 and 4, I witnessed a very extraordinary sign in the heavens...The facts are as follows: I was passing the Mill by the water-works reservoir. On the gallery I observed the miller uttering exclamations of surprise, and looking earnestly towards the west. On inquiring what took his attention so much, he said: 'Look, sir, I never saw such a sight in my life!'

"On turning in the direction towards which he was looking, the west, I also was astounded—numberless black disks in groups and scattered were passing rapidly through the air. He said his attention was directed to it by his little girl, who called to him in the Mill, saying, 'Look, father, here are a lot of balloons coming!'

"They continued for more than 20 minutes, the time I stayed. In passing in front of the sun they appeared like large cannon shot. Several groups passed over my head, disappearing suddenly, and leaving puffs of grayish brown vapor very much like smoke.

"I am, sir, your truly, James E. Beveridge, Darland, Chatham."

Source: Letter to the editor of Chatham News and Symonds' *Monthly Meteorological Magazine* (Dec. 1867): 8.

473.

8 June 1868, Radcliffe Observatory, Oxford, England
Unknown astronomical body

Astronomers recorded, at 9:50 P.M., a luminous object that moved quickly across the sky, stopped, changed course to the west, then to the south, where it hovered for four minutes before heading north.

Source: "Remarkable meteor" *English Mechanic* 7 (July 10, 1868): 351.

474.

25 July 1868, Parrammatta, NSW, Australia ⓘ
Flying ark, spirit voices, strange formulas, abduction

The following account, based upon a transcript of a manuscript that has never been located, must be taken with great caution. It purports to tell the story of Mr. Frederick William Birmingham, an engineer and local council alderman, who saw what he described as an "Ark" as he was standing under the verandah of his rented cottage in Duck's Lane and looked up to the sky:

"While looking at it (…) I said to myself aloud 'Well that is a beautiful vessel', I had no sooner ended the sentence than I was made aware that I was not alone, for, to my right hand and a little to the rear of my frontage a distinct voice said, slowly, 'That's a machine to go through the air'– in a little time I replied 'it appears to me more like a vessel for going upon the water, but, at all events, it's the loveliest thing I ever saw.' During this part of the conversation the machine made three courses: the first a level, the second a rapid backward descent, and the third left descent, but with a forward and curved easterly movement."

Birmingham's description goes on:

"The machine then quite stopped the forward motion and descended some twenty feet or so as gently as a feather on to the grass [and] showed its bottom partially, its side fully, and a half front section or view, its peculiar shapings are well impressed upon my mind and the colour seemed to blend with faint, flitting shades of steel blue, below, and appearing tremulous and like what one might term, magnified scales on a large fish, the latter being as it were flying in the air. The machine has not the shape of anything that has life."

This observation led to a classic contact or abduction scenario:

"Shortly after my declaring it was the loveliest thing I ever saw–the spirit said to me 'Have you a desire or do you wish to enter upon it?' I replied Yes, – 'then come'– said the spirit, thereupon we were lifted off the grass and gently carried through the air and onto the upper part of the machine, which was about 20 yards distant from where we were standing – (the spirit appeared like a neutral tint shade and the shape of man in his usual frock dress). While I stood on the machine the spirit moved to a cylinder pointing and indicating its purposes by downward motion of hand then made sign (that another and similar, was beyond and back of the Pilot house – as I term a part of the machine) which former I could not from my position see – the spirit then went further to the right two steps or so and went down in the machine to his waist returned to me and while passing on one side going to the rear of the machine the spirit – *en passant* – and making a sign, pointing, said 'step in' and I partly turned in the direction indicated to me I saw steps (three, I think) steep ones. I stepped down into the – let me call it – 'Pilot house' which had a floor about three and one half feet lower than the first or upper floor it was enclosed at the sides, end and top and only open in front, and nothing was in the Pilot house that I could discern but a table with passage all around it, and this table or bench seemed covered all round its sides and top alike a solid or at all events a thing about five feet long or so, and 3 1/2 broad and 2 1/2 feet high covered like with oil skin or something of that sort, or perhaps iron covered with rubber cloth tightly–the side spaces round it were about 2 feet wide and everything appeared very strong, the sides I noticed (when about 'stepping in') were extremely thick, about six inches – and I wondered why they were so strong in a machine to go through the air.

"I was now alone in the machine at the rear end of the tablet or table resting my fore fingers and thumbs on its edge looking vacantly with downcast eyes upon the table and repenting like at my saying yes – when the spirit previous to my entering upon it had spoken to me – I felt miserably queer – just like one undertaking a billet or post he knows nothing of, so I remained for some considerable time, when I was aroused as it were from my reverie by the voice of the spirit on my right hand (and his hand resting upon the table with several printed paper within it) who said 'here are some papers for your guidance.'"

Associated with this sighting, and with the papers that contained formulas to make a flying machine, the witness later experienced paranormal phenomena. Prior to the observation of the "Ark" itself he had had a vision of faces in the sky. Some time later he experienced poltergeist phenomena when the latch of a gate kept raising itself in full view without visible cause. In April 1872 he observed three clouds of very peculiar shape, which flew away quickly. He took this observation to be another divine instruction, the meaning of which he could not decipher.

Source: The document containing this report has an interesting history. It is known as the *Memorandum Book of Fred Wm. Birmingham, the Engineer to the Council of Parramatta* and subtitled *Aerial Machine*. Researcher Bill Chalker has traced its post-1940s whereabouts and spoken to some of the people involved. It seems it was originally written in ink in a small black imitation-leather book, which came into the possession of a teacher named Wallace Haywood, a resident of Parramatta. In the 1940s he passed the book to a Mrs. N. de Launte, a qualified nurse who was looking after his wife, and she finally gave it to ufologist Tasman V. Homan in the 1950s. Homan made a transcription of the book, including the sketches contained in it, with the help of four other people. A copy of this 15-page typed version was discovered among the papers of an astrologer called June Marsden. Mr. Fred Phillips, then honorary president of the Sydney-based UFO Investigation Centre (UFOIC), showed this to Bill Chalker in 1975. The original (if it exists?) has not been found, hence our reservations.

475.

8 June 1869, Fort Wayne, Indiana, USA: Circle of fire

About 2:30 A.M. an object "larger than the moon when full" was observed in the western sky for half an hour.

"It was of a bright red color, and at intervals of a few minutes, darted forth on every side bright rays like the straws of a broom, and from the ends of these were sent out sparks like those of a Roman candle. Suddenly this would cease, and only the circle or ball of fire remained, when again the rays would blaze out around the whole circumference of the central ball."

Two witnesses, including the doorman of Pike's Opera House, watched the phenomenon as it went down behind Mount Davidson, following the motion of the stars and "still blazing and sputtering forth sparks and jets of fire."

Source: "Singular Celestial Phenomenon," *Fort Wayne Daily Democrat* (Fort Wayne, Indiana), 9 June 1869.

476.

7 August 1869, Adamstown, Pennsylvania
A silvery object lands

At noon a luminous object was seen to descend from the sky to a dry, swampless area 200 yards north of the village, which is situated in Lancaster County. "It was square and became a column about 3 or 4 feet in height and about 2 feet in thickness." The object reflected sunlight "like a column of burnished silver" but after 10 minutes it disappeared. Several people gathered at the spot where it had rested but there were no landing traces.

Source: *Reading Eagle* (Pennsylvania), 14 August 1869.

477. 

7 August 1869, Ottumwa, Iowa
Astronomer's sighting

About 25 minutes before the totality of the solar eclipse, Professor Zentmayer observed some bright objects crossing from one cusp to the other of the solar crescent.

Each object took two seconds to make the crossing. The points were well-defined and must have been miles away from the telescope, given their sharpness.

Other sources indicate that similar objects were seen at the same time by Professor Swift in Mattoon, Illinois and in Shelbyville, Kentucky by Alvan Clark Jr., George W. Dean and professor Winlock, showing the objects were not local insects or seeds picked up by the wind.

Source: Henry Morton, "Solar eclipse—August 7, 1869," *Journal of the Franklin Institute*, S. 3, 58 (whole series, vol. 88): 200-16, at 213-4; Henry Morton, "Apparence d'une pluie météorique," *Cosmos: Les Mondes* 21: 241-3; "Meteors observed during a total eclipse of the Sun," *Popular Astronomy* 2 (March 1895): 332-3.

478. 

Spring 1870, Alen, Norway ⓘ
Flying object with occupant

The grandmother of Lars Lillevold saw a flying object in the sky. "Somebody" aboard the object beckoned to her.

Source: J. S. Krogh, *The Hessdalen Report* (CENAP Rept, 1985), 11.

479. 

22 March 1870, Atlantic Ocean: Circle with five arms

On a very clear day, with the wind blowing from the north-northeast, Captain Frederick William Banner and his crew of the American bark "Lady of the Lake" observed an object at 6:30 P.M. in the south-southeast. The ship was located about halfway between Senegal and Natal, Brazil, at latitude 5.47 N and longitude 27.52 W.

The object was described as "a circular cloud" light gray in color with a semicircle near the center and four arm-like appendages reaching from the center to the edge of the circle. Banner noted that "from the center to about 6 degrees beyond the circle was a fifth ray, broader and more distinct than the others, with a curved end."

The object moved to the northeast, much lower than the cloud cover. It was last seen at 7:20 P.M. about 30 degrees above the horizon.

Source: *Quarterly Journal of the Meteorological Society*, vol.1, new series, No.6 (April 1873): 157.

480.

15 August 1870, Dunbar, Scotland
Hovering ball of light

About 8:45 P.M. a bright sparkling ball of light tinged with blue appeared about 45 degrees above the northern horizon.

"From the head or ball there issued a tail of the same bright colour…pointing in a north-easterly direction. A remarkable circumstance was that it appeared quite motionless and stationary. By-and-by, however, a second tail seemed to branch off from about the middle of the first one, at an angle of 45 degrees, thus giving the tail of the figure a cleft or forked appearance.

"This second tail seemed to come and go, being occasionally detached for a few seconds, sometimes being lost to sight altogether... The phenomenon lasted with little variation for fully 20 minutes, and then proceeded very slowly in a south-westerly direction.

"No noise or explosion of any kind was heard during its passage. It attracted a great deal of attention, and was witnessed with a great deal of excitement by the inhabitants of the villages to the west."

Source: *The Scotsman* (Edinburgh), 17 Aug. 1870, 2. Two days later the same paper added that the object had been seen by people on the Greenock Esplanade.

481.

26 September 1870, Berlin, Germany: Slow transit

"As I was last night examining the constellation Lyra through my 4½-inch achromatic, with a power of 46, I observed a luminous object, with a distinct comet-like tail, pass slowly through the field of my glass, apparently starting from Vega and falling in the direction of Epsilon Lyræ. The hour by my watch was 12:15, Berlin time. The time occupied by this object in its transit across the disc of the glass was about 30 seconds, but before it had reached its edge it disappeared suddenly from view. I at first thought it was a falling star, but on reflection it appeared to me that a falling star would never have remained so long visible in the telescopic field."

Source: Mr. Barbazon's letter to the editor of *London Times*, 30 Sept. 1870, 9.

482.

2 August 1871, Marseille, France
Magnificent red object

Camille Flammarion reports an observation made by Mr. Coggia, also the discoverer of a comet. At 10:43 P.M. he observed a long-duration "bolide" that could not have been a classical meteor, given its slow rate of progression in the sky. He described it as "a magnificent red object." It moved eastward, slowly, clocked for no less than nine minutes. It stopped, moved north, and was stationary again at 10:59. It turned eastward again and was lost to sight at 11:03 P.M.

Source: Coggia. "Observation d'un bolide, faite à Observatoire de Marseille le 1er août," *Comptes Rendus* 73 (1871): 397-8.

483.

31 August 1872, Rome, Italy: Slow sky object

French astronomer and author Camille Flammarion notes another observation of a slow-moving object that could not have been a meteor, given its trajectory.

Source: Flammarion, *Bradytes*, op. cit., 135.

484.

2 June 1873, Paris, France: Three round objects ⓘ

Astronomers from Paris observatory are reported to have observed three round bodies evolving slowly at an altitude estimated at 80 km, leaving no trail.

Source: *Le Journal du Ciel*, unknown date and issue number.

485.

Ca. 30 August 1873, Brussels and Ste-Gudule, Belgium Starlike mystery

At 8 P.M. an object was seen rising above the horizon in a clear sky. It was starlike, mounted higher and higher for two minutes, and then disappeared suddenly.

Source: "Le météore de Bruxelles," *Nature* 2 (Paris, Sept. 13 1873): 239.

486.

30 November 1873, Poissy, France: Slow transit

Several observers tracked a maneuvering object, red like Mars in color. It was in the sky for 10 minutes. They first saw it in the North above the Big Dipper, then it approached the star 'gamma' without touching it, moved away on several curves and disappeared in the west. Camille Flammarion and Mr. Vinot, editor of *Journal du Ciel*, later made inquiries that confirmed the sighting.

Source: Flammarion, *Bradytes*, op. cit., 159.

487.

24 April 1874, Prague, Czechoslovakia
Dazzling white object in front of the moon

Professor Schafarik observed "an object of so peculiar a character that I do not know what to make of it."

He was observing the three-quarter moon at about 3:30 in the afternoon in bright sunshine, using a 4-inch achromatic telescope by Dancer with power 66, field 34 minutes of arc, when: "I was surprised by the apparition, on the disc of the moon, of a dazzling white star, which travelled slowly from E.S.E. to W.N.W and after leaving the bright disc, shone on the deep blue sky like Sirius or Vega in daylight and fine air. The star was quite sharp and without a perceptible diameter."

Source: "Telescopic Meteors" in *The Astronomical Register* 273, (September 1885): 205-211. Professor Schafarik discusses the frequency and appearance of telescopic meteors, which he places into four classes, for which he hypothesizes various explanations, ranging from faint shooting stars at the limit of the atmosphere to such mundane objects as birds, bats, the pappus of various seeds and "convolutions of gossamer." He was genuinely puzzled, however, by the above observation.

488.

Ca. 17 February 1875, Pwllhi, Caernarvonshire, Wales
Eight lights on erratic trajectories

A reader of the Field newspaper reports that eight lights were seen at once, at an estimated distance of 8 miles, moving in "horizontal, perpendicular and zigzag directions. Sometimes they were a light blue colour, then like the bright light of a carriage lamp, then almost like an electric light, and going out altogether, in a few minutes [they] would appear dimly again, and come up as before."

Source: *Notes and Queries*, 17 April 1875. *London Times* of 5 October 1877 gives the name of the witness as Mr. Picton Jones.

489.

About 12 January 1876, Sheridan, Pennsylvania
Gliding light and human figure

A man who was riding home on horseback at night had trouble controlling his terrified horse when they were faced with a bright light on the bank of a creek. Very bright at first, the light decreased in intensity, appearing to recede in the process. Urging the animal to move forward, the witness saw the light again in a field, borne by what appeared to be a human figure clothed in white that glided along the ground. When it came within 100 yards the horse dashed forward, almost unseating the witness.

Source: *Reading Eagle* (Pennsylvania), quoted in the *St. Louis Democrat* (Missouri), 17 January 1876.

490.

4 April 1876, Peckleloh, Germany: Planetary mystery

Mr. Weber, of Berlin, observed what he believed to be an intra-mercurial planetoid. The time of the observation was 4:25 A.M. Berlin Mean Time. Astronomer Wolf reported this sighting to Le Verrier in August 1876.

Source: "Les Planètes entre le Soleil et Mercure" *Année Scientifique et Industrielle* 20 (1876) : 6-11, at 7.

491.

17 March 1877, Gunnersbury, England
Long-lasting red celestial unknown

A large red "star" was witnessed in the sky about 8:55 P.M. in the constellation of Serpens. It seemed to be brighter than Arcturus. After no less than 10 minutes it began to "increase and diminish in magnitude two or three times," giving the impression that it was flashing, after which it disappeared.

Source: *Nature* 15 (March 1877): 451.

492.

23 March 1877, Vence, France
Luminous balls emerge from a cloud-like formation

A number of lights appeared in the sky, described as balls of fire of dazzling brightness. They emerged from a cloud about a degree in diameter and moved relatively slowly. They were visible more than an hour, moving northward.

Source: "Eclairs en boule observés à Vence, en Provence" in *Année Scientifique et Industrielle* 21 (1877): 45-6.

493.

7 September 1877, Bloomington, Indiana, USA
Five luminous objects in the sky, stationary

Mr. John Graham "had his attention arrested by a sudden light in the heavens, and upon looking up he saw a stationary meteor between Aquila and Anser et Vulpecula, about right ascension 295°, declination 15°N. It increased in brightness for a second or more, and disappeared within less than half a degree east of the point in which it was first seen. Immediately after the extinction of the first, three others, separated by intervals of three or four seconds, appeared and vanished in the same place; with the exception that one disappeared about as much west of the radiant as the first did to the east of it. Mr. Graham's curiosity was excited and he continued to watch till, after an interval of a few minutes, a fifth meteor, corresponding in appearance to the preceding, was seen in the same place. The meteors resembled stars of the first magnitude."

A possible interpretation of this observation would be an exceedingly unlikely train of identical meteorites falling directly in the direction of the observer over a period of several minutes.

Source: *Scientific American*, 29 Sept. 1877, New Series, 37: 193.

494.

22 January 1878, Dallas, Texas, USA ⓘ
Dark object: the very first "flying saucer?"

Mr. John Martin, a farmer who lived some six miles north of town, was out hunting in the morning when his attention was directed to a dark object in the northern sky.

"The peculiar shape, and the velocity with which the object seemed to approach, riveted his attention, and he strained his eyes to discover its character. When first noticed it appeared to be about the size of an orange, which continued to grow in size."

Going through space at a wonderful speed, it came directly overhead and Mr. Martin compared it to "a large saucer."

First mentioned in the UFO literature by Major Donald Keyhoe, this use of the word "saucer" by John Martin has triggered many debates among researchers. John Martin's use of the term seems to relate to the size of the object rather than its shape, just as Kenneth Arnold, in 1947, would use it to describe the motion of the crescent-shaped objects he witnessed. It is interesting, nonetheless, that "saucer" should be the word that came to the mind of these men when they tried to describe what they saw. The possibility that Martin observed a balloon must also be considered. The site is more likely to be Dallas than Denison.

Source: Denison *Daily News* for 25 January 1878, 1.

495.

c. 1st February 1878, Osceola Township, Iowa, USA
Strange light in the road

A newspaper reported that a strange phenomenon had occurred in Osceola Township, Iowa, one evening the evening the week before. A young man "well known in the community" was crossing the fields when his attention was attracted by a light moving along the road some way ahead. It was "much larger than a lantern," and it came nearer:

"When the light reached a point in the road nearly opposite him it stopped and came directly toward him with great velocity, until it was within a few feet of him when it stopped. The observer describes it as about the size of a half bushel and of intense brightness. It then rose in the air a distance of several rods and then began to descend where the gentleman stood. He says that he is not usually easily frightened, but he could not account for the strange sight and he retraced his steps to the house he had just left."

The light followed him up to his neighbor's house, where the witness told of what he had seen. Two men there offered to accompany him home. They started out but the light had apparently disappeared. Then, suddenly, "it again made its appearance and was distinctly seen by all three." This time it did not approach as closely as before, but would disappear and reappear in an entirely different direction and at a distance from where it was last seen.

The article finishes with the statement that the light was also seen by others in the neighborhood, none of whom could explain the strange occurrence.

Source: "Strange Phenomenon," *Ackley Enterprise*, Iowa, February 8, 1878. The report was originally published in the Hampton Chronicle but as there are no precise details, our date of February 1st is only an estimation.

496.

29 July 1878, Rawlins, Wyoming, USA
Unidentified planetoid observed by two astronomers

Professor Watson has observed a shining object at a considerable distance from the sun during the total eclipse. A confirmatory observation was made by Professor Swift of Denver, Colorado. Astronomer Lockyer commented: "There is little doubt that an Intra-Mercurial planet has been discovered by Professor Watson."

Source: Lewis Swift, "Discovery of Vulcan," *Nature* 18 (19 September 1878): 539. Also *The Observatory* 2 (1878): 161-2 and J. Norman Lockyer: "The Eclipse" *Nature* 18 (29 August 1878): 457-62, at 461.

497.

30 July 1878, Edwardsville, Kansas, USA
Unknown light rushes down the train tracks

Mr. Timmons, "one of the most substantial farmers and reliable men in Wyancotte county," reports that "the section men on the K. P. road, on my farm, seeing the storm coming up very fast, got their hand-car on the track and started full speed for Edwardsville. They had run but a little ways when the entire crowd, at the same time, saw coming around the curve of Edwardsville what they supposed to be a locomotive at full speed.

"They jumped down and took their car off the track as fast as possible when they saw it was not a locomotive. Whatever it was came down the track giving off a volume of dense smoke with occasional flashes resembling a head light in the centre of smoke. It came three-fourths of a mile from where they first saw it, then turned off the track at a pile of cordwood, went round it once, then went off in a southwesterly direction, through a thick wood. The section men came running to my house evidently much frightened and bewildered by what they saw."

Note: globular lightning may have produced this effect, as the ball of plasma could have been guided by the train tracks until it grounded itself. The duration of the phenomenon, however, makes it most unusual.

Source: *Atchison Globe* (Kansas), 7 August 1878.

498.

11 August 1878, McKeesport, Pennsylvania, USA
Planetoid passing in front of Jupiter

Two amateur astronomers, Messrs. Gemill and Wampler, observed an unusual celestial object using a 5-inch telescope. At 10:05 P.M. they noticed a dark round spot on the eastern margin of the disc of Jupiter. It moved west, just above the northern belt, parallel with the planet's equator, and passed off the face at 1:24 A.M. on 12 August, having crossed the disk in 3 hours and 19 minutes.

The object appeared as a perfect sphere, much larger than any of Jupiter's satellites. It was well-defined and sharp, most intensely black. The observers commented "it was neither a satellite nor the shadow of one, because all four satellites were in full view all the time."

Note: Jupiter has other satellites that were unknown at the time, but they are much smaller than the four satellites in question, and could not explain the effect observed.

Source: *The Indiana Progress* (Indiana, Pennsylvania), 22 August 1878.

499.

12 April 1879, Manhattanville, New York, USA
Unexplained astronomical phenomenon

"Upon the evening, Mr. Henry Harrison was searching for Brorsen's comet, when he saw an object that was moving so rapidly that it could not have been a comet. He called a friend to look, and his observation was confirmed. At 2 A.M. this object was still visible (...) Mr. Harrison disclaims sensationalism, which he seems to find unworthy, and gives technical details: he says the object was seen by Mr. J. Spencer Devoe, of Manhattanville."

Source: "A Curious astronomical phenomenon" in *Scientific American* n. s., 40: 294; *New York Tribune*, 17 April 1879, 2, c.3; also Henry Harrison's entry in *Scientific American Supplement* 7 (21 June 1879): 2884-5.

500.

10 October 1879, Dubuque, Iowa, USA
Large unexplained airship overhead

"People who were up at a very early hour this morning were astonished at seeing what appeared to be a large balloon going over the city. It was seen by quite a number of persons in different parts of the city, and was visible for an hour."

The object disappeared on the horizon, moving in a southwesterly direction. It is noteworthy that an employee of the *Times* named Thomas Lloyd saw this balloon as it was very high in the southeast and traveled south slowly, rising and falling in its course.

A real balloon (the "Pathfinder") piloted by professor John Wise had taken off from the town of Louisiana, Missouri in this period, but it had fallen into lake Michigan some ten days before, and could not have been the cause of the sighting.

It is noteworthy that the last eight sightings in the Chronology come from the United States, and that the last one is a report of an unknown "airship" flying slowly over a city. But that, as journalists of the nineteenth century liked to say, "is another story."

Source: *The Inner Ocean* (Chicago, Illinois), 11 October 1879.

Epilogue to Part I-F

Three aspects stand out when we review 19th century reports of unusual aerial phenomena:

- *Meteors and meteorites* were reported with greater enthusiasm and in more abundance than in previous times. The press, exploiting people's interests and concerns for the sake of sales, began a love story with these still-mysterious astronomical phenomena, and it lasted till the end of the century. Given the interest in meteorites in the name of scientific progress, the public was encouraged to report the latest observations at their local newspaper offices.

- *Claims of extraterrestrial encounters* were first made in this century. Such claims were not published until the 1850s on, but tall tales involving the alleged inhabitants of the moon, allegedly spied through telescopes, had become popular decades earlier. It was not long before stories of aliens on the moon would turn into stories of aliens on visits to Earth. Indeed, by the late 19th century the press would be full of articles speculating on what extraterrestrials drank and ate, on the average height of Venusians and on whether airships had already made the journey across interstellar space to meet us. As early as 1847 the Mormons, followed by the Jehovah's Witnesses, began to discuss interplanetary travel and speculated on which physical planet God inhabited. We have generally avoided including examples here for want of truly convincing cases.

- *Reports of UFO crashes* were first claimed in the nineteenth century. Though examples can be found in the literary efforts of earlier generations, allegedly *factual* reports had never been published before. That most of these cases were probably hoaxes is not in doubt, hence our avoidance of them, but they do allow us a glimpse of the world's new mentality.

We stopped our Chronology before 1880 because the world was about to change radically and irreversibly, with increasingly common access to novel forms of energy and transportation: The Suez Canal was opened; John D. Rockefeller founded Standard Oil; Most significantly, the first mobile gas engine was demonstrated by Siegfried Marcus, and other engineers rushed to make plans for new vehicles based on the internal combustion engine, which had been demonstrated as early as 1860 by Lenoir.

With the introduction of automobiles, the telegraph and an oil-based economy, the basic structure of the modern world was established. In science, the first measurements of the speed of light were accomplished, and the kinetic theory of gases published. In technology, high tension induction coils, cast-iron frame buildings, Bunsen gas burners, and Singer's sewing machines were developed.

Most importantly for our purpose, the social context was revolutionized by changing standards in journalism, the increasing demand for escapism and instant news and the renewed fascination for the exotic and the unknown.

In 1876 Italian priest Pietro Secchi announced that he had discovered "canals" on Mars. When astronomer Antonio Schiaparelli confirmed this observation in 1877, and Asaph Hall discovered the two satellites of Mars the same year at the Naval Observatory, it led to much renewed speculation about life on other planets, which in turn tended to color reports of unexplained aerial objects and inspired today's fascination with the extraterrestrial theory, to the exclusion of any other hypothesis about these phenomena.

It is not for lack of data that we decided to stop this chronology when we did. The end of the nineteenth century would see an extraordinary burst of sightings, popularized by the new media and amplified by the growth of urban centers, the greater ease of travel and the vast extension of the railroads and the telegraph. A catalogue of unexplained aerial phenomena beyond our chronology would deal with history's first major "wave" of reports about 1885 and with an even bigger one from the fall of 1896 to 1897. The records of that era, now known as "the Airship Wave," if they are ever analyzed and published, will dwarf the present book.

PART II

Myths, Legends, and Chariots of the Gods

In our effort to understand how certain recurring themes linked to unexplained aerial phenomena have evolved and spread throughout human history, we have tested many claims for the sake of accuracy. Naturally, given the hoary age of these accounts, it was not possible for us to measure the truth or falsity of every story compiled. However, in the process of analysis, we have uncovered many spurious items that cloud the literature of the field. Some of them deserve special documentation.

Descriptions of unexplained objects or phenomena in the sky are found in the records of the earliest civilizations that used some form of writing. Several serious authors, such as Alexander Kazantsev and professor Agrest in Russia or Aimé Michel in France, have suggested that some prehistoric rock carvings and primitive statues were indicative of contact with non-human visitors from the sky.

Less cautious or less scholarly writers such as Erich von Daniken and Zecharia Sitchin have expanded this notion into the popular theme of Ancient Astronauts, where it is assumed that the Earth was either visited or colonized by beings from another planet. Some forms of the Ancient Astronaut theory quote the Bible and other ancient texts in support of the notion that these beings intermarried with primitive earthlings or modified them genetically to produce modern humans.

Indeed, the literature of earlier centuries is rich in legends involving beings flying in the heavens, sometimes alongside humans as witnesses or as participants in their warfare or their lovemaking. Although such accounts are too vague in date and circumstances to be included in our chronology, they cannot be ignored in any study of the history of the field. Having said this, the reliability of these accounts must be critically challenged, either because they were the product of poetic imagination, because they were fabrications used in blatant support of political or religious movements of the time, or because they were invented by opportunistic authors and popularized by overly credulous readers.

Other accounts in the literature were genuine historical events that were misinterpreted in good faith by observers at the time, and propped up later as "evidence" for various theories, some of which still flourish in contemporary works. Accordingly, we have classified the stories we have rejected from the main chronology, under four major categories.

Deceptive story, hoax, fictional account or tall tale.

These accounts may be deliberately couched as true happenings, or they may have been lifted from their fictional context by later retelling as true facts. History is full of examples where a simple rumor gave rise to major movements while truthful accounts were only reconstructed much later.

Religious vision.

The real (or imagined) arrival of beings and artifacts from outside Earth has had enormous impact on human societies, and the evidence can be considered from many points of view. Theology has been shaped by a belief in sky-dwelling divinities. If mysterious craft are seen in the sky and stones fall from the clouds, what can Man's position in the scheme of things be?

Religious visions have their own characteristics, and we do not feel qualified to judge their relevance to the overall problem. While they may represent true happenings for large groups of believers, these accounts are not amenable to scientific study in the same sense as the observation of an objective phenomenon.

It has not escaped our notice that a genuine paranormal phenomenon may come to us dressed up as a religious vision, either because the witnesses interpreted it in such terms, or because the standards of the society around them demanded such an interpretation. Therefore we do not exclude reports purely on such a basis.

Natural astronomical phenomenon.

Throughout history, mankind has anxiously observed the heavens for signs of future events. The sky has answered with a bewildering series of displays, such as comets and meteors, which we recognize today as natural phenomena. Ancient accounts of such "wonders in the sky" provide precious information for today's astronomers, in the form of accurate data on the periodicity of comets, and the frequency of meteor showers, to give only two obvious examples. Auroral displays (aurora borealis) are frequently the occasion for historical amazement, and rightly so: The natural mechanism for such phenomena was not fully understood by physicists until the present (21st) century.

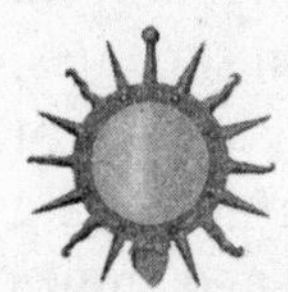

Optical illusion or atmospheric effect.

These deserve a category apart. Here we deal with sincere witnesses faced with spectacular sky displays such as luminous crosses, multiple suns, multiples moons, or fantastic mirages. The mechanism behind such displays has only become understood in recent centuries, and new discoveries are still being made today about the properties of the atmosphere, lightning, tornadoes (often seen by terrified witnesses as sky serpents or dragons), the propagation of light through the air, and yes, even swamp gas!

It would be most interesting to compile an exhaustive list of events reported in the ancient literature under the general topic of sky phenomena. Some authors such as William Corliss have published catalogues of scientific anomalies that give fascinating compilations for comets, meteors, globular lightning, or *aurorae borealis*. Such work was not within our scope, but we needed to tell the reader why certain well-known incidents had been excluded from our main chronology.

The following list, selected among hundreds of items, makes interesting and sometimes comical reading. It illustrates the vagaries of the human mind, indeed even the scientific mind, as it tries to come to grips with phenomena beyond its understanding.

400 million years ago, Kentucky, USA
Crashed saucer, strange alien bodies

One of the most common recurring themes in the literature of this field is that of a flying machine that comes down from the sky and crashes, along with its extraterrestrial occupants. Far from being unique to Roswell, crashes of alien artifacts constitute a standard story, complete with descriptions of small cadavers and mysterious writing on the recovered craft. The present case, however, which is little known even within the paranormal community, must set some sort of record in terms of the extremely ancient date of the alleged incident.

In January 1969, the American periodical *Beyond Magazine* published a curious article about an alleged extraterrestrial fossil found in Kentucky. "Reader Melvin R. Gray of 417 South 5th St., Louisville, Kentucky, 40202," wrote columnist Brad Steiger, "has discovered a stone which has what he considers very suspicious indentations." Mr. Gray's examination of the stone led him to conclude that it contained fossilized remains of tiny humanoid creatures and "what may at one time have been a tiny flying saucer no larger than our present day washbasins or dishpans."

No photographs illustrated the article but Gray described the stones as looking like "a small chunk of meteor." In order to get a better idea of what the "beings" looked like he made plaster, fiberglass, and aluminum castings from the rock. He reported:

> "The fossilized creatures themselves are humanoid in appearance, looking very much like ourselves, and approximately three inches tall. (…) The stone looks rather cindery as if it may have hurtled through a long trail of space, melting as it went and finally splashing into some river or lake before it

> was entirely consumed, leaving … a fossil-like imprint for a permanent record to tell the world … that we had visitors to our earth … who had met with some terrible calamity."

Steiger himself was not entirely convinced. While acknowledging that, with the aid of a magnifying glass, he could make out the outline of "a tiny human pilot sitting in a bucket-type seat" on the casts that Gray sent him, he wondered whether it was merely a trick of nature? There was no reason to think Gray had made the "fossil" himself.

A second article was published about Melvin Gray's fossil in Ray Palmer's magazine *Flying Saucers*. In "A Fossilized Alien Spaceship and its Occupants," Executive Director of the Kentucky-based National UFO Research and Investigation Committee, Buffard Ratliff, wrote that, after reading about Gray in *Beyond*, he contacted Mr. Gray and obtained the fossil.

Gray told Ratliff that he and his wife had come across the stone while cutting the grass in their back yard. He then examined the artifact carefully "for a period of approximately seven months and made several discoveries that led him to believe it might possibly be from outer space." Ratliff and Gray were able to find "seven very small creatures…in or on the fossilized stone." Three of the creatures were ape-like in appearance. The other four were humanoid. All were approximately three inches in height, vertebrates, and very strong for their size.

Ratliff and Gray concluded that the three tiny ape-like creatures "could very well be humanoids in special space suits," and that these beings were in a separate section of the craft they labeled "B." However they were quick to point out that one of the humanoids was also in that section, as opposed to section "A" of the spacecraft. As the two sections seemed to be divided, "apparently where the spaceship is fitted together," this "indicates intelligent construction and design by intelligent beings."

After some more interpretation the researchers arrived at a breathtaking theory: the craft had come from outer space and crashed into a large body of water during the last ice age. The water extinguished the fire, but the craft sank to the bottom and became encased in sand and clay, becoming a fossil. There it lay dormant for some 400,000,000 years till Melvin Gray almost trod on it as he was mowing the lawn in his back yard in Louisville, Kentucky.

It is evident from the information presented in *Beyond* and *Flying Saucers* that neither Gray nor Ratliff were able to present any basis for their incredible theory other than their own imaginative interpretation of the rough exterior of the stone. What reference material the investigators used, and exactly what tests were carried out was not explained.

12,000 years ago
Granite disks tell a story about Alien vehicles

This is yet another case of crashed spacecraft leaving mysterious material covered with alien writing, a recurrent theme in contemporary ufology. In July 1962 a German magazine called *Das Vegetarische Universum* [The Vegetarian Universe] published an article about a strange finding made in the mountains between China and Tibet. It is a tale that regularly turns up in UFO contact lists, in books, magazines and on the Internet. It therefore deserves our attention. The author, Reinhardt Wegemann, reports that:

> "In the borderland between Tibet and China lies the cave area of the high mountains of Baian-Kara-Ula. Here the strange discovery of hieroglyphic writing tablets was made 25 years ago. Several thousand years ago, record-shaped plates were sawed out of the hardest granite rock, with untraceable and completely unknown appliances."

Wegeman went on to state that 716 rock plates had been recovered, each resembling records, with a hole in the centre and a groove spiralling to the outer edge. He stated that it took two decades for archaeologists and linguists to decipher the script, the content of which so stunned the Academy of Prehistory in Beijing that they forbade its publication. However, one of the researchers, Professor Tsum Um-nui, is said to have discussed this matter with a small group of colleagues and decided to release a report without official consent. The archaeologists reportedly came to the conclusion that "The grooved writing tells of vehicles from the air, which must have arrived 12,000 years ago. In one place it says literally that the Dropa came down from the clouds with their air gliders. Ten times the men, women and children of the Kham hid in the caves until dawn. Afterwards they understood the signs and saw that the Dropa came with peaceful intentions…"

The story adds that the aerial fleet was destroyed on landing and that graves of small humans of the Dropa and Kham race with thin bodies and unusually large heads can be found in the caves, along with star maps carved on the rock walls. Furthermore,

> "Rock particles were scraped off one of the writing plates and were sent for analysis to Moscow. A sensational discovery was made: The grooved plates are strongly cobalt and metallic. When a whole plate was tested with an oscillograph, a surprising rhythm of oscillation showed up, as though, once 'loaded,' the plates with the grooved writing would have somehow served as electrical conductors."

The critical analysis of this amazing tale yields certain surprises. Oddly enough, very few who support the reality of it know anything about its origins. Whole books and articles have been written on or around this story with absolutely no reference to its author or first publication. This absence of detail is what gives the impression that its origins are "shrouded in mystery," and therefore leaves its reality status open to all sorts of theories. Unfortunately, after several years inquiring through colleagues and journalist friends, and conducting thorough searches of newspaper archives, no German writer by the name of Reinhardt Wegemann could be found.

Fig. 38: The Dropa hoax

In July 1964 the same article was published again, as if new, in the German UFO magazine *UFO-Nachrichten*. Here, "Wegemann" made no mention of the fact that his report was now 'old news,' and added no new revelations about the discs.

From this moment on, the "Dropa" would become famous all over the world. The French/Belgian UFO organization BUFOI referred to them in March 1965, and in 1966 Wegemann's article was translated into Russian and published by the Soviet journal *Neman*. A year later, Dr. Vyatcheslav Zaitzev wrote about the discs of Baian Kara Ula for the first edition of the Soviet magazine *Sputnik*. Owing to the enormous distribution of this publication, many have erroneously cited Zaitzev as the original source of the story.

News of the Dropa was published for the first time in the United States on 26 February 1967. A journalist of the *Los*

Angeles Herald-Examiner, using the article from *Sputnik*, compared the finding of the cave drawings to a star map allegedly seen by UFO abductee, Betty Hill. By this time Reinhardt Wegemann had been forgotten and the tale's origins completely obscured. Over the next three decades details would be lost, others invented, and the spelling of the name of the Dropa tribe would become increasingly exotic: Dzopa, Dhzopa, Dzohpa, Dhropa, and so on.

The first skeptical enquiries began in 1973, when the director of the British magazine *Flying Saucer Review*, Gordon Creighton, a serious scholar of the field, reported he could find no record of any archaeological expedition to Baian Kara Ula in 1938. Creighton also pointed out that the name of the mountains was more usually written "Bayan Khara Uula," Mongol words meaning "the good black mountains" ("Bayan Har Shan" in Chinese), and that there were no records of any archaeologist named Chi Pu Tei. Likewise, all attempts to trace Tsum Um Nui or his report have failed.

In 1979 a further twist came in a book called *Sungods in Exile*, edited by one David Agamon, who declared it to be the posthumous work of a British scientist called Karyl Robin-Evans. The work describes an expedition to Baian Kara Ula led by Robin-Evans in 1947 with the aim to gather information about a disc that had been purchased in India or Nepal by a colleague of his in Oxford, a Polish scholar named Sergei Lolladoff. According to Agamon, the expedition met with a tribe of dwarves in a remote valley in the region and these beings, the Dropa, told him that their ancestors had come from a planet in the Sirius system and had been trapped on the earth in the year 1014 AD due to a mechanical problem with their spacecraft.

Years later, Agamon (using his real name, Gamon) confessed in letters to the editor of *Fortean Times* that *Sungods in Exile* was a hoax and none of the characters in it were real. Even so, photographs taken by Gamon of a fake "Dropa disc" are still believed by many to be authentic, giving rise to rumors and speculation. Meanwhile, the real Dzopa people of Tibet live in blissful ignorance of the whole affair.

Circa 4780 BC: The fiery Vimanas of King Citraketu

The earliest dated story we are able to find about flying devices of non-human origin comes from the ancient literature of India. For instance the *Bhagavata Purana*, also known as *Srimad Bhagavatam*, a text that is part of Hindu literature, states that while Indian King Citraketu was traveling in outer space on a "brilliantly effulgent ship given to him by Lord Vishnu," he saw Lord Shiva: "The arrows released by Lord Shiva appeared like fiery beams emanating from the sun globe and covered the three residential ships, which could then no longer be seen." (*Srimad Bhagavatam*, Sixth Canto, Part 3). If the reference to this particular King is trustworthy, the event would have taken place about 4,780 BC.

The Vedic literature, including India's national epic, the *Mahabharata*, a poem of vast length and complexity, contains many descriptions of flying machines generally called Vimanas. Another text, the *Ramayana*, which can be loosely translated as 'the travels of Rama,' tells of two-storied celestial chariots with many windows that roar off into the sky until they appear like comets. Sanskrit books describe at length these chariots, "powered by winged lighting...it was a ship that soared into the air, flying to both the solar and stellar regions."

There are no physical remains of ancient Indian aircraft technology but references to ancient flying machines are commonplace in the Indian texts. Several popular epics describe their use in warfare. Depending on one's point of view, either it contains some of the earliest known science fiction (a sort of Indian *Star Wars*) or it records conflict between beings with weapons as powerful and advanced as anything used today.

Fig. 39: Flying Vimana at Ellora caves, India

It is a curious fact that the yantras (Sanskrit for "machines") described in later Indian texts were less powerful than those mentioned in greater and older works. Does this imply a gradual departure from fantasy towards realism? Some have proposed the change reflects a loss of knowledge. Richard L. Thompson writes: "Some ascribe this to the fantastic imagination of ancient writers or their modern redactors. But it could also be explained by a progressive loss of knowledge as ancient Indian civilization became weakened by corruption and was repeatedly overrun by foreign invaders. It has been argued that guns, cannons, and other firearms were known in ancient India and that the knowledge gradually declined and passed away toward the beginning of the Christian era." (*Alien Identities*, San Diego: Govardhan Hill, 1993, 258.)

Circa 2637 BC, China
Relativity and the Emperor's dragon

According to an article circulating on the Internet, the legendary First Emperor Huang-Ti (the "Yellow Emperor", who instituted the calendar that survives in China to this day for festival dates, and is said to be the ancestor of all Han Chinese) had a "dragon" named Changhuan, that could move through space at enormous velocities. One ancient writing mentioned that it "originated in the land where suns are born," and was over 3,000 years old. Its enormous speed had an effect on the movement of time, affecting the ageing process, a surprising early reference to the relativity of time, 4,400 years before Albert Einstein.

Huang-Ti is said to have manufactured 12 gigantic mirrors of unknown nature and used them "following the Moon," as well as miraculous tripods about 4 meters high. The legends of ancient China said that the "tripods" depicted "dragons, flying in the clouds."

We have not been able to verify these statements or to consult the sources listed, which appear to come from the late scientist and orientalist Igor Lissevich (magazine "Asia and Africa Today", 1974, No. 11, in Russian). Lissevich also presented his scientific findings at the 1975 Zelenchuk SETI Symposium ("Problem of SETI", Moscow 1981, in Russian). Igor Lissevich knew Chinese and was a reliable source. The original references are quoted as "Records of the foremost deeds of Huang-Ti the Great" and "Glorification of the three tripods of Huang-Ti" written by Zao Ji. The Yellow Emperor Huang-Ti is said by tradition to have reigned from 2698 BC to 2598 BC.

Circa 2357 BC
Japan's "divine man" and his luminous monster

Entries in various UFO lists mention that "According to Tau-se from an ancient manuscript called *Sey-to-ki*, during the time of Emperor Ton-Yo, in the year of "Mon-Sham" a divine man descended from the sky, using a "monster that was emitting light" (spacecraft?). The people called this man "the master." He received the name *Tan-kun* (Sandalwood God) and his country was called Peson."

The source for this item is *Space Visitors in Ancient Japan* by Mikhail Rosenshpitz in *Unbelievable World* No. 8, August 2004. When we tried to research this item it was found to contain spurious information and we could not locate anything supporting it, unfortunately a frequent situation with both online cases and UFO books.

In this case, our first goal was to obtain a copy of the original article by Rosenshpitz. This proved more complicated than we expected because it turned out that *Unbelievable World* did not exist. The correct Russian name of the magazine was *Neveroyatnyi Mir*, a paranormal news journal distributed in the Ukraine. Steering away from sensationalist press wherever possible, we decided this was not a source we could use.

Observant readers may have noticed that the year given is also spurious, because even legendary Japanese rulers date back only as far as Emperor Jimmu, who supposedly founded Japan in March 585 BC! No real or mythical Japanese emperor had a name resembling Ton-Yo. After playing with different spellings we realized the account must be from Chinese tradition, not Japanese. Indeed, Tau-se was just an unusual transcription of Tao-se, or "Tao Teacher." The information may refer to the legendary Lord Yao or Tangyao, who supposedly reigned between 2357 and 2258 BC However, as we could find nothing resembling the story of the "divine man" and the light-emitting monster in the Chinese literature available to us, we quickly lost faith in the account.

Circa 2208 BC
A Chinese Emperor flies away from danger

The emperor of China is said to have flown in an aerial machine and descended back to earth. (reference: Hervey – Winkler catalog, published by FUFOR – the Fund for UFO Research). It turns out that this is a story about Emperor Shun, who supposedly reigned between 2258 and 2208 BC. However the actual incident has nothing to do with the 'aerial machine' account.

In the *Shi Ji* (Historical Records) Sima Qian relates that Shun's father Gu Sou wanted to kill him. Finding him at the top of a granary tower, he set fire to it. Shun escaped by assembling a pile of large conical straw hats together and leaping down! See *The Shorter Science and Civilization in China* Vol. 4, by Colin A. Ronan: Cambridge University Press 1994, 290.

Circa 1900 BC
Egypt, the death star and a gold serpent

The first ancient reference to an unidentified object from the sky in relation to strange beings is found in an authentic Egyptian papyrus generally considered to belong to the twelfth dynasty, 1991 to 1802 BC. The text, known as *The Tale of the Shipwreck*, was discovered by chance in 1880 by Golenischeff in the Ermitage Museum of Saint Petersburg and is now on display at a Moscow Museum. It tells how the lone survivor of a shipwreck was carried by the waves to a mysterious tropical island that nobody had seen before. The ruler of the island was a giant, glowing, human-headed serpent, "his body overlaid with gold, and his color as that of true lapis-lazuli." This being seemed pleased to meet the unfortunate sailor and invited him to his home as a guest.

Egyptologist G. Maspero, very much an authority in his day, translated the extract in *Les Contes de l'Egypte Ancienne* (4th Edition, Paris 1911):

> *We are seventy-five Serpents in number, my children and my brothers, not mentioning the young girl who was brought to me by the magic art. Because when a star fell, those who were in the fire with her came out and the young girl appeared; and I was not amongst the beings of the flame, I was not amongst them, else I would be dead, but I found her among the corpses, alone.*

What 'star' is the serpent-being referring to? Unfortunately no details are given in the papyrus. Was the 'star' a meteorite, as most scholars suggest? It seems possible but it cannot be proved. There was no word to describe meteorites in the Egyptian hieroglyphic system, so the word 'star' ("seba") could be used as a wild card for any kind of luminous phenomenon travelling in the sky.

Analysis of the tale reveals that it already contains imagery that would become the framework of 'encounter' stories for the next three and a half thousand years. The island, which the text actually says will sink into the sea again like fabled Atlantis, would be replaced by what is nowadays called a 'window area.' The reptile king would hardly change at all over time, as humanoid serpents and "reptilian beings" are a staple element in mythology and UFO lore all over the world. And wherever supernatural beings dwell in folklore, mysterious lights, or crashing objects are never far away.

Similar stories come from ancient China: the dragon king had his palace on an island in the ocean. This island was said to vanish and reappear regularly, confusing sailors and giving rise to many strange beliefs. "Sometimes," writes Donald Mackenzie, "a red light burns above the island at night. It is seen many miles distant, and its vivid rays may be reflected in the heavens." A Japanese story describes the island as "a glowing red mass resembling the rising sun."

Circa 1766 BC, China: Feathered guests from the sky

"The Xian were immortals capable of flight under their own divine power. They were said to be feathered, and a term that has been used for Taoist priests is *yu ke*, meaning 'feathered guest'. The *fei tian*, which might be translated as 'flying immortals', also add to the numbers of airborne beings in the Chinese mythological corpus."

"The Chinese tales of *fei che*, flying vehicles, exhibit the first understanding, perhaps, that humans would fly only with some kind of technological apparatus."

Source: Dr. Benjamin B. Olshin, *Mechanical Mythology: Private Descriptions of Flying Machines as Found in Early Chinese, Korean, Indian, and Other Texts* (from extensive quotes available online).

Circa 1515 BC: Egypt: The infamous Tulli papyrus

Shiny objects "brighter than the Sun" flew south and left a foul odor, according to an ancient Egyptian document found among the papers of Alberto Tulli, a director of the Egyptian museum at the Vatican.

The text appeared in 1953, in Issue 41 of *Doubt*, journal of the Fortean Society, when novelist and co-founder of the Society, Tiffany Thayer (1902-1959), published the hieroglyphic translation of what would soon be known as the

"Tulli Papyrus." Accompanying the transcription was a letter from its translator, an amateur Egyptologist of Russian-Italian descent, Boris de Rachewiltz. This letter explained that the papyrus had been passed on to Tulli's brother Gustavo, a priest. Rachewiltz had been sent the hieroglyphic transcription for translation.

Rachewiltz explained to *Doubt* that the papyrus had been longer, and indeed we must assume the unpublished part referred to an incident during the reign of Thutmosis III because the fragment we have provides no sign of this. Several versions of the translation have been published, but the following is the first, as it appeared in *Doubt*:

> "In the year 22, third month of winter, sixth hour of the day (…) The scribes of the House of Life found it was a circle of fire that was coming in the sky. (Though) it had no head, the breath of its mouth (had) a foul odour. Its body one 'rod' long and one 'rod' large. It had no voice. Their hearts become confused through it: then they laid themselves on their bellies (…) They went to the King (…?) to report it. His majesty ordered (…) has been examined (…) as to all which is written in the papyrus-rolls of the House of Life. His Majesty was meditating upon what happened. Now, after some days had passed over those things, Lo! They were more numerous than anything. They were shining in the sky more than the sun to the limits of the four supports of heaven. (…) Powerful was the position of the fire circles. The army of the king looked on and His Majesty was in the midst of it. It was after supper. Thereupon they (i.e. the fire circles) went up higher directed to South. Fishes and volatiles fell down from the sky. (It was) a marvel that never occurred since the foundation of this Land! Caused His Majesty to be brought incense to pacify the hearth (…to write?) what happened in the book of

the House of Life (...to be remembered?) for Eternity."

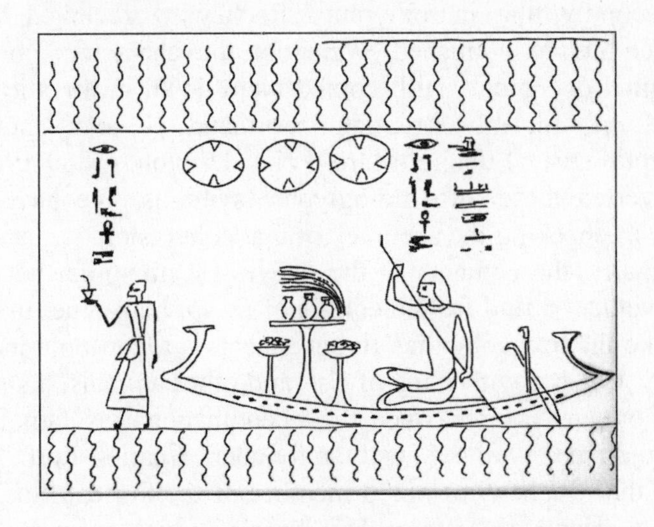

Fig. 40: The Tulli papyrus hoax

If the Tulli papyrus is authentic, the objects it describes must indeed be classified as UFOs. Their shape, luminosity and silent movement in the sky are familiar. The text quickly became a classic in books of the genre, used first by George Adamski and Desmond Leslie in their *Flying Saucers Have Landed* (1953) and later by popular writers such as Harold Wilkins in *Flying Saucers Uncensored* (1956). We could devote many pages to the evolution of this document during its first 50 years of life in ufology. However, space allows us only to outline the reasons we have not included it in the main body of this book.

First of all, the whereabouts of the Tulli Papyrus are completely unknown. Not even Boris de Rachewiltz knew where the original was kept. Later correspondence with Rachewiltz revealed he had only ever received the Egyptologist's personal notes, not the papyrus itself, and that even Albert Tulli had only made his transcription during a visit to the house of an antiquarian in Cairo in

1934. Reportedly, the papyrus had been too expensive for him to purchase at the time.

Secondly, the hieroglyphics Rachewiltz received, and which *Doubt* published, were not the characters on the original document. Tulli copied them down in hieratic – a kind of ancient Egyptian shorthand – and another Egyptologist, Etienne Marie-Felix Drioton (1889-1961), converted these into hieroglyphic symbols. We have no way of checking the accuracy of that conversion.

Finally, the contents of the papyrus seem somewhat too convenient a find for the editors of *Doubt* magazine. In one stroke the text combines flying saucers – a hot topic in the early 1950s – with rains of fish and other animals, a staple of Fortean research since the phenomenon was famously popularized by the Society's founder, Charles Fort. The fact that Rachewiltz was a member of the Fortean Society and a friend of Tiffany Thayer, also gives us cause for questioning the document's authenticity.

Circa 1447 BC, Nile Valley, Egypt
Moses and the blue object

Moses is seen brandishing a rod and triggering rains of blood, in an ancient illustration that shows a complex flying object. This picture is taken from the *Ashkenazi Haggadah*, in a section showing the Plagues of Egypt (Exodus 5-9). Reference: The British Library, Add. Ms. 14762.

The artist has represented an astonishing blue device hovering in the sky. The picture shows an object with four circular structures or openings, surrounded by flames and what appear to be bloody explosions. The hand of God at the end of a reddish-brown sleeve is extended below it, pointing to the assembled–and somewhat astonished – Hebrews. A flame is burning atop a nearby column.

The illustration represents a well-known event, the Seventh Plague sent against the Egyptians. The arm in the picture is

Fig. 41: Moses and the blue object

textually referred to as "God's outstretched arm." Here is the relevant Old Testament passage, as used by Jews today:

> *Shemot (Exodus) 9:23-24: And Moshe stretched out his rod towards heaven: and HaShem sent thunder and hail; and the fire rained down upon the ground; and HaShem rained hail upon the land of Mitzrayim. So there was hail and fire flaring up amidst the hail, very grievous, such as there was none like it in all the land of Mitzrayim since it became a nation.*

The image is from a 15th century manuscript composed and illuminated by Joel ben Simeon. That is, an illustration created 3,000 years after the event. It shows the fire (red and yellow) and hail (grey) sent by God. While the text itself does not mention any flying object in the sky, the artist, possibly influenced by tradition, has felt it necessary to display God's manifestation in the form of something resembling a flying machine.

Ca. 1440 BC, Elim, Sinai Desert, Arabic Peninsula
Manna Machine

The Zohar, a sacred Jewish text, describes a device miraculously providing food for the Hebrews as they flee:

"There are three upper heads; two, and one which contains them. The dew of the white head drops into the skull of the Small-faced One and there is it stored. And those parts which are found in the beard, they are shaped and lead downwards in many directions. In his lower eyes there are a left and a right eye, and these two have two colors, except when they are seen in the white light of the upper eye."

Modern writers have speculated that in this puzzling text the ancient Hebrews, who lacked a technical vocabulary, used anatomical analogies to describe a complex flying machine that generated food to sustain the crowd as it moved through the desert.

Fig. 42: The Manna Machine

The Bible never tells us exactly what manna was and where it came from, but there are many Old Testament passages which describe its physical qualities and conditions associated with its appearance. The Bible's first reference to manna is in the *Book of Exodus* as the children of Israel are fleeing from Egypt and follow Moses into the wilderness. After six weeks of wandering, they begin complaining to Moses that they are tired and hungry. What happens next is truly extraordinary:

"Then said the LORD unto Moses, 'Behold, I will rain bread from heaven for you; and the people shall go out and gather a certain rate

every day, that I may prove them, whether they will walk in my law or not (16:4).'

And when the dew that lay was gone up, behold, upon the face of the wilderness there lay a small round thing, as small as the hoar frost on the ground (16: 14). And when the children of Israel saw it, they said one to another: It is manna, for they knew not what it was. And Moses said unto them, 'This is the bread which the Lord hath given you to eat.'"

Before 1200 BC
Mesopotamia's dark meteors and standing fireballs

At the dawn of recorded history, Mesopotamian tablets deserve a mention, if only because a popular author, Zechariah Sitchin, has offered an interpretation of some passages in terms of visits by astronauts from other planets (notably in *The Twelfth Planet*, Avon Books 1978).

It is a fact that some cuneiform literature deals with interesting celestial anomalies. Assyriologists have recognized these writings to be astromantic in nature, that is, texts explaining how to forecast the future by watching meteoric phenomena, as opposed to astrology, which deals with the movements of the planets. Some of these records are from 1200 BC or earlier, and were written in Hittite, but it is thought that they were copied from older Akkadian originals, not yet located.

The vast majority of these texts described phenomena that can be explained today as the natural observation of meteors, fireballs, and comets. The scribes did not generally report on specific incidents that had occurred but rather provided meanings to particular kinds of sightings. A handful of cuneiform references to sky phenomena have puzzled archaeologists and astronomers. For example, the following text: "If a fireball moves across the Wagon-Star and stands..." seems to describe a meteor that stays motionless in the sky. The word "sallummu" has been translated as "fireball" but it very literally could have been anything bright passing through the sky that stood still. One possibility is that 'sallummu' was a meteor train that remained visible for some time across the face of the Wagon Star (Ursa Major), but the original text is not clear enough to reach a conclusion.

Other texts mention a more complex picture: "Two great stars flashed one after the other in the middle watch." (R. Campbell Thompson, *The Reports of the Magicians and Astrologers of Nineveh and Babylon*. Luzac & Co., London 1900, 202)

The "dark meteors" were among other mysterious objects described in cuneiform. For example: "If a meteor comes from above the Wagon Star and is dark and passes at the right of the man: that man will see injury." Since meteorites passing overhead are necessarily luminous or fiery rather than dark it is tempting to retain such quotes as indication of an exceptional phenomenon, but the context is so vague that it is ultimately anyone's guess. To the extent that no date is associated with the observation, we have not retained these cases in our chronology.

Circa 852 BC, Bath, England
Did this English King crash a Druid Airship?

According to writer John Michell (1967), King Bladud is said to have been killed at Troja Nova, in the London area, when riding a "druid airship" that crashed into the temple of Apollo.

Although Bladud is legendary, the story of his flight has some factual basis. The trouble is, the legend doesn't mention a druid airship but only chicken feathers! Bladud, or Blaiddyd, was the legendary founder of Bath. His son was King Lear, whose story William Shakespeare famously adapted for the theater. There is no evidence that Bladud actually existed before Geoffrey of Monmouth named him in his *History of the Kings of Britain* (1135), where he is said to have been the tenth ruler in line from Brutus.

This legend has enjoyed centuries of embellishments. Usually it is said that young Bladud contracted leprosy in Athens and was banished from his native kingdom on return to Britain. Years went by, during which he lived as a swineherd in Swainswick, on the outskirts of present-day Bath, where he discovered by accident that the mud of a bog in the marshy ground could cure skin diseases in pigs. The prince wallowed in the mud himself to see what effect it had on his own lesions, and eventually managed to cure himself completely. He then returned to Court, where he was welcomed. On his father's death Bladud became king himself. He founded the city of Bath and there built the temple of Aqua Sullis, dedicated to Minerva, goddess of healing.

According to this legend, Bladud practiced magical arts, such as necromancy, and this led him to conduct an experiment. He constructed some wings from chicken feathers, and attempted to fly towards (or from) the Temple of Apollo in New Troy, present-day London. Unfortunately he fell and broke his neck. How this quaint legend entered UFO databases is a complete mystery to us! See also Fabyan, *The Chronicles* (1516) f. viii and H. C. Levis, *The British King Who Tried to Fly* (London: 1919).

479 BC, Athens, Greece
What was the flying object at the battle of Salamis?

A "horn-shaped object" is said to have flown over during the battle of Salamis (Salamine) near Athens, Greece, between the Greeks and the Persians. The term "horned star" has often been used to describe comets.

Upon consulting *Cometography: A Catalogue of Comets* by Gary W. Kronk and Brian G. Marsden (Cambridge University Press, 1999, 154) we find that interestingly, these authors do include the Salamis observation in their catalogue, calling it a Cerastes type comet.

Pliny writes in his *Natural History* (Book II, Chapter XXII) that "There are stars that suddenly come to birth in the heaven itself; of these there are several kinds. The Greeks call them 'comets,' in our language 'long-haired stars,' because they have a blood-red shock of what looks like shaggy hair at their top. The Greeks also give the name of 'bearded stars' to those from whose lower part spreads a mane resembling a long beard. 'Javelin-stars' quiver like a dart; these are a very terrible portent."

336 BC, Venice, Italy
Alexander the Great, and the UFO that shook Venice

According to a letter he wrote to Aristotle, Alexander the Great seized Venice when an object came down from the sky, shot a beam, and crumbled the walls. This is an entirely spurious account: We traced the story and discovered it was about the use of gunpowder, not an unexplained flying object. The Aristotle letter is a known medieval forgery, only famous because Dante mentions it.

330 BC: Sur (Tyre), Lebanon: Great silver shields

Two strange craft were seen to dive repeatedly at Alexander's army. They looked like great silver shields that went back up into the sky over the Macedonian camp. These "flying shields" flew in triangular formation, led by a large object, while the others were smaller by almost half. In all there were five. They circled slowly over Tyre while thousands of warriors on both sides stood and watched them in astonishment. Suddenly from the largest "shield" came a lightning flash that struck the walls, which crumbled. Unfortunately, no book about Alexander the Great contains the account. The story came from American writer Frank Edwards in 1959, who provided no reference.

Fig. 43: An interpretation of Alexander's "silver shields"

In 1966 Italian writer Fenoglio (the man who invented the Alençon story and many others) embellished the tale. He did not give a reference either, except to say that Alexander's historian recorded the event. The problem is that Alexander's historian was Callisthenes, whose "Deeds of Alexander" are lost.

Fenoglio also states 19th century historian Gustavo Droysen "intentionally does not cite it, believing it to be a fantasy of the Macedonian soldiers." In 1970 Gordon Creighton referred to the sighting in *Flying Saucer Review*. He mentioned Edwards' name but added details from Fenoglio. He stated that he did not know an original source.

In 1976 another English researcher, W. Raymond Drake, interpreted Edwards' and Fenoglio's versions as two separate events, one in Venice and one in Tyre. He misread the article of 1966 and understood Fenoglio had actually quoted from Droysen.

Until some original source can be located, we are left with the suggestion that Alexander's army at Tyre simply witnessed fiery projectiles, some sort of flaming weapon.

Ca. 300 BC, Kun-Lun Mountains, China
The Charioteer of the Moon

According to Professor Couliano, Taoist K'u Yuan (or Qu Yuan) flew with Wang-Shu,"the charioteer of the Moon," in a chariot drawn by dragons.

Fig. 44: Qu Yuan, from the Ming Dynasty

The lady charioteer crops up in many florid poems, like this one, *The Sorrow of Separation*, by K'u Yuan (340-278 BC). It gave rise to the idea that he traveled with the moon entity, but in fact it is just a well-known poem, full of metaphor. The verse most relevant to our reference is this:

I could, for the time being, roam leisurely and remain carefree.
I let the moon-driver Wangshu act as leader
And told the wind-god Feilian to follow closely.
A phoenix acted as vanguard.
The thunder god told me that luggage was not ready.
I ordered the phoenix to fly fast, day and night.
A cyclone gathered around, leading clouds to welcome me.

133 BC, Amiterno, 70 Roman miles NE of Rome, Italy
Sun at night

This case is not to be confused with the sighting at Amiterno in 218 BC mentioned in our Chronology.

The literature of the UFO field often quotes this later event, with the mention that "Under the consulate of Africanus and Fabius, the sun was seen shining at midnight." In reality this object probably was a comet, as recorded by Lucius Annaeus Seneca and Marcus Junianus Justinus in their books. According to Seneca, during the reign of Attalus III, King of Pergamum, "a comet appeared, of moderate size at first. Then it rose up and spread out and went all the way to the equator, so that its vast extent equaled the region of the sky which is called the Milky Way" (*Quaestiones Naturales*).

Justinus noted that at the birth of Mithradates VI Eupator "a comet burned so brightly for 70 days that the entire sky seemed to be on fire. In its greatness it filled a quarter of the heavens, and with its brilliance it outshone the sun, while its rising and setting each took a period of four hours." (Historiae Philippicae). While there is no proof both writers were referring to the same event, the reign of Attalus III has been dated from 137 to 132 BC, while Mithradates VI Eupator was born between 133 and 131 BC. The descriptions do seem to coincide. Furthermore, a Chinese document, the *Han shu*, records that a spectacular comet was seen between August and September 134 BC lasting 30 days, while the twelfth century text, *T'ung kien kang mu*, notes the reign changed due to a comet seen in 133 BC.

85 BC, Roman Empire: Burning shield

"In the consulship of Lucius Valerius and Gaius Marius a burning shield scattering sparks ran across the sky at sunset from west to east." (Pliny: *Natural History* Book II, Chapter XXXIV.)

The description matches the behavior of an ordinary meteor. There is also a date problem with this item. The consulship in 85 BC was held by Cornelius Cinna III and Papirius Carbo I, not the people mentioned. The only consulship held by two people named as in Pliny's claim, C. Marius VI and L. Valerius Flaccus, was in 100 BC. Several sources were checked, including *Who's Who in the Roman World* by John Hazel (Routledge UK, 2002, 110). Valerius did, however, become a suffect consul in 86 BC, which may be the source of the error.

72 BC, Phrygia, near Otryae, Turkey
A falling meteorite stops a battle

At the time of the war between Lucullus and Mithridates, "Marius, whom Sertorius had sent out of Spain to

Mithridates with forces under him, stepping out and challenging him, prepared for battle. In the very instant before joining battle, without any perceptible alteration preceding, on a sudden the sky opened, and a large luminous body fell down in the midst between the armies, in shape like a hogshead, but in color like melted silver, insomuch that both armies in alarm withdrew. This wonderful prodigy happened in Phrygia, near Otryæ." (*Plutarch's Lives*: *Lucullus,* translated by John Dryden, 1683.) There is no reason to believe this object was anything but a natural phenomenon.

68 BC, China: A "Guest Star" hovers over the land

The Han Shu, composed in 100 AD, reports that a "guest star" was observed during "first watch of the night" on July 23rd 68 BC. It "stayed between the left and the right star of Chio, pointing southeastward and measuring about 2 degrees. Its color was white." (Han Shu p.46)

While this particular phenomenon has yet to be identified, it must be noted that the Han Shu uses the same wording to describe the appearance of comet Swift-Tuttle in August of the same year. The Han Shu records many such sightings with great precision, and modern astronomers correlate them to records of comets.

62 BC, Roman Empire: Flashes of fire

Dio Cassius, in his *Roman History* (c. 229 AD), writes "in the west flashes of fire darted up into heaven." There are many such reports in records of that period, and they are often quoted in UFO chronologies. However they are so vague that they could describe several natural phenomena, including meteors, comets or auroral displays.

July 43 BC: The Comet of Murtine, Croatia

Pliny the Elder tells that Augustus wrote "On the very days of my games, a comet was visible over the course of seven days, in the northern region of the heavens. It rose at about the eleventh hour of the day and was bright and plainly seen from all lands. The common people believed that this star signified the soul of Caesar had been received among the spirits of the immortal gods. On this account, it was added as an adornment to the head of the statue of Caesar that I, not long afterwards, dedicated in the Forum."

Gaius Suetonius Tranquillus included Augustus' account in his book *The Lives of the Caesars*, and Seneca also mentions it in *Quaestiones Naturales*, stating the phenomenon appeared at the 11th hour of the day. Plutarch, Siculus and Obsequens all wrote about it, as well as Servius in his commentaries on Virgil's *Eclogue and Aeneid* in the 4th century. Servius relates the phenomenon was observed in the daytime and lasted for three days, but he was writing long after the event. Astronomers believe there could be a link with a comet recorded in the *Han Shu* for May and June 43 BC.

Circa 5 BC, Galilee: The Star of Bethlehem

The birth of Christ presents us with a remarkable, if controversial, report of an extraordinary star. It is often cited in the UFO literature as an example of the relationship between signs in the sky, celestial beings and human reproduction. A very bright object was reportedly seen in the sky, leading the three Magi to the place where Jesus was born. Many possible explanations have been advanced for this "Star of Bethlehem," thought by some to be Venus, or a nova. The lack of a precise date or even year for the birth of Jesus makes it difficult to reach a definite conclusion about the nature of this celestial object.

Fig. 45: The Star of Bethlehem

The New Testament describes an encounter between a young woman named Mary and an entity from the sky, described as an angel: "The angel Gabriel was sent by God to a town in Galilee called Nazareth, to a virgin betrothed to a man named Joseph, of the House of David; and the virgin's name was Mary. He went in and said to her, 'Rejoice, so highly favored! The Lord is with you.'"

> "She was deeply disturbed by these words and asked herself what this greeting could mean, but the

> angel said to her, 'Mary, do not be afraid; you have won God's favor. Listen! You are to conceive and bear a Son, and you must name him Jesus. He will be great and will be called Son of the Most High. The Lord God will give him the throne of his ancestor David; he will rule over the House of Jacob forever and his reign will have no end." (Luke 1:26-38)

This scene has inspired numerous painters, who consistently show a light above Mary, sometimes associated with a bird, and the angel speaking to her. We will return to the question of artistic representation of such events in Part III. Interestingly, the thirteenth-century *Golden Legend* of Jacques de Voragine, an authoritative source on the lives of the Saints and the chronology of Catholic feasts, questions the nature of the star:

"Some say that it was the Holy Ghost, Who had taken this form in order to guide the Magi. Others think it was an angel who also appeared to the shepherds. Still others, with whom we agree, are of the opinion that it was a heavenly body newly created, and that once it had fulfilled its mission, it was absorbed once more into the matter of the universe."

Ca. 28 AD, Jerusalem: Judas enters a luminous cloud

"Look, you have been told everything," Jesus says to Judas after whispering secrets to his friend and most loyal follower. "Lift up your eyes and look at the cloud and the light within it and the stars surrounding it. The star that leads the way is your star." Judas lifted his eyes and saw the luminous cloud, and he entered it.

Source: *The Gospel of Judas*, a newly-deciphered Coptic codex released in 2006 by the National Geographic Society, Washington, D.C.

Circa 249, Britain: A terrible bloody sword in the air

"At his [Decius'] coming to the Throne it Rained Blood in divers parts of this Kingdom, and a Terrible Bloody Sword was seen in the Air for three Nights, a little after Sunset." (Britton, C.E., *A Meteorological Chronology to A.D. 1450*, London: H.M.S.O., 1937, 13.) Note that Trajan Decius was emperor from 249 to 251. We conclude this refers to a classic comet.

312, near Verona, Italy
A pagan Emperor sees a cross in the sky

Emperor Constantine and others see a luminous cross in the sky. The emperor establishes Christianity in Rome.

Fig. 46: The vision of Emperor Constantine

Note that luminous crosses in the sky are not very unusual. They are caused by refraction phenomena similar to what one may observe by looking at a bright light through the mesh of a screen door. In this particular case the emperor is said to have seen luminous writing that read "In hoc signo vinces" (*You will win through this Sign*), which would eliminate the optical phenomenon as a simple explanation. However the writing may have been seen (or even heard) in a vision rather than an actual observation in the sky.

In his remarkable book, *L'Atmosphère et les Grands Phénomènes de la Nature* (Paris: Hachette 1905), Camille Flammarion gives many examples of similar phenomena caused by reflexions or refraction due to ice crystals or water droplets in the atmosphere. Whether the cross was a natural phenomenon or not, this case is so important that it deserves special comments. There are two main sources for the story. One comes from Eusebius, the other from Lactantius. They do not provide an exact location, and contain contradictions. (The later sources are historically not very valuable.)

Fig. 47: Cross-shaped atmospheric illusions

Eusebius described the same events in the life of Constantine in two separate books. In the earliest of these he does not refer to the cross in the sky, literally as if it had never happened. In the later book he tells that before Constantine marched to Rome to battle Maxentius, he and all his soldiers witnessed a cross in the sky. That night, when asleep, the emperor had a dream in which Christ instructed him to make a copy of the sign they had seen, for use in all future battles. Lactantius, on the other hand, did not mention the cross in the sky at all. Instead he wrote that Constantine, while in the vicinity of Rome and before the battle with Maxentius, was simply instructed in a dream to use a special symbol, not forever but in the battle at hand.

Eusebius wrote that the symbol consisted of a cross with the Christ monogram (the chi-roh) at the top. Lactantius wrote that the symbol was itself the chi-roh.

Constantine had already experienced a vision in 310 AD, as the appearance of the pagan god, Apollo, whom the emperor worshipped as a sun god. It is odd that he should have met gods from two opposing religions in the space of two years. The inference is that Constantine's only real vision was in 310 AD, and that he or someone else recycled it for the conversion story of 312 AD. Constantine had already used the sign of the cross (even the Latin cross) on his coins before his conversion, in reference to the Sun.

353, Antioch (Antalaya), Syria
Amazing luminous cross

As Constantius the Victorious, proclaimed Gallus as Caesar, he saw a cross in the form of a column of light appear in the western sky over Antioch.

"After these things, the Emperor Constantius having created Gallus his kinsman Caesar, and given him his own name, sent him to Antioch in Syria, providing thus for the guarding of the eastern parts. When Gallus was entering this city, the Savior's sign appeared in the East: for a pillar in the form of a cross seen in the heavens gave occasion of great amazement to the spectators." (Ecc. Hist. 2.28.2)

Source: Chronicle of Michael the Syrian, Jacobite patriarch of Antioch: 1166-1199 A.D.), who took this story from Socrates (*Ecclesiastical History*, Book II).

384, Roman Empire: A pillar in the sky

"A terrible sign appeared in the sky, shaped like a pillar (*columna*). It was in the time of the Roman Emperor, Theodosius." A number of atmospheric phenomena, or a comet, can produce this effect.

Source: *Lycosthenes,* op. cit., 279.

393 (or 390), Roman Empire
Brilliant globes, swarming like bees

A brilliant globe is observed close to Venus. Many others join in, "swarming like bees." The first reference for this item is Bougard, *Inforespace* no. 22 (August 1975) p.34, quoting Lycosthenes *Prodigiorum ac Ostentorum Chronicon.*

Further research shows that the primary source is historian Philostorgius, in his *Ecclesiasticae Historiae* (composed in 425 AD). He wrote that after Theodosius I returned to Rome following a victory over Clemens Maximus, there appeared "a new and strange star…which announced the coming of very great calamities upon the world." It was first seen at midnight in the east and was said to be "large and bright, and in brilliance it was not much inferior to the morning star (…) A concourse of stars gathered around it on every side, like a swarm of bees."

Later "the light of all the stars mingled together" and the object took on the appearance of a huge double-edged sword. It lasted for 40 days. Our conclusion: the object probably was a comet, which was actually visible in 390 AD.

Circa 523, Kent, Britain
Weird sky phenomena, drops of blood

"In his time strange sights were seen of Dragons, Lions and other furious wild Beasts Fighting in the Air. In the West of Kent it Rained Wheat, and soon after great Drops of Blood, upon which ensued extream Dearth ..."

Scholar C. E. Britton comments: "Vaguely allocated to the reigh of Octa (ca. 513-533). Legendary."

Here we have a further challenge, since nothing indicates that these events were correlated at all, or even happened at te same time. (Britton, C.E., *A Meteorological Chronology to A.D. 1450*, London: H.M.S.O., 1937.)

553, Clonfert, Ireland
Saint Brendan flies up into the sky

Another early instance of what would be called "abduction" today took place when "Brennain of Birra was seen ascending in a chariot into the sky this year."

This refers to Saint Brendan ("Bréanainn") of Clonfert (ca. 484 – ca. 578), an early Irish monastic saint sometimes believed to have sailed to America.

Source: *The Annals of the Four Masters*, historical chronicles compiled in the 17th century by four friars of the Abbey of Donegal in

Bundrowes, near Bundoran. They are also known as *The Annals of the Kingdom of Ireland.*

577, France
Mock suns, and a glittering star in the moon

"Thereafter, on the night of the third day of the Ides of November, while we were celebrating the vigil of the Holy Martin, there appeared to us a great wonder. A glittering star was seen to shine in the center of the Moon; above and below the Moon appeared other stars all near to it, and round about it was the circle which is wont to portend rain. We know not what these things signified.

"And often in this year we saw the Moon darkened, and before Christmastide there was a loud thunder. Moreover, there appeared around the sun the meteors which the country people also call suns, such as those described by me as visible before the calamity in Auvergne.

"It was declared that the sea had risen beyond its usual bounds, and many other signs were seen."

Here again, the sightings are consistent with natural phenomena.

Source: Gregory of Tours, *History of the Franks*, Volume 23, trans. O. M. Dalton (Oxford, 1927), 198.

584, France, exact location unknown: A battle of lights

Many witnesses. Brilliant rays of light hitting one another in the sky.

Given the lack of details, some natural phenomena (such as an aurora borealis) could produce this effect. However we cannot completely reject the case on the basis of the information given, and the source is impeccable.

Source: Gregory of Tours, *History of the Franks*, op. cit.

September 585, France, exact location unknown
Domes in the sky

"The kind of domes people are used to seeing" cross the sky rapidly. Here again we seem to have a tantalizing hint that widespread rumors existed about very unusual aerial objects, of round shape. However, a thorough check of the chronicles of Grégoire de Tours fails to disclose such a description: is there a translation error here, on the part of a usually reliable author ?

Source: *Inforespace* 22 (August 1975): 35. M. Bougard quotes Gregory of Tours, *History of the Franks*, op. cit.

610, Medina, Saudi Arabia
Angel apparition to Islam's prophet

This event concerns the apparition of angel Gabriel to 40-year old Mohammed, Islam's prophet. Muhammad received his first revelation on the mountain of Hira outside Mecca, while he searched for solitude. At the time of the contact with the angel he first experienced great pain, and feared that he was going to die.

The first fraction of the Koran Muhammad received is believed to be the beginning of Sura 96:

"*Recite in the name of your Lord, who created mankind from clots of blood. Recite, and your Lord will be bountiful.*

He who has taught by the pen taught mankind what was not known."

One could argue that this event does not include any description of an aerial phenomenon, and thus does not belong in this compilation. However it does involve an entity from the sky (an angel) and an episode of contact with transmission of a message that has had a major impact on men's beliefs – an impact that continues to this day. In

Fig. 48: The apparition to Mohammed

that sense it epitomizes the complexity of "contact" claims that are an integral part of the phenomenon, both in a social sense and in the larger scope of the societal significance of the relationship between men and the higher worlds in which they believe.

14 January 616, China
A fireball kills 10 people among rebels

The History of the Sui Dynasty, 581-618 records a spectacular fireball that fell into a rebel's camp, partly destroying it. This is interesting because many UFO accounts deal with enemies being frightened or persecuted by mysterious lights – such as when the Christians fought the Turks. This narrative leaves no doubt that the phenomenon had natural causes.

The Chinese document states, "A large shooting star like a bushel fell onto the rebel Lu Ming-yueh's camp. It destroyed his wall-attacking tower and crushed to death more than 10 people."

Source: Kevin Yau, Paul Weissman and Donald Yeomans, "Meteorite falls in China and some related human casualty events," *Meteoritics* 29 (1994): 867.

After 618, China
Capture of a celestial ship – It flies away!

The Dong Tien Ji (*Peeping on the Sky*) says: "In the Tang Dynasty a celestial ship, over 50 feet long, was found and placed in the Ling De Hall. The ship gave out a metallic sound when struck, and was of very hard material which was rustproof.

Li Deyu, the Tang Prime Minister, cut over a foot of a slender, long stick of the ship and carved it into a figure of a Taoist priest. The Taoist figurine flew away and then returned. In the years of Emperor Daming, the figurine disappeared and the ship also flew away."

Source: Paul Dong, *China's Major Mysteries: Paranormal Phenomena and the Unexplained in the people's Republic of China* (China books, 2000), 68-9.

637, Japan: The Barking of the celestial dog

A great star floated from East to West and there was a noise, like that of thunder. The people of that day said it was the sound of the falling star. Others said that it was earth-thunder. Hereupon the Buddhist Priest, Bin, said, "It is not the falling star but the Celestial Dog, the sound of whose barking is like thunder."

Source: *Nihongi or Chronicles of Japan* (London: G. Allen & Unwin, 1956). Quoted by W. Raymond Drake in *Gods and Spacemen in the Ancient East* (London: Sphere, 1973), 104.

24 March 639, Japan, exact location unknown
Noisy star

A big star flew from east to west with a roar like thunder. Min, a Buddhist priest, said it was star *Amagitune*, which is said to mean, "Fox lives in the sky."

In spite of the reported sound, we would argue this was a meteor. For a long time, scientists discounted reports of sounds in connection with meteors, because the speed of sound is so slow compared to light that any sound should only be audible well after the passage of the meteor. Only recently was it realized that the perception of sound can be created inside the skull of the witness by microwaves propagating at the same speed as the light itself.

Source: Morihiro Saitho, *Nihon-Tenmonshiriyou*, Chapter 7, "Meteor, The messenger from space."

640, Faremoutiers-en-Brie, France
A Virtuous Virgin is taken to Heaven

"In the year of our Lord 640, Eadbald, king of Kent, departed this life, and left his kingdom to his son Earconbert, who governed it most nobly 24 years and some months. His daughter Earcongota, as became the offspring of such a parent, was a most virtuous virgin, serving God in a monastery in the country of the Franks, built by a most noble abbess, named Fara, at a place called Brie. Many wonderful works and miracles of this virgin, dedicated to God, are to this day related by the inhabitants of that place; but for us it shall suffice to say something briefly of her departure out of this world to the heavenly kingdom. The day of her summoning drawing near, (…) she let (others) know that her death was at hand, as she had learnt by revelation, which she said she had received in this manner:

"She had seen a band of men, clothed in white, come into the monastery, and being asked by her what they wanted, and what they did there, they answered they had been sent thither to carry away with them the gold coin that had been brought thither from Kent. Towards the close of that same night, as morning began to dawn, leaving the darkness of this world, she departed to the light of heaven. Many of the brethren of that monastery who were in other houses, declared they had then plainly heard choirs of singing angels, and, as it were, the sound of a multitude entering the monastery. Whereupon going out immediately to see what it might be, *they beheld a great light coming down from heaven, which bore that holy soul, set loose from the bonds of the flesh, to the eternal joys of the celestial country.* They also tell of other miracles that were wrought that night in the same monastery by the power of God."

Source: Bede the Venerable, *Ecclesiastical History of England*, trans. A. M. Sellar (London: George Bell & Sons, 1907).

Circa 685, Lindsey, England
Miracles from Heaven chase the Devil away

How a light from Heaven stood all night over King Oswald's relics, and how those possessed with devils were healed by them: "I think we ought not to pass over in silence the miracles and signs from Heaven that were shown when King Oswald's bones were found, and translated into the church where they are now preserved.

"It was revealed by a sign from Heaven with how much reverence they ought to be received by all the faithful; for all that night, *a pillar of light, reaching from the wagon up to heaven, was visible in almost every part of the province of Lindsey*. Hereupon, in the morning, the brethren of that monastery who had refused it the day before, began themselves earnestly to pray that those holy relics, beloved of God, might be laid among them. (…) Then they poured out the water in which they had washed the bones, in a corner of the cemetery. From that time, the very earth which received that holy water had the power of saving grace in casting out devils from the bodies of persons possessed.

"Lastly, there came to visit her (the queen) a certain venerable abbess, who is still living, called Ethelhild, the sister of the holy men, Ethelwinand Aldwin, the first of whom was bishop in the province of Lindsey, the other abbot of the monastery of Peartaneu; not far from which was the monastery of Ethelhild. When this lady was come, in a conversation between her and the queen, the discourse turning upon Oswald, she said, *that she also had that night seen the light over his relics reaching up to heaven*. The queen thereupon added, that the very dust of the pavement on which the water that washed the bones had been poured out, had already healed many sick persons.

"The abbess thereupon desired that some of that health-bringing dust might be given her, and, receiving it, she tied it up in a cloth, and, putting it into a casket, returned home.

"Some time after, when she was in her monastery, there came to it a guest, who was wont to be grievously tormented with an unclean spirit at night; he being hospitably entertained, when he had gone to bed after supper, was suddenly seized by the Devil, and began to cry out, to gnash his teeth, to foam at the mouth, and to writhe and distort his limbs. (…) When no hope appeared of easing him in his ravings, the abbess bethought herself of the dust, and immediately bade her handmaiden go and fetch her the casket in which it was. As soon as she came with it, as she had been bidden, and was entering the hall of the house, in the inner part whereof the possessed person was writhing in torment, he suddenly became silent, and laid down his head, as if he had been falling asleep, stretching out all his limbs to rest. 'Silence fell upon all and intent they gazed,' anxiously waiting to see the end of the matter. And after about the space of an hour the man that had been tormented sat up, and fetching a deep sigh, said, 'Now I am whole, for I am restored to my senses.'"

Source: Bede the Venerable, *Ecclesiastical History of England*, op. cit.

April 750, Córdoba, Spain: Three suns, a sickle of fire

"In the nones of April, on Sunday during the first, second and almost the third hours, all the citizens of Cordoba saw three suns which shone and twinkled in a wonderful way preceded by a sickle of fire and emerald; and, from its appearance, by order of God, his angels devastated all the inhabitants of Spain with intolerable hunger."

Source: *Cronica Mozarabe* of the year 754 (or "Continuatio Hispana de San Isidoro").

778: Notre-Dame de Sabart (Ariège, France)
Luminous virgin

Tradition states that the Sabart sanctuary, near Tarascon, dates from the Great Charlemagne. The emperor dedicated a chapel to Notre-Dame in recognition of her help in his fight against the Saracens. The sanctuary is the site of a pilgrimage on September 8th. According to legend, a luminous virgin was unearthed at this place by two heifers led by Charlemagne himself. The chapel is adorned with a wonderful stained glass window dating from the thirteenth century, the oldest such window in the Midi.

Source: René Alleau, *Guide de la France Mystérieuse* (Paris: Tchou, 1964).

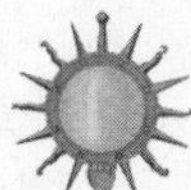

793, Northumbria, England: Fiery Dragons, Evil Men

According to the Anglo Saxon chronicle, "Here in this year, dire portents appeared over Northumbria, and sorely terrified the people. They consisted of immense whirlwinds and flashes of lightning, and fiery dragons were seen flying in the air.

A great famine immediately followed those signs, and a little after that in the same year, on 8 June, the ravages of heathen men miserably destroyed God's church on the island of Lindisfarne, with plunder and manslaughter."

These descriptions are consistent with electrical storms, possibly associated with tornadoes.

Source: G. P. Cubbin, ed., *The Anglo Saxon Chronicle, A Collaborative Edition*, vol. 6 (Cambridge: Boydell & Brewer, 1996), 17.

819, Clent, Shropshire, England: Beam of light

A column of white light projects a beam towards a thorn tree where rested the head of murdered King Kenelm.

The connection with aerial phenomena is very tenuous indeed, yet this event is quoted in the literature of the field as if it was unidentified.

Source: Delair J Bernard, *UFO Register* (1971), quoting the Chronicles of Richard of Cirencester.

About 1000, Europe: Flying cross

A rare book entitled *Aragón reyno de Cristo y dote de María SS. ma fundado sobre la columna immobile de Nuestra Señora de su Ciudad de Zaragoza*, published in Zaragoza in 1739, mentions all kinds of celestial prodigies associated with religious images over the centuries. Among these, two incidents of flying crosses are recorded. Neither case is dated but they occurred at some time between the 10th and the 11th centuries. One of these was the "Miraculous appearance of the Sacred Cross over the carrasca in the Royal Field of the Town of Aynsa," when Captain Garci-Ximenez, soon to become the first king of Sobrarbe, conquered the Moors of the town, "freeing it from Muslim tyranny."

Among so many fears, Garci-Ximenez turned his eyes to the Sky, asking the God of Armies for help, and, as a presage of victory, He gave him a marvellous sign, a Sacred Red Cross that appeared over a carrasca. The sight of this spurred Garci-Ximenez on, as if he had heard the voice that Constantine the Great heard from the sky: "With this sign of the Holy Cross you will overcome."

And, of course, they overcame. The second reference to a cross in the sky, under the heading "Prodigious apparition of the Holy Cross over the valley of Arahones in the ancient Kingdom of Sobrarbe." On this occasion, when the Christian troops, led by Iñigo Arista, try to reconquer a place called Campo del Rey, in Aragues, they are surrounded by the Moors and ask Heaven for help. A cross then appears in the air, giving the Christians the spirit they need to fight on.

29 June 1033, England: An eclipse and the Antichrist

"Just as the superstitious idolatries of Antichrist were arrived at their height by overspreading the Christian world, upon June 29 (which is by some called St. Peter's day) at six o'clock in the morning, a terrible eclipse of the sun happened, in which he became like sapphire; so that it made men's countenances look pale, as if they had been dead; and every thing in the air seemed of a saffron colour."

There is no unusual aerial phenomenon here, only an assumed connection between a religious incident (the arrival of the Antichrist) and a solar eclipse.

Source: John Howie, *An Alarm unto a secure generation...*, (Glasgow: John Bryce, 1780).

1066, River Setoml near Kiev, Ukraine
Red star and little man

The initial story we found stated that "in 1065" a dwarf-like entity was pulled out of the river by a fisherman and thrown back, while local residents observed a strange sign in the sky – a huge star with blood red beams of light.

"This phenomenon lasted for seven straight days. It was seen only during the evening. Around the same time a child-like dwarf type entity was found by fishermen in the river Setomi (this river does not exist at present). The dwarf was pulled out of the river in a net. The fishermen kept watch over the strange entity until late afternoon and then threw it back into the river out of fear and repugnancy. The dwarf like entity was very strange with a very wrinkled face and other "shameless" details on his face and body."

Review of this case disclosed several problems: First, the year itself was incorrect: in the "*Povest Vremennyh Let*" (Tale of Bygone Years) the date of incident is given as 1066. Second, the "star with red beams" was none other than Halley's Comet! Its nearest approach to the Earth was March 27, 1066. The "monster" was only a deformed child who was dropped in the river Setoml (not Setomi) by his mother. His body was accidentally found by the fishermen (the manuscript clearly stated this). This child even had its genitalia on the face! Russian manuscripts often stated that the birth of "monstrous" children was an omen or a curse for all peoples.

Source: *Povest Vremennyh Let* (an ancient manuscript), and Dmitri Lavrov in *Ukrainian News* 18 Feb. 1998. Further research by Mikhail Gershtein, Magonia Exchange (Magoniax) Project.

1 March 1095, Piacenza, Italy
Blue luminous dove, a great cross

In the public square, in front of the church of Saint Maria di Campagna, there was a meeting of the most powerful figures of the century. It was March 1st, 1095, and the preparations for the first crusade were under discussion. According to the legend, during the assembly, just as Countess Matilde di Canossa began to speak, a blue, luminous dove descended from the sky. Later, when these powerful leaders officially announced the launch of the crusade, the clouds opened miraculously, the public square was bathed in a powerful light, and a great cross appeared in the sky, identical to that which appeared before Emperor Constantine, with the words "In hoc signo vinces" written upon it. Probably just a bit of Christian brand management at the time of the crusades…

Source: P. Carpi, *Magia e segreti dell'Emilia-Romagna* (Borelli: Modena, 1988), 114.

Eleventh-century Europe: Astronauts in trouble

A new genre of folktales developed in Medieval Europe between the 11th and the 13th centuries. In these stories, a member of the aerial crew of a cloud ship runs into trouble as he descends to retrieve a lost spear or loosen a trapped anchor. These wonderful tales were told for a long time. Although it is likely they all derived from the same original source, certain details were slightly altered in each retelling.

Aside from the issue of whether actual UFOs were seen during this period, which we discussed in Part I of this book, tales of 'cloud ships' were retold and embroidered to support the argument that mysterious beings traversed the sky with the ease that humans travel over the sea in ships.

There was an almost universal belief that the world was composed of three levels or 'decks': the earth, the heavens and the marine kingdoms under the sea, between which it was not impossible to travel in the right conditions or by following certain instructions. For this reason, surreal stories of celestial sailing vessels dropping anchors upon the earth or divers from above drowning in our air became believable urban myths in medieval times. Consider the following account from Bishop Patrick's *Hiberno-Latin Mirabilia* (1074-84 AD):

There was once a king of the Scots at a show
With a great throng, thousands in fair array.
Suddenly they see a ship sail past in the air,
And from the ship a man then cast a spear after a fish;
The spear struck the ground,
and he, swimming, plucked it out.
Who can hear this wonder
and not praise the Lord of Thunder?

Other Irish documents, such as the *Book of Glendalough*, composed in 1130, repeated the same story of the airship and the fish in every detail. However, the *Book of Leinster* (ca.1170), while stating that it occurred at the royal fair of Tailtiu, speaks of *three* ships in the sky, and alleges that King Domhnall, son of Murchad, was among the witnesses. This is interesting: Domhnall was the 161st Monarch of Ireland, reigning between 738 and 758 AD, and a report that a flying ship was seen in the sky in that period does in fact exist. The *Annals of Ulster*, which covers the years 431 to 1588, states, albeit with no reference to the fair or to the king, that as early as the year 749, "Ships with their crews were seen in the air above Cluain Moccu Nóis."

Another, much later work, *The Annals of the Four Masters*, a series of historical chronicles compiled between 1632 and 1636 by four friars of the Abbey of Donegal in Bundrowes, near Bundoran, states that, "Ships with their crews were seen in the air" in 743. As this book contains numerous errors we are more inclined to take the date of 749 as the correct one. Was Domhnall, as opposed to Congalach, the royal witness in the original version of the airship sighting?

We see these stories as interesting forerunners of the ufology era, with a series of episodes in which the pattern shows either aerial voyagers in trouble, or airship operators who adopt a posture conveying a message, such as an outstretched arm.

Source: Aubrey Gwynn, ed., *The Writings of Bishop Patrick 1074-1084* (Dublin: Dublin Institute for Advanced Studies, 1955), cited in *Anchors in a three-decker world,* Miceal Ross, Folklore Annual 1998. Note that the term "Scots" here refers to the Irish.

1211, Britain: Death of a sky visitor
But where is the Alien body?

In later retellings of the above story the motive behind the sailor's descent is a trapped anchor, thus substituting the traditional spear for an object more suited to navigation.

Gervase of Tilbury collected a similar tale in his work, *Otia Imperialia* (1211 AD):

"As people were coming out of church in Britain, on a dark cloudy day, they saw a ship's anchor fastened in a heap of stones, with its cable reaching up from it into the clouds. Presently they saw the cable strained, as if the crew was trying to pull it up, but it still stuck fast. Voices were then heard above the clouds, apparently in clamorous debate, and a sailor came down the cable. As soon as he touched the ground the crowd gathered around him, and he died, like a man drowned at sea, suffocated by our damp thick atmosphere. An hour afterwards, his shipmates cut the cable and sailed away; and the anchor they left behind was made into fastenings and ornaments for the church door, in memory of this wondrous event."

It is not reported whether the dead sailor's body is shipped home in the airship, or whether the deceased is given a Christian burial on earth. In either case, this would be the first account of an aerial navigator that dies in an accident on our planet, some seven centuries before Roswell.

1250, Cloena (Clonmacnoise)
A ship with occupants, captured anchor

Some forty years later, the story was repeated by the anonymous author of an influential book written in Old

Norse. The *Kongs Skuggsjo*, better known by its Latin name, the *Speculum Regale* ["the king's mirror"], was written around 1250 AD. The event took place in Clonmacnoise.

"There happened something once in the borough called Cloena, which will also seem marvellous. In this town there is a church dedicated to the memory of a saint named Kiranus. One Sunday while the populace was at church hearing mass, it befell that an anchor was dropped from the sky as if thrown from a ship; for a rope was attached to it, and one of the flukes of the anchor got caught in the arch above the church door. The people all rushed out of the church and marvelled much as their eyes followed the rope upward.

"They saw a ship with men on board floating before the anchor cable; and soon they saw a man leap overboard and dive down to the anchor as if to release it. The movements of his hands and feet and all his actions appeared like those of a man swimming in the water. When he came down to the anchor, he tried to loosen it, but the people immediately rushed up and attempted to seize him. In this church where the anchor was caught, there is a bishop's throne.

"The bishop was present when this occurred and forbade his people to hold the man; for, said he, it might prove fatal as when one is held under water. As soon as the man was released, he hurried back up to the ship; and when he was up the crew cut the rope and the ship sailed away out of sight. But the anchor has remained in the church since then as a testimony to this event."

The strong Christian overtones are noticeable in this version. Here it is not a king but a bishop who is present during the event, and the action occurs in the air above a church. The fact that the diver is allowed to return to his ship unharmed is another moralistic touch.

According to folklorist John Carey, the move from spears to anchors was due in part to the popularity of another legend of the same period, in which the crew are aboard a ship actually sailing in the sea, not in the sky. In this version, the anchor gets stuck in an underwater monastery, to be freed by a blind boy who swims down and finds himself in a subaquatic world.

Return of the celestial diver seven centuries later… in a Texas hoax!

In April 1897, in the middle of a wave of mysterious airship sightings, some American newspapers published two British folktales from Gervase of Tilbury. One of these was none other than the legend of the anchor and the church that we have cited above.

This article, entitled "A Sea Above the Clouds: Extraordinary Superstition Once Prevalent in England," published first in the *Boston Post*, must have impressed one reader at least, for a couple of weeks later an anonymous writer wove yet another airship yarn from it.

Anchor of the Airship.

Said to Be on Exhibition at Merkel, Attracting Much Attention.

Merkel, Texas, April 26 – Some parties returning from church last night noticed a heavy object dragging along with a rope attached. They followed it until in crossing the railroad, it caught on a rail. On looking up they saw what they supposed was the airship. It was not near enough to get an idea of the dimensions. A light could be seen protruding from several windows; one bright light in front like the headlight of a locomotive. After some 10 minutes a man was seen descending the rope; he came near enough to be plainly seen.

He wore a light-blue sailor suit, was small in size. He stopped when he discovered parties at the anchor and cut the ropes below him and sailed off in a northeast direction.

The anchor is now on exhibition at the blacksmith shop of Elliott and Miller and is attracting the attention of hundreds of people.

In this updated American version, railroad tracks replace the tombstone where the anchor gets caught, and the pilot – dressed, naturally enough, in a sailor's suit – returns to his craft safe and sound, but there can be no mistaking the origin of the tale.

1122, London, England
Another flying ship loses its anchor

Mr. Page, associate national correspondent of the French Antiquary Society, reports a story told by a 12th century monk from Limousin named Geoffroy de Vigeois.

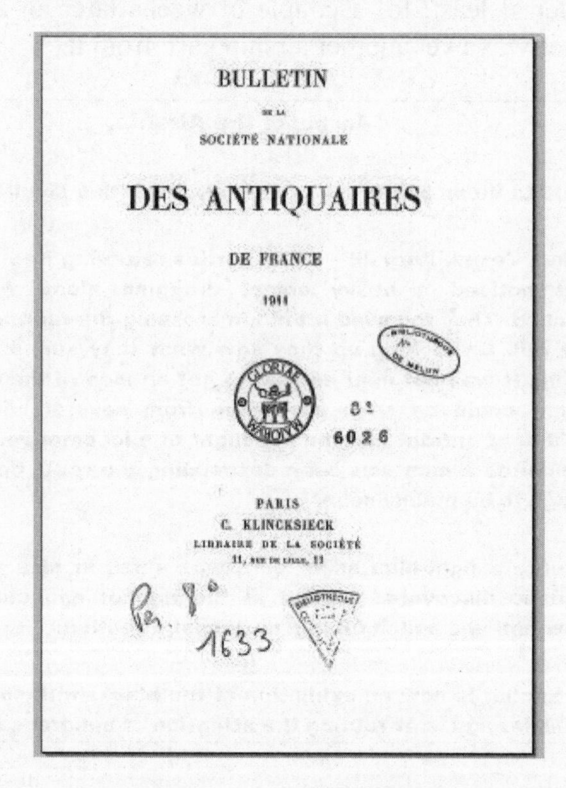

BULLETIN

DE LA

SOCIÉTÉ NATIONALE

DES ANTIQUAIRES

DE FRANCE

1911

8°
6026

PARIS
C. KLINCKSIECK
LIBRAIRE DE LA SOCIÉTÉ
11, RUE DE LILLE, 11

Per 8°
1633

Fig. 49: Bulletin des Antiquaires

The event concerns a flying ship "navis sursum in aëre," which landed in the middle of London. The inhabitants rushed on the anchor of this ship, and the passengers were forced to cut the rope in order to take to the air again.

"Mr. le Comte H.-F. Delaborde reminds us that Leonardo da Vinci has studied aerial navigation, and was not the first. He assumes that the event may have originated with a simple mirage. The legend must have grown as it traveled, as often happens."

Source: Geoffroi du Vigeois, *Chronica*, A.D. MCXXII, éd. Philippe Labbe, *Nova Bibliotheca manuscripta* (Parisiis, 1657), II, 299-300; *Bulletin de la Société des Antiquaires de France* (1911) : 102-103.

1141, Bingen, Germany
Figures within fiery flying disks

Sainte Hildegard, 41, reports: "Heaven was opened and a fiery light of exceeding brilliance permeated my whole brain"

Hildegard of Bingen (1098-1179) was a remarkable woman, a pioneer in many fields. At a time when few women wrote, Hildegard produced major works of theology and visionary texts. Kristina Lerman, writing on the UCSB (Santa Barbara) website, observed: "When few women were accorded respect, she was consulted by and advised bishops, popes, and kings.

"She used the curative powers of natural objects for healing, and wrote treatises about natural history and medicinal uses of plants, animals, trees and stones. She is the first composer whose biography is known. She founded a vibrant convent, where her musical plays were performed. Revival of interest in this extraordinary woman of the middle ages was initiated by musicologists and historians of science and religion."

As a girl, Hildegard started to have visions of luminous objects at the age of three. She soon realized she was unique in this ability and hid this gift for many years.

Fig. 50: Visionary painting by Sainte Hildegard

However, in 1141, Hildegard had a vision of God that gave her instant understanding of the meaning of the religious texts, and commanded her to write down everything she would observe:

"And it came to pass ... when I was 42 years and 7 months old, that the heavens were opened and a blinding light of exceptional brilliance flowed through my entire brain. And so it kindled my whole heart and breast like a flame, not burning but warming..."

It is now generally agreed that Sainte Hildegard suffered from migraine, and that her experiences were a result of this condition. The way she describes her visions, the precursors and the debilitating aftereffects, points to classic symptoms of migraine sufferers. Although a number of visual hallucinations may occur, the more common ones described are the "scotomata" that often follow perceptions of phosphenes in the visual field. Scintillating scotomata are also associated with areas of total blindness in the visual field, something Hildegard might have been describing when she spoke of points of intense light, and also the "extinguished stars."

Migraine attacks are usually followed by sickness, paralysis, blindness – all reported by Hildegard, and when they pass, by a period of rebound and feeling better than before, a euphoria also described by her. Among the strangest and most intense symptoms of migraine aura, are the occurrences of feelings of sudden familiarity and certitude...or its opposite. Such states are experienced, momentarily and occasionally, by everyone; their occurrence in migraine auras is marked by their overwhelming intensity and relatively long duration. As Kristina Lerman notes, "It is a tribute to the remarkable spirit and the intellectual powers of this woman that she was able to turn a debilitating illness into the word of God, and create so much with it."

September 1157, Germany
Three suns, three moons

Lunar halo, mock moons, sundogs and crosses of light are represented in this medieval book.

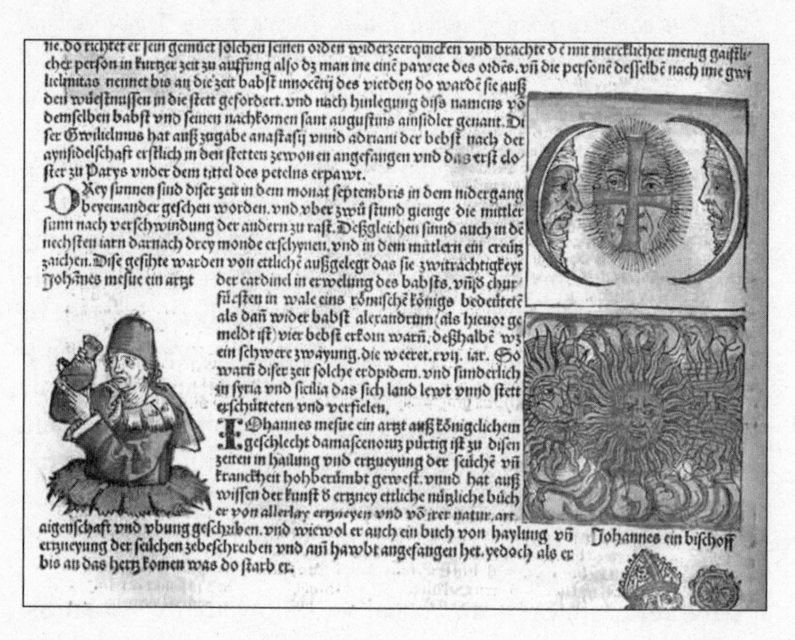

ne. do richtet er ſein gemüet ſolchen ſeinen orden widerzeerquicken vnd brachte de mit mercklicher menig gaiſtlicher perſon in kurtzer zeit zu auffung alſo dz man ime einē pawere des ordēs. vñ die perſonē deſſelbē nach ime gwilielmitas nennet bis an die zeit babſt innocētij des vierden do wardē ſie auß den wüeſtnuſſen in die ſtett gefordert. vnd nach hinlegung diſs namens vō demſelben babſt vnd ſeinen nachkomen ſant auguſtins ainſidler genant. Diſer Gwilielmus hat auß zugabe anaſtaſij vnnd adriani der bebſt nach der aynſidelſchaft erſtlich in den ſtetten zewonen angefangen vnd das erſt cloſter zu Parys vnder dem tittel des petelns erpawt.

DRey ſunnen ſind diſer zeit in dem monat ſeptembris in dem nidergang beyeinander geſehen worden. vnd vber zwū ſtund gienge die mittler ſunn nach verſchwindung der andern zu raſt. Deßgleichen ſinnd auch in dē nechſten iarn darnach drey monde erſchynen. vnd in dem mittlern ein creütz zaichen. Diſe geſihte warden von etlichē außgelegt das ſie zwitrachtigkeyt der cardinel in erwelung des babſts. vñ̄ d chur fürſten in wale eins römiſchē königs bedeütetē als dañ wider babſt alexandrum (als hievor gemeldt iſt) vier bebſt erkorn warñ. deßhalbē wz ein ſchwere zwayung. die weeret. xvij. iar. So warñ diſer zeit ſolche erdpidem. vnd ſunderlich in ſyria vnd ſicilia das ſich land lewt vnnd ſtett erſchütteten vnd verfielen.

Johānes meſue ein artzt

IOhannes meſue ein artzt auß königlichem geſchlecht damaſcenorū pürtig iſt zu diſen zeiten in hailung vnd ertzueyung der ſeüchē vñ kranckheit hohberümbt geweſt. vnnd hat auß wiſſen der kunſt d ertzney etliche nützliche büch er von allerlay ertzneyen vnd vō irer natur. art. aigenſchaft vnd vbung geſchriben. vnd wiewol er auch ein buch von hailung vñ ertzneyung der ſeüchen zebeſchreiben vnd anī hawbt angefangen het. yedoch als er bis an das hertz komen was do ſtarb er.

Johannes ein biſchoff

Fig. 51: Phenomena in Germany

The text reads:

"In the month of September, there were seen three suns in a clear sky [and more than two hours after the disappearance of the other two, the middle sun disappeared as well—Schedel]. And a few days later, in the same month, three moons [were seen], and in the moon that stood in the middle, a white cross. Whereupon the Doctors and the most skillful searchers of natural things, being sent for from the universities of Paris, Bononia [Bologna] and Venice, did interpret the prognostication, signifying that there should arise a discord between the Cardinals in choosing the Pope [...]

"There were seen many signs in the sky towards the North, as it were fiery torches and the likeness of reddish human blood. Neither did these wonders deceive them, for King Suenus [Sven III of Denmark] spoiled the country of the Wagians and all places were spoiled by war."

Source: Hartmann Schedel, *Liber Chronicarum* (Nuremberg, 1493), fol. 203v; Lycosthenes, op. cit., 413-414.

1173, Northern Ireland: A mass of fire in the air

On the night the bishop of all Northern Ireland died, "the night was illumined from nocturns until cockcrow, and the ground was all in flames; and a large mass of fire ascended over the town, and proceeded towards the southeast; and all persons arose from their beds, imagining that it was day."

Again, a classic description of auroral displays.

Source: *Annals of Loch Ce* (Millwood, NY: Kraus Reprint, 1965), 149.

1290, England
The disk that flew over Byland Abbey

In 1254, "the perfect form and likeness of a mighty great ship," was said to have been seen in the sky by "certain monks of St. Albans," in England. In the classic flying saucer book *Flying Saucers Have Landed,* Desmond Leslie and George Adamski published the translation of a document called the "Ampleforth Abbey manuscript," allegedly a 13th century document. They had come across the story in a letter to *The Times* on February 9th 1953 that ran as follows:

"Sir – Reports of "flying saucers" usually evoke a small crop of cynical replies that far more sensational objects were seen towards the end of the last century, &c. While going through some early manuscripts pertaining to Byland Abbey, in Yorkshire, I came across material for this sort of criticism which is surely unsurpassed. A document dated circa 1290 mentions a round flat silver object like a discus which flew over the monastery exciting "maximum terrorem" among the brethren.

I am, Sir, yours faithfully, A. X. Chumley, Ampleforth College, York"

The controversial fragment itself was in Latin and read as follows:

"…took the sheep from Wilfred and roast them in the feast of SS. Simon and Jude. But when Henry the Abbot was about to say grace, John, one of the brethren, came in and said there was a great portent outside. Then they all went out and LO! a large round silver thing like a disk flew slowly over them, and excited the greatest terror. Whereat Henry the Abbott immediately cried that Wilfred was an adulterer…"

The story came apart when two boys confessed to having written the passage as a joke. In January 2002 one of us (C.A.) contacted the archivist at Ampleforth Abbey, who prefers not to be named, in order to discover the identities and motives of the hoaxers. He replied that he had been at school with them himself. One of the boys had been killed in an accident in the mid-1950s, he said, while "the other half is a distinguished academic, now in retirement," who preferred to remain anonymous.

"It was done on purpose in order to bring out the folly of the credulous," he added. As for the name "Chumley," which had been attached to the letter in *The Times*, it "was a known local name, spelt more usually as Cholmondly."

When asked if he knew how the surviving hoaxer felt about the fuss made by the prank – a prank that was (and still is) cited by ufologists the world over, the archivist replied: "I think he finds it rather tiresome. Consider to what extent you wish to dwell–or rather be pursued about–the japes of your youth!"

January 1319, Mozhcharyk, Russia
Fiery columns, pillars of fire

At the time when Prince Mikhail of Tver was murdered by his brother and a wicked Tatar, "many believers and even infidels at that time saw two clouds which came over the body of blessed Prince Mikhail, and they came together and parted and they shone as the sun. These people told us of it with tears, and giving many oaths. The body was sent to Mozhcharyk.

"At that time there were Russian merchants present who wanted to place the body in a church and cover it with a saintly cover; but they were not permitted to do so and the body was put in a barn, under guard. Then others living there saw at night a fiery column extending from the earth to the sky. Others saw a rainbow which bent over the barn in which the body lay.

"From thence the body was taken to the town of Bezdezh, and when they drew near the town, many people in the town saw a vision; around the sledge there was a multitude of people with candles and others on horseback with lanterns, riding in the air. And so they brought the body to the town, but the body was not placed in the church, but only in the yard. Two of the guards lay down in the sledge above the body but they were seized by great fear and were thrown out of the sledge and pushed afar off. When they arose and recovered, they went and confessed what had happened to the priests who were there. I heard it from them and I have written it exactly."

Another source reports that during the course of January, at night over most areas of Russia numerous witnesses observed "fiery pillars," similar to those sighted in 1111, which extended from the ground toward the sky. Some people also sighted a "heavenly arc." Yet others saw horse-like flying entities, equipped with "lanterns."

These descriptions suggest a widespread natural phenomenon such as an aurora borealis, which could be seen over large portions of Russia. The mechanism of the aurora's fantastic luminous displays was a mystery until the 20th century, when physicists understood the behavior of charged particles from the Sun caught in the Earth's magnetic field. These particles emit light of vivid colors in wide, undulating curtain-like swatches that can be likened to pillars, rainbows or arches.

Source: *The Nikonian Chronicle*, trans. S. A. Zenkovsky, vol. 3 (Princeton, NJ: Kingston Press, 1984-1989), 110-112

6 November 1331, Florence, Italy: Miraculous cross

An old chronicler records the following observation:

"That evening a miraculous sign appeared in the sky: a vermilion cross over the palace of the Priors. Each bar was more than a palm and a half wide; one line appeared to be more than forty feet high, and the transverse was a little less. The cross remained for as long as it takes a horse to run two laps. The people who saw this – and I saw it clearly – could understand that God was firmly set against our tormented city."

Here we have a clear example of a religious interpretation given to a natural phenomenon that puzzled observers of the time.

Source: Dino Compagni, *Chronicle of Florence*, trans. Daniel E. Bornstein (Philadelphia: Univ. of Pennsylvania Press, 1986), 47-48.

1503, Freiburg im Breisgau, Germany: Flying spear

A painting illustrates an observation of a flying spear in the sky, watched by a monk in prayer. The Latin manuscript calls the object is "the Lance of Christ." The description of flying spears is a common reference to meteors and bolides in the atmosphere.

Fig. 52: Freiburg meteor

The manuscript was never printed.

Source: Jakob Mennel, *Über Wunderzeichen* (*De Signis Portentis Prodigiis*) (1503). Current location: Österreichische Nationalbibliothek, Vienna.

20 April 1535, Stockholm, Sweden: Five fiery disks

Five sun-like disks were seen in the sky. Swedish reformer and scholar Olaus Petri (1493-1552) had a painting made by Urban to memorialize the event. The object's trajectories were drawn up by Dutch painter Jacob Matham. The Rijksmuseum in Amsterdam has a copper plate showing Matham at work on this drawing. Controversy about the event lasted for a century. The painting can be seen now in Stockholm Cathedral.

Our research indicates that the "controversy" in question didn't have to do with the sighting or its depiction but with its interpretation by the common people, who took it as a bad omen after the king converted to Protestantism. Perceiving the painting as a threat to his power, the king had Petri arrested.

The painting itself is beautiful. It clearly shows solar parhelia due to atmospheric effects.

Source: Christiane Piens, *Les Ovni du Passé* (Belgium: Marabout, 1977).

25 October 1593, Manila, Philippines
Involuntary desertion

According to Father Gaspar de San Agustin, on the morning of Monday, October 25th 1593, a Spanish soldier suddenly appeared in the Main Square in Mexico City. He belonged to a regimental unit stationed in Manila in the Philippines, some 9,000 miles away. The soldier, whose name is not stated, was not merely lost: he was unable to explain how he had reached the other side of the Pacific Ocean! Following his arrest he was ordered by the Inquisition to return to Manila.

This story has been published in a number of versions since Father Gaspar de Agustin included it in his 1698 book *Conquista de las Islas Filipinas*. Here is the original paragraph, in English:

"It is worthy of reflection that on the same day that the tragedy of Gómez Pérez occurred, the art of Satan had already made it knowledge in Mexico. With [Satan's help] some women inclined to performing such acts transported a soldier, who was in the sentry box on the walls of Manila, to the main square in Mexico City. This was carried out without the soldier's even being aware of it, and in

the morning he was found walking about the square in Mexico with his weapons, asking everyone who passed by to give him their name. But the Holy Inquisition of the city ordered him to return to these Islands, where many who knew him assured me of the truth of this event."

Many details were added later. It is often claimed, for example, that to prove that he had actually been in Manila the night before, the soldier told Mexican authorities that His Excellency the Governor of the Philippines, Gómez Pérez Dasmariñas, had just been assassinated with a blow to the head. In this version two months pass before a galleon arrives in Acapulco bringing an important message from Manila: Governor Dasmarinas has indeed been murdered by the Chinese during a military mission against the Molucas Islands. The date of the assassination is given as October 24th. However, no such incident is described in Father Gaspar's book.

The name of the soldier is sometimes said to be Gil Pérez and we are told he has been charged with desertion. These claims are also later accretions. The origin of the story is not known, and there are legitimate doubts about its authenticity. Father Gaspar cites Antonio de Morga's *Sucesos de las Islas Filipinas* as his source but the story does not appear there, so we are left wondering whether the priest invented it himself. The legend is mentioned by UFO writers (such as Morris Jessup) to support theories of abduction or teleportation.

Mid-17th century, France: A controversial "jeton," a flying disk?

The token shown below, which experts believe was struck around 1656, shows a disk with light or a beam emanating from it. This object ("jeton" in French) has been mentioned in numerous books and magazine articles as a commemoration of an event in which a flying object was observed in the clouds, passing over a wide landscape.

The object is the size of a U.S. quarter or a one-euro coin and similar to thousands of other tokens that were produced in Europe around that time for religious and educational purposes. Rather than a flying saucer, it is likely to represent the "Shield of Numa." In fact jetons bearing very similar images have been found dating to mid-17th century France.

Fig. 53: French *jeton*

It is said that on March 1st, 707 BC, during the outbreak of a plague, the second king of Rome, Numa Pompilius, witnessed the fall of an oval shield from the sky during a ceremony.

Somewhat astonished, he sought advice from the nymph Egeria and the nine Muses, who assured him that Jupiter had dropped it as a sign of his benevolence. The pestilence soon came to an end, so the grateful king had eleven identical copies made by an armourer, and those were used in dances and celebrations every year.

Around 60 AD the Roman poet Lucanus composed a long work in which he proposed an explanation for the event, suggesting that a stormy wind had whisked the shield out of a soldier's hand and sent it spinning across the sky. The story was long remembered in the literature and is likely depicted on this token, where the inscription reads: *Oportunus Adest*, "it arrives in time." The same "flying disk" design appears on many jetons from different periods.

November 1661, Chaldan Monastery, Tibet
Flying double hat

This case was mentioned by researcher W. Raymond Drake in 1975, based on the diary of a Jesuit father, Albert d'Orville:

D´Orville, a Belgian, wrote about a fascinating sighting at Lhasa, Tibet: "1661 November. My attention was attracted by something moving about in the heavens. I thought it was some unknown species of bird which lived in that country, when the thing on approaching took an aspect of a double Chinese-hat (the classical conical straw-hats) and flew rotating silently as if borne on invisible wings of the wind. It was surely a prodigy, an enchantment. That thing passed above the city, and as if it wished to be admired, it completed two circles, then surrounded by mist it vanished, and no matter how one strained its eyes it could no longer be seen.

"I asked myself whether the altitude where I was had not played some trick, however perceiving a lama not far away I asked whether he had seen it. After assenting by nodding his head, he said to me, 'My Son, what you have seen is not magic. Beings from other worlds have for centuries sailed the seas of space, they brought intellectual illumination to the first people populating Earth, they banished all violence and taught men to love one another, but these

teachings are like seed scattered on stone, which does not germinate. These Beings, all light, are well received by us and often descend near our monasteries teaching us and revealing things lost for centuries during the cataclysms which have changed the aspect of the world.'"

This would be a most interesting event if it had happened. Unfortunately the diary of Albert d'Orville does not seem to exist. We have come to the conclusion that the case is likely to be a hoax, first mentioned by a man named Alberto Fenoglio in a 1966 magazine entitled *Non è Magia*. It is quoted as authentic by several contemporary ufological writers and is widely reproduced all over Internet sites.

1663, Montréal, Québec, Canada
The language of Heaven

This case is based on a relation of what occurred in the Mission of the Fathers of the Society of Jesus in the country of New France, from the summer of 1662 to the summer of 1663:

Heaven and Earth have spoken to us many times during the past year, and that in a language both kind and mysterious, which threw us at the same time into fear and admiration. The Heavens began with Phenomena of great beauty, and the Earth followed with violent upheavals, which made it very evident to us that these mute and brilliant aërial voices were not, after all, mere empty words, since they presaged convulsions that were to make us shudder while making the Earth tremble.

As early as last autumn we saw fiery Serpents, intertwined in the form of the Caduceus, flying through mid-air, borne on wings of flame. Over Québec we beheld a great Ball of fire, which illumined the night almost with the splendor of day—had not our pleasure in beholding it been mingled with fear, caused by its emission of sparks in all directions. This same Meteor appeared over Montréal, but seemed to issue from the Moon's bosom, with a noise like that of Cannon or Thunder; and, after traveling three leagues in the air, it finally vanished behind the great mountain whose name that Island bears.

But what seemed to us most extraordinary was the appearance of three Suns. Toward eight o'clock in the morning, on a beautiful day last Winter, a light and almost imperceptible mist arose from our great river, and, when struck by the Sun's first rays, became transparent,—retaining, however, sufficient substance to bear the two Images cast upon it by that Luminary. These three Suns were almost in a straight line, apparently several 'toises' distant from one another, the real one in the middle, and the others, one on each side. All three were crowned by a Rainbow, the colors of which were not definitely fixed; it now appeared iris-hued, and now of a luminous white, as if an exceedingly strong light had been at a short distance underneath.

This spectacle was of almost two hours' duration upon its first appearance, on the seventh of January, 1663; while upon its second, on the 14th of the same month, it did not last so long, but only until the Rainbow hues gradually fading away, the two Suns at the sides also vanished, leaving the central one, as it were, victorious.

These are classic descriptions of what we recognize today as natural atmospheric phenomena.

Source: *The Jesuit relations and allied documents: travels and explorations of the Jesuit missionaries in New France*, 1610-1791: the original French, Latin, and Italian texts, with English translations and notes. Reuben Gold Thwaites, 1853-1913. (Cleveland: Burrows, 1899.)

19 January 1665, Québec, Canada
Fireballs, preceded by explosions

"About a quarter to six in the evening, there was heard to come from beneath the ground a report so loud as to be taken for a cannon-shot. This sound was heard by persons distant three and four leagues from one another; while our Savages, knowing that the cannon is not fired toward evening, except to give warning of the appearance of the Iroquois, left the woods where they were, and came all through the night to ask us why we had fired such a terrible cannon shot.

"About seven minutes after this report, there appeared over Québec a ball of fire which merely passed by, coming from the mountains toward the North and emitting so bright a light that houses two leagues from Québec was seen in broad day. In the course of the year there were seen several other similar fireballs, not only at Québec, but below Tadoussac, and on the way to Three Rivers."

Source: *The Jesuit relations and allied documents: travels and explorations of the Jesuit missionaries in New France*, 1610-1791: the original French, Latin, and Italian texts, with English translations and notes. Reuben Gold Thwaites, 1853-1913. (Cleveland: Burrows, 1899.)

31 March 1676, Florence, Tuscany, Italy: Fiery globe

An Italian magazine called *Alata Quaderni* (No. 1, Feb 1979), mentions an ancient source which says "there

Fig. 54: Phenomenon in Tuscany

appeared in the Tuscan sky a luminous thing in the shape of a disc or bag of grain or sheaf." However, an authentic document describes it as a globe.

The object was undoubtedly a meteor, seen over much of Italy. It is mentioned by Father Louis Cotte in his *Traité de Météorologie* (Paris, 1784, p. 83) as "a luminous globe that crossed the Adriatic Sea," and was seen all the way from Livorno to Corsica.

Source of the illustration: *Notizie Diverse di Firenze*—anno 1676, 406-407, found by a researcher in the Magonia Internet group. The original belongs to the Marquis Alessandro Loteringhi della Stufa, Calcione castle in Arezzo (Italy).

1685, Hatfield, Yorkshire, England: Fearful unknowns

Objects turning in the sky, frightening men and cattle. One destroyed some trees, fell into a river.

This phenomenon, like the following in Rutland, may relate to a weather anomaly, such as a mini-tornado, but its duration and behavior do place it within the literature of unusual aerial phenomena.

Source: *Philosophical Transactions* 281 (1702):1248, and 284 (1702): 1331.

24 March 1718, Island of Lethy, India
Globe of fire with residue

A globe of fire appeared to drop a load of gelatinous substance. The strange object came out of the sky and touched the ground on Lethy Island in the East Indies. Witnesses who approached the site found a "jelly-like mass, silvery and scaly."

Until the nature of "shooting stars" was understood in the 19th century, people thought that any blobs of jelly-like material found in their fields were related to meteors.

Source: *A Catalogue of Meteorites and Fireballs, from A.D. 2 to A.D. 1860*, compiled by R. P. Greg, Esq., F. G. S, for the British Association for the Advancement of Science in 1860.

19 October 1726, Ath and Liège, Belgium
Circles of light

Large glowing clouds collide: fireballs and circles of light: "About 7:30 at night that Saturday, large clouds of fire arose from our horizon, pushed by the southern wind towards the north. They seemed to collide in an extraordinary manner, yet soundlessly. The sky, although calm, was all afire. Among these volumes of fire one could notice large luminous circles, open at the bottom, pushing one another like the waters of an agitated sea. About 11 o'clock these phenomena passed over the city, continuing long into the night, creating panic and wonder everywhere."

This case and the next one are part of a local "wave" of terrifying observations probably caused by a spectacular auroral display over Northern Europe.

Source: Gilles Joseph de Boussu, *L'histoire de la ville de Ath* (Mons, 1750).

19 October 1726, Echilleuses, France
"Pyramids" in the sky and unexplained rays of color

Pyramids in formation in the sky, with red and blue rays, seen in Echilleuses and at Villefranche du Rouergue.

The incident was recorded by a witness as follows: "About 8 hours of the evening an extraordinary sign in the stars; it looked like sorts of pyramids that made spears sometimes red, sometimes blue in color, and seemed to move in ranks like an army in the air." The spectacle, probably caused by the same aurora noted above, seemed so horrible that parishes in the region rang their bells. It lasted until 11 P.M.

Source: Departmental archives of Loiret, courtesy of Merrs. Franck Marie and Pierre-Valéry Archassal.

1743, Holyhead near Peibio, Anglesey, Wales
Ships in the sky

Mr. Morris, an experienced mining engineer, master of many languages and eminent antiquarian, had a report from Anglesey. This was made by a farmer named William John Lewis whose steading lay near Peibio, a little place only a stone's throw from Holyhead.

"Plowing" (as it was written) "with his servant boy in ye fields", he saw bearing down upon him a ship of 90 tons, rigged like a ketch, with its fore-tack at the cat-head and its pennant and antient flying. The day was described as indifferent and cloudy, but the detail of the ship could be clearly seen. It was "coming from ye mountains of Snowdon", not by sailing on the waves around Holy Island, but moving "about a Quarter of a mile High from ye Ground".

The farmer called his wife. She ran from the farmhouse in time to see the ship in the sky retreating, its pennant lowered to the deck and all sails furled. It was steering stern foremost, making for whence it had come, the mountains of Snowdonia.

Mr. Morris hastened to Holyhead and interviewed first the wife and then the husband, separately. Neither had any doubt about the circumstances. The wife had not acquaintance with sea terms, but was quite sure of what she had seen; her only doubt was what the neighbours might think if she allowed Mr. Morris to publish the affair. He found the husband at an inn, visiting Holyhead on farm business. He had no doubt that the man was sober and sincere, with no trace of the "melancolick" disposition that might have led him to exaggerate or imagine.

The ship had been plain to see, exact in every detail; the keel could be observed from below; the sails were distended with the wind; when the foresail was lowered it hung in a natural way over bow. In the end a cloud hid the vessel from sight, but not before the farmer, his wife, and his boy had had their observation supported by a flock of birds that assembled to examine the phenomenon and flew round it from all directions. When the vessel began its backward journey, the birds with one accord flew from it northwards in the opposite direction.

What finally persuaded Mr. Morris was the way in which the farmer – William John Lewis – assured him that he had seen another such ship exactly ten years earlier in much the same place, and that, ten years before then again, he had seen just such another. The ships were in each case very like the old packet-boats that plied between Holyhead and Ireland; the very ropes of the rigging could be counted one by one.

He concluded: "Since the hill at Holyhead is the only height in Anglesey to face the distant loftiness of Snowdon, some trick of refraction may have been responsible for picking up vessels plying the Menai Straits and setting them, pennant and antient and all, to steer the skies above Peibio."

Source: Wynford Vaughan-Thomas and Alun Llewellyn, *The Shell Guide to Wales* (Michael Joseph Ltd, 1969). The account is found under 'Holyhead' in the Gazetteer section of the book.

31 August 1743, Castel Nuovo, Italy
A light follows Casanova

Casanova saw a "pyramid-shaped flame" 50 cm high, 1.2 m above ground, 3 meters away. It followed him all day:

"An hour after I had left Castel Nuovo, the atmosphere being calm and the sky clear, I perceived on my right, and within ten paces of me, a pyramidal flame about two feet long and four or five feet above the ground. This apparition surprised me, because it seemed to accompany me. Anxious to examine it, I endeavoured to get nearer to it, but the more I advanced towards if, the further it went from me. It would stop when I stood still, and when the road along which I was travelling happened to be lined with trees, I no longer saw it, but it was sure to reappear as soon as I reached a portion of the road without trees. I several times retraced my steps purposely, but, every time I did so, the flame disappeared, and would not shew itself again until I proceeded towards Rome. This extraordinary beacon left me when daylight chased darkness from the sky."

Casanova's reaction is interesting. First he dismisses the event as a skeptical scientist: "What a splendid field for ignorant superstition, if there had been any witnesses to that phenomenon, and if I had chanced to make a great name in Rome! History is full of such trifles, and the world is full of people who attach great importance to them in spite of the so-called light of science." But then he adds, more humbly:

"I must candidly confess that, although somewhat versed in physics, the sight of that small meteor gave me singular

Fig. 55: Giacomo Casanova

ideas." And he concludes with the same words as so many witnesses of unusual phenomena: "I was prudent enough not to mention the circumstances to anyone."

Source: Giacomo Casanova, *The Memoirs of Jacques Casanova de Seingalt*, trans. Arthur Machen (New York: Putnam's, 1959), Vol. I, 222.

15 September 1749, Rutland, England
Watering intruder

An object created a sprout that roared, took water from a river, shot light beams to the ground, and broke rocks. Although this case sounds similar to that of Hartfield in Yorkshire, the two locations are separated by a fair distance.

The weather was calm, warm and cloudy with some showers. The witnesses described "great smoke with the likeness of fire" either as a single flash or as multiple arrows darting down to the ground, whose "whirling, breaks, roar and smoke frightened both Man and Beast."

The phenomenon went down the hill, took up water from the river Welland, and ran over fields and trees, tearing branches. The Royal Society correspondent reports: "I saw it pass from Pilton over Lyndon lordship, like a black smoky Cloud with bright Breaks; an odd whitling Motion, and a roaring Noise, like a distant Wind, or a great Flock of Sheep galloping along on hard Ground…"

Source: "An Account of an extraordinary Meteor, which resembled a Water-Spout, communicated to the President, by Tho. Barker, esq." Read on Dec. 14, 1749. *Philosophical Transactions* (Nov-Dec. 1749), no. 493.

September 1768, near Leipzig, Germany
Goethe's unknown lights

On the way to the University at Leipzig, 16-year old Goethe and two companions see a bright "tube" at ground level with blinding small lights jumping around. The trip was difficult, under steady rain. The travelers had to get out of the carriage to help the horses in steep slopes. During one of these walking sections, Goethe noticed something unusual:

Fig. 56: Goethe

"Suddenly, in a ravine on the right side of the road, I beheld a sort of amphitheater, marvellously illuminated. Within a space shaped like a pipe an incalculable number of small lights were shining, stacked like steps one on top of the other. They were so bright that the eye was blinded. But what was the most troubling in this sight was that the lights were not fixed, they jumped this way and that, going up and down and in all directions. Most of them, however, remained stable and radiated."

"It is with the greatest reluctance that I consented, when I was called, to move away from this spectacle that I would have desired to examine closer. The postillon, when I interrogated him, stated that he had never known of such a phenomenon, but in the neighborhood there was an old quarry, the hole of which was filled with water. It remains to be known whether it was a pandemonium of elementals or an assembly of luminous creatures, I would be unable to decide."

While the great writer and philosopher is more likely to have observed a display of spontaneously burning methane (marsh gas) than the dance of the Fairies, his observation is interesting.

Source: Goethe's autobiography, 6th book. As published in *The Autobiography of Goethe. Truth and Poetry: from my own life.* Translated by J. Oxenford. Vol. 1 (London, 1867), 203.

18 August 1783, Greenwich, England
Ten Balls of Light

"At 11 minutes after nine in the evening, a very singular phenomenon was seen at Greenwich. It being rather dark, of a sudden an uncommon light appeared, without any cause visible, for full two minutes; this phenomenon, coming from the N.N.W. perfectly horizontal in its course, and without any vibration, continued to the S.S.E. It passed over Greenwich, and near the Royal Observatory, till the elevated trees in the park took it from the sight. *Though it was transitory, the motion was not rapid, for you could distinctly discover its form, colour, &c. Its duration was near two minutes, during which there was no variation in its lustre*. Its magnitude and animated effect, made it appear near our earth. Two bright balls parallel to each other, led the way, the apparent diameter of which appeared to be about two feet, and were followed by an expulsion of eight others, not elliptical, seeming gradually to mutilate, for the last was small.

Fig. 57: The Greenwich train of meteors

"Between each ball, a luminous serrated body extended, and at the last a blaze issued, and terminated in a point. Minute particles dilated from the whole. While this luminary was passing, the atmosphere was exceedingly bright; but immediately after it became dark, though the moon was up.

"The balls were partially bright, as imagination can suggest; the intermediate spaces, not so exquisite in their colourings. The balls were tinted first by a pure bright light, then followed a tender yellow, mixed with azure, red, green, &c. which, with a coalition of bolder tints, and a reflection from the other balls, gave the most beautiful rotundity and variation of colours, that the human eye could be charmed with.

"The sudden illumination of the atmosphere, the form, and singular transition of this bright luminary, rendered much to make it awful; nevertheless the amazing vivid appearance of the different balls, and other rich, connective parts, not very easy to delineate, gave an effect equal to the rainbow, in the full zenith of its glory. It appeared also almost all over the island of Great Britain nearly at the same time, as well as in France, Flanders, &c."

Although this event is often cited in UFO compilations, this was undoubtedly a meteor, first seen over the Shetland isles with the apparent size of 1/3 the moon, equal to twice the full moon over Kent. It seemed to burst into two over Lincolnshire. When it passed over Windsor it was about 60 miles up, traveling 20 miles a second. It was heard to explode over York some minutes later. The phenomenon was observed from as far away as Ireland and Burgundy.

Source: "Singular Phenomenon," *The Annual Register* (Aug. 18, 1783): 214.

17 July 1790, Alençon, France
Crashed UFO, the pilot escapes!

This is another account in the long series of "crashed UFO with occupant" stories. At 5:00 in the morning, several farmers saw a huge globe in the sky, surrounded by flames. They first took it to be a balloon that had caught fire, but its speed and the strange whistling sound coming from it led them to think otherwise. The globe descended slowly, touching the top of a hill, where it tore up the plants along the slope. The flames from the object set fire to the small trees and the grass. Fortunately, the locals managed to stop the fire from spreading.

In his report on the incident, police inspector Liabeuf wrote that the sphere was still hot in the evening. It showed no signs of damage despite the heat. "It stirred up so much curiosity that people came from all directions to see it."

After some time, a much unexpected thing happened. A door burst open in the sphere and a human came out!

"This person was dressed in a very strange fashion. He wore a suit which clung to his body, and when he saw all this crowd he said a few words which could not be understood, and ran to take flight in the woods."

The peasants drew back from the sphere instinctively – which was fortunate for them, because the object exploded, throwing pieces everywhere. A search was undertaken to find the mysterious visitor but he was never discovered.

The Alençon incident has been included in many anthologies of UFO reports, dozens of books, and has become one of the best-known "folkloric" cases in the field. The reader may feel a little disappointed, therefore, though perhaps not very surprised, to discover that the event never really occurred.

The earliest reference to this case comes from an article published by Italian author Alberto Fenoglio, whom we've already met in connection with the supposed ufological deeds of Alexander the Great. A writer known to have invented some UFO reports in his time, Fenoglio seems to have created the story about Inspector Liabeuf for a purportedly serious article about sightings in ancient history, published in the Italian magazine *Clypeus*. This article was widely distributed and translated into several languages. The truth of the matter finally came to light in 1975 when Italian researcher Edoardo Russo conducted an investigation into Fenoglio's claims. In spite of this, books and magazine articles presenting the story of the Alençon 'crash' as a genuine case continue to be published in good faith every year in many countries.

Three or four historical cases may have inspired Fenoglio to compose a story dated June 17th, 1790. For instance, on

July 24th, 1790, an incident occurred in the municipality of La Grange de Juillac, France, involving several black "stones from heaven" that fell with a hissing noise before hundreds of witnesses.

On April 26th, 1803, at 1:00 P.M., a fireball was seen over Caen, Pont-Audemer and near Alençon. Up to 3,000 stones are said to have fallen amid detonations, one of which weighed 17 lbs. (*Astronomie populaire*, Paris, 1840. Tome IV, 225.)

More famously, precisely a year before the date given by Fenoglio, a fireball witnessed near the city of Worms, in the Rhineland, led to the writing of a controversial book. The canon of Trier, Worms and Spires cathedrals, Johann Friedrich Hugo von Dalberg (1760-1812) saw a meteorite from his family's country house and was told by neighbours that it had crashed nearby.

Dalberg went on to write *Über Meteor-Cultus der Alten, vorzüglich in Bezug auf Steine,die vom Himmel gefallen* (On the Meteor Cult of the Ancients, Especially with Regard to Stones Fallen from the Sky), published in 1811, a book suggesting that meteorites originated in space, where they defied gravity and waited for an opportunity to drop. "These Air-stones have from the start an inner, electrical life," he wrote, "and can consequently stay floating, so long as they are surrounded by the neutral-electric ether...As blazing spheres, sometimes exploding in the upper air, sometimes on their descent, they plunge down towards the heavenly body into whose spherical electrical atmosphere they are drawn." Did Fenoglio envision one of these plunging down at Alençon?

March 1796: Don region, Russia
The Devil and the brawling Cossack

According to writer Peter Kolosimo, the inhabitants of a Russian village in the Don region were surprised to find a large metal ball in one of their fields. The ball measured ten feet in diameter. People from everywhere flocked to see it, wondering where it had come from. Clearly it had not been delivered by road, as there were no wheel tracks to be seen anywhere in the vicinity. It could only have fallen from the sky, they thought. Except for a regular pattern of circles etched into its surface, the ball was as smooth as marble.

The village folk tried to move it but their effort was useless: it would not budge an inch. Then a man named Pushkin arrived. Pushkin was a drunkard and a gambler, even a heretic, and everyone looked down on his ways. But despite his faults, he was also known to be very courageous. They led him to the spot: “He drew his saber, spurred his horse toward it, he cursed it and defied it,” the legend says. “Whether it came from heaven or hell he challenged it to fight back.”

The man struck the object with his sword again and again. Suddenly the crowd around him began to howl with terror: one of the circles on the ball had opened up, revealing a single inhuman eye!

Pushkin sneered and carried on with his blows against the object. He struck it so hard, in fact, that the blade of his saber snapped off.

The peasants fled in fear. When they looked behind them they saw the drunkard and his steed were suddenly becoming transparent, fading into the air like ghosts. They could still faintly hear Pushkin’s voice, cussing angrily, but even this quickly faded away. “The villagers were not unduly perturbed by this," it is said. "The devil had gotten his own back with the brawling Cossack."

Two days passed: nothing was seen or heard of Pushkin. Then to everyone's surprise both he and his trusty horse staggered back into the village as if half asleep. He seemed calm enough, but he soon flew into a rage and began to howl that he was going to put an end to the unholy globe and set fire to it and the woods and everything around it.

Hearing this, everybody in the village trailed along after him to watch the spectacle, but he never could take his revenge on the mysterious metal ball, for "all that was there to be seen was his sorry mortification. The ball was no longer there."

Unlike the case of the crash at Alençon, we have been unable to prove that this tale is a modern hoax. However, not one Russian specialist we have approached had ever heard of the story, and the general consensus is that it originated as a fictional tale.

Early 19th century, Penrhynisaf, North Wales
Three hours' missing time

The Rev. R. Jones's mother, when a young unmarried woman, is said to have started one evening towards her home, accompanied by a servant man, David Williams, called on account of his great strength and stature, Dafydd Fawr, Big David, who was carrying a flitch of bacon. The night was dark, but calm. Williams walked in the rear of his young mistress, and she, thinking he was following, went straight home. But three hours passed before David appeared.

Interrogated as to the cause of his delay, he said he had only been about three minutes behind her. Told that she had arrived three hours ahead, David would not believe it. At length, he was convinced that he was wrong in his timing, and he proceeded to account for his lagging behind:

He had observed, he said, a brilliant meteor passing through the air, followed by a ring or hoop of fire, and within this hoop stood a man and woman of small size, handsomely dressed. With one arm they embraced each other, and with the other they took hold of the hoop, and their feet rested on the concave surface of the ring. When the hoop reached the earth they jumped out of it, and proceeded to make a circle on the ground. As soon as this was done, a large number of men and women appeared, and to the sweetest music that ear ever heard commenced dancing round and round the circle. The sight was so entrancing that the man stayed, as he thought, a few minutes to witness the scene. The ground all around was lit up by a subdued light, and he observed every movement of these beings.

By and by the meteor which had at first attracted his attention appeared again, and then the fiery hoop came to view, and when it reached the spot where the dancing was, the lady and gentleman who had arrived in it jumped into the hoop, and disappeared in the same manner in which they had reached the place. Immediately after their departure the Fairies vanished from sight. The man found himself alone in darkness, and he proceeded homewards.

Unfortunately, we have found no original document to authenticate the circumstances of the story, or even the year of the event, so it has to remain as an interesting fable.

Source: Elias Owen, *Welsh Folk-Lore, A Collection of the Folk-Tales and Legends of North Wales*, (1896 edition). Facsimile reprint by Llanerch Publishers, Felinfach, Wales 1996, 93-4.

22 February 1803, Hara-Yadori, near Tokyo, Japan
Female visitor

A saucer-shaped “ship” of iron and glass floated ashore. It was 6 meters wide and carried a young woman with very white skin. The episode began when a group of fishermen and villagers saw a ‘boat’ just off the shore of Hara-yadori in the territory of Ogasawara etchuu-no-kami. **(1)**

People approached the object in their own small boats and managed to tow it to the beach. The object was round. The upper half was composed of glass-fitted windows with lattice, shielded by a kind of putty, and the lower hemisphere consisted of metal plates. Through the glass dome the witnesses could see letters written in an unknown language and a bottle containing a liquid, perhaps water.

Fig. 58: Japanese object and occupant

The villagers arrested the girl and tried to decide what to do with her. One of the villagers, who had heard of a similar case that had happened at another beach not far from there, suggested that the woman was possibly a foreign princess, exiled by her father because of an extramarital love affair. The box, he said, may even contain her lover's head. If this was so, it would be a political problem, and that would imply some sort of cost: "We may be ordered to spend a lot of money to investigate this woman and boat. Since there is a precedent for casting this kind of boat back out to sea, we had better put her inside the boat and send it away. From a humanitarian viewpoint, this treatment is cruel for her. However, this treatment would be her destiny." Backing their decision with such straightforward logic, they forced the visitor back into the domed object, pushed it out, and it drifted out of sight.

This is not the only version of the story but it is probably the earliest. It comes from the Japanese *Toen-Shosetsu*, a compilation of stories written in 1825 by various authors, including Bakin Takizawa, a Japanese novelist. There was even a reproduction of a sketch of the object, showing something like a typical round, domed flying saucer.

A second version of the story was published in 1844 in a book called *Ume no Chiri*, written by Nagahashi Matajirou. This version said the incident took place on 24 March 1803. The beach was now named Haratono-hama. The girl was 1.5 meters tall and her dress was strange, made of an unknown material. Her skin was white as snow. She spoke to the astonished crowd in a language they were at a loss to interpret. She also had a strange cup of a design unknown to the witnesses.

Was there a precedent for a tale of this kind in Japan? Kazuo Tanaka explains that the report seems to be based on a variety of Japanese folklore known as *Utsuro-fune* or *Utsubo-fune*, a series of stories handed down over generations that preserved "the ancient national memory of Japanese immigration." In these tales, a founding member of a family, usually a noblewoman, would be said to have come across the sea by boat. If the tale was believable it could raise one's family to a higher social status. In the lyrics of one folkloric song from Kyushu Island we find references to "a daughter of a nobleman" who was sent to sea in a boat with glass windows. It even mentions that "the food in the boat was delicious cake." **(2)**

Could the whole report, then, have been a fiction based on much older hearsay? Tanaka draws this conclusion: no official document of the period mentions the incident of the woman in the round boat, and there are no references to beaches called Haratono-hama or Hara-yadori, which would be suspicious omissions if the story were true. On the other hand, the erudite ufologist Junji Numakawa has pointed out that the name of the beach could easily have changed over time. If the beach was originally named Kyochi-gama, as he postulates, this could be meaningful, as 'gama' or 'kama' means pot or cauldron, and the pot-like recipients used at the time were not unlike the craft in which the mysterious woman arrived. **(3)**

Sources: **(1)** Kazuo Tanaka *Did a Close Encounter of the Third Kind Occur on a Japanese Beach in 1803?* Skeptical Inquirer Volume 24, Number 4 July/August 2000. Masaru Mori, *The Female Alien in a Hollow Vessel*, Fortean Times No. 48, Spring 1987, 48-50 and *The UFO Criticism by J. N. from Japan*, Vol. 1, No. 1, January 2001. The latter is the English version of a privately published newsletter (UFO Hihyo) written and distributed by Tokyo-based researcher Junji Numakawa.

(2) This song was collected by the great Japanese folklorist Yanagida Kunio (1875-1962) and reproduced in a paper of his titled *The Story of Utsubo-fune* in 1925.

(3) *The UFO Criticism by J. N. from Japan*, Vol. 1, No. 1, January 2001.

15 December 1813, Connecticut, USA
First Congressional Hearings!

This episode took place during the American war against England. When unexplained "blue lights" were seen repeatedly over the harbor, they were interpreted as treacherous signals intended for the British fleet. The lights were "thrown up, like rockets, from Long Point," and distinctly seen, but never identified. "The gentleman from whom we receive this information plainly saw the lights, and states, that they were answered by three heavy guns from the ships of the enemy (England), at intervals of about ten minutes; that he was further informed, by an officer from Fort Trumbull, that the lights were continued during the whole night."

Considerable emotion was stirred up in the newspapers and in Congress when a letter from Commodore Decatur to the Secretary of the Navy, dated 20 Dec. 1813, confirmed the sightings: "These signals have been REPEATED, and have been seen by twenty persons at least in this squadron, there are men in New London who have the hardihood to

Fig. 59: Commodore Stephen Decatur

affect to disbelieve it, and the effrontery to avow their disbelief."

A heated debate followed at the House of Representatives:

"Mr. Law, of Conn., after some remarks, in too low a voice to be heard by the reporter, called the attention of the House to the story about certain blue lights which had been put in circulation, and had received countenance by the publication of an official letter from one of our naval officers to the head of a department. The motion Mr. Law was about to make, was not, he said, induced by any belief that the report was correct; for he could not believe that his native town contained in its bosom men so abandoned as to light those torches (...) He could not believe, under these circumstances, that these lights were exhibited as represented, but that some delusion must have existed on the subject.

It was proper, he conceived, that the fact should be enquired into, and placed on its proper footing, as it had been alluded to frequently in this House. With this view he offered the following resolution:

> *Resolved, That a committee be appointed to enquire whether any treasonable correspondence has been held, or information by means of blue lights or signals by fire given from the shore at or near the harbor of New-London in the state of Connecticut, to the blockading squadron off that harbor, whereby the enemy might learn the state, condition or movements of the American ships under the command of commodore Decatur now in that port; and that the committee be authorised to take evidence by deposition or otherwise as they may deem necessary, and report thereon to this House.*

Mr. Mosely supported the motion. Mr. Grundy was willing such an enquiry should be made, if the matter were referred to the naval committee. Mr. Fisk said he was sorry to hear a wish expressed by the gentleman from Tennessee for the proposed enquiry. When he looked at the principle of this motion, he trembled at the consequences of its adoption. What was the principle? It was nothing more than a proposition to exercise, through a committee of this House, the inquisitorial power to enquiry whether treason has been committed in a particular instance. He hoped no such precedent would receive the sanction of the House. Mr. Eppes and Mr. Jackson of Virginia opposed it. Mr. Roberts moved to lay it on the table. Mr. Calhoun thought it a matter too diminutive to engage the House."

The unidentified lights were never explained.

Source: *Proceedings of Congress, House of Representatives* Monday, January 24, 1814. Also Niles' *Weekly Register* (Baltimore, MD), Vol. 5:121 (December 25, 1813) and Vol. 6:133 (March 19, 1814).

Early spring 1820, Manchester, New York, USA
Two glorious entities

Joseph Smith, founder of the Mormon religion, writes: "It was on the morning of a beautiful, clear day...I kneeled down and began to offer up the desires of my heart to God...I saw a pillar of light exactly over my head...

"When the light rested upon me I saw two Personages, whose brightness and glory defy all description...One of them spake unto me, calling me by name and said, pointing to the other 'This is My Beloved Son. Hear Him!'...

"I asked the Personages who stood above me in the light, which of all the sects was right, (for at this time it had never entered into my heart that all were wrong) and which I should join.

"I was answered that I must join none of them, for they were all wrong...

"I soon found, however, that my telling the story had excited a great deal of prejudice against me among professors [believers] of religion, and was the cause of great persecution, which continued to increase; and though I was an obscure boy, only between fourteen and fifteen years of age...yet men of high standing would take notice sufficient to excite the public mind against me, and create a bitter persecution; and this was common among all the sects all united to persecute me."

Source: *The Pearl of Great Price*, by Joseph Smith – History 1:5-8, 14-19, 22.

21 September 1823, Palmyra, New York, USA
Golden apparition

In this next episode, Joseph Smith was shocked to see a light appear in his room, and a human figure within the light: "Indeed the first sight was as though the house was filled with consuming fire. The appearance produced a shock that affected the whole body. In a moment a personage stood before me surrounded with a glory yet greater than that by which I was already surrounded…"

The figure, whose feet did not touch the floor, revealed itself as "angel Moroni" and gave Smith specific instructions. The scene repeated itself three times during the night. After the third time Smith was surprised to hear the cock crow and to find that daylight was approaching, "so that our interviews must have occupied the whole of that night." The next day Smith found himself so exhausted that he couldn't work in any useful way in his normal chores. His father thought he was sick and told him to go home. On the way he fell to the ground when trying to cross a fence, and remained unconscious. The angel appeared to him once more and told him to reveal his instructions. Smith went on to found the Mormon religion.

By placing the case in this section on Myths, we do not mean to state that the story was invented or that no such event took place. Millions of people today do take the report at face value. Ufologists, on the other hand, might claim that it represents a typical "bedroom visitation" type of alien contact. We do believe that it is unrelated to the events we seek to study in our Chronology of unexplained aerial phenomena.

Source: Francis Kirkham, *Concerning the Origin of the Book of Mormon* (Salt Lake, 1937), and Smith's own account in publications of the Church of Latter Day Saints.

Epilogue

This section gives only a tiny sample of the hundreds of items we have extracted from the literature, for which we found evidence of natural explanations or strong indications of a mischievous or deluded author. A complete listing of such cases is impossible to contemplate, since it would have to include every comet, every meteor shower, every atmospheric illusion and every tornado ever mentioned in books or broadsheets down through the centuries. At the time, many of these phenomena were taken as omens of disaster or as manifestations of the divine realm.

In passing, one must note that in spite of such fanciful interpretations the reliability and accuracy of the observations was good enough for us, in the 21st century, to reconstruct the nature of phenomena that are known to today's science, but were a complete, often terrifying enigma to ancient witnesses.

Most importantly, this compilation of mythical or legendary material demonstrates the powerful impact of this imagery, not only in folklore (including contemporary folklore) but in spiritual beliefs and mainstream history. Religious tradition in every part of the world is replete with allusions to celestial phenomena that inspired chroniclers to invoke moral principles and warnings to humanity in writing that has survived through the ages.

This work also shows that the dominant narratives in today's literature on "extraterrestrial" encounters – complete with saucer crashes, strange writing and abduction by non-human entities – were already present in widely-reported stories that predate the industrial revolution.

PART III

Sources and Methods

Anyone attempting to review the historical and social impact of unexplained aerial phenomena immediately faces two difficult challenges: (1) where to find reliable information that can be further investigated and verified? And (2) how to select suitable material for presentation without biasing the reader towards pre-established conclusions?

A fact-driven study

It is natural to begin with available sources in literature and on the Internet. Until now, mainstream believers in extraterrestrial visitors have actively discouraged such research, because it seemed obvious to them that the phenomenon was of recent vintage. Thus Budd Hopkins, a contemporary authority on alien abductions, has sharply criticized one such compilation as "an odd confluence of UFO case studies, free-wheeling speculation, and folklore *of obviously uncertain authenticity* (our emphasis)."

Indeed, most UFO books begin with the blunt affirmation that the "flying saucer era" started on 24 June 1947 when Kenneth Arnold spotted several objects apparently flying in formation over Mount Rainier, Washington, implying that any cases before that date are irrelevant. Many specialists, such as Jerome Clark, see no indication that the phenomenon existed before the mid-19th century.

This attitude is driven by ideology rather than data: If the UFO phenomenon did start in the summer of 1947 with Arnold and (shortly thereafter) the infamous Roswell crash, then one is justified to claim that it originates with visitors from space who have spotted our atomic explosions and decided to come to Earth to investigate – and perhaps save us from ourselves: A seductive view, but one that is contradicted by the mass of previous cases. While it is true that the amount of available data went through a sharp rise about 200 years ago, we have seen that this was due more to progress in the publishing and dissemination of news around the world than to a dramatic increase in the actual frequency of incidents.

At the other end of the spectrum are the devotees of the Ancient Astronaut theory, who claim that contact with extraterrestrials was established very early in the history of the human race, or even, as the Raëlian cult argues, that we are the product of E.T. experimentation or inter-breeding. As we have noted, they find support for their view in many religious traditions and, indeed, in ancient writing and the Bible itself, which alludes to sexual intercourse between the Nephilim (gods from Heaven) and the daughters of men.

When we began this project we took a different approach: one that is strictly fact-driven, rather than belief-driven. Leaving ideologies aside, we were striving to compile a list of documented sightings, with as little reference as possible to a particular theory – although naturally the beliefs of the witnesses and those around them had to be noted as a factor in the way the story was transmitted to us.

One of us (JV) had long collected items from the literature and from folklore in an effort to find out whether the phenomenon of unidentified aerial objects had an identifiable "start date" in history or followed recognizable patterns in time. Beginning in the mid-sixties, he compiled and published computer catalogs of reports culled from books and newspapers around the world.

In a similar vein, Chris Aubeck began to re-examine the totality of the available literature, taking advantage of the new search capabilities of the Internet to leverage the information available in books. As an online collaboration group, the Magoniax Project he initiated in 2003 with fellow researcher Rod Brock was thus able to track down journals and obscure sources in several languages to assemble the largest collection of such stories in the world today.

The two authors began collaborating through web-based software to merge their files and catalogs of that period. As a result, we were often able to go back to original sources rather than citing popular trade books or contemporary compilations.

Classical Sources

Contrary to common opinion, we are not dealing here with "folklore of obviously uncertain authenticity," as abductionist Budd Hopkins once asserted. Early sources are plentiful, sometimes officially certified and verified, detailed and quite distinguished, even in the remote classical era. An early researcher, Raymond W. Drake, remarked (in *Flying Saucers* No. 39, December, 1964):

"The Romans worshipped the Gods for a thousand years; their augurs prophesied the future from signs in the skies. Julius Obsequens recorded 63 celestial phenomena, Livy 30, Pliny 26, Dio Cassius 14, Cicero 9, confirming their psychological impact on the educated Roman mind. Lycosthenes writing in AD 1552 collated 59 heavenly portents in Roman times."

No less a political and classical authority than Cicero mentions the topic in his writing (*De Re Publica* 1.19.31), where a character named Laelius scolds the young Tubero for his fascination with a celestial phenomenon – a vision of a double Sun in the sky – reported to the Roman Senate. Another Roman philosopher, Seneca, on the other hand (in

Questiones Naturales 7.1.1) sides with the young man because man's imagination gets dulled by the endless repetition of ordinary phenomena. It takes an exceptional sighting, a "sweet spectacle," to bring back our feeling of wonderment before the beauties of nature, he argued.

There appeared in the Elemente diuers tokens and ſtrange Comets.
Theodoric king of the the *Oſtrogotes* iſſuing out of *Meſſea* and going into *Italy*, met twice with Ordoaces King of *Italy* and put him to flight, and cloſing him in at *Papia*, did there beſiege him thrée yeares. When Eoricus king of the *Vigotes* was dead, Alaricus hys ſonne did ſuccéede him in the kingdome.

In *Minias* a riuer in *Spaine* fiſhes were taken, in whoſe ſcales the coyne of that yeare was as it had bin engrauen, the ſame yere Feltheus and Feua kings of the *Rugians* went about to thruſt the *Herules* out of the Country, whom Odoaſer vanquiſhed and ſlewe, and almoſt defaced the nation of the *Rugians*. — Minies. 499

When Deuterus a Biſhop of the *Arians* at *Conſtantinople* Baptiſed one Barba, & diſtinguiſhing the trinitie amiſſe ſaid. I Baptiſe thee Barba in the name of the Father through the ſonne in the holy Ghoſt (*Baptizo te Barba in nomine patris per filium in ſpiritu ſancto*) the water vaniſhed away. — 504

Moreouer in thoſe dayes, there appeared in *Affrica* Gods reuengement vpon an *Arian* called Olympius, who whilſt he waſhed his bodye in bathing water, belching out certaine vnworthye and blaſphemous words, touching the holy beliefe and the trinitie, there came ſodainely a fierie Darte — 506

Fig. 60: A fragment from the English translation of Lycosthenes, whose book bore the full title "*The Doome warning all Men to the Iudgemente wherein are contayned for the most parte all the Straunge Prodigies hapned in the Worlde, with diuers Secrete Figures of Reuelations tending to mannes stayed conuersion towardes God: in maner of a generall Chronicle, gathered out of sundrie approued authors by St. Batman professor in diuinite*, by Konrad Lykosthenes, 1518-1561." (Imprinted by Ralphe Nubery assigned by Henry Bynneman. Cum priuilegio Regal, London 1581).

In many historical periods the phenomenon was taken very seriously indeed. It should be remembered that it was the Roman custom to report every year to the Consuls anything that could be interpreted as a portent, as the Consuls wanted to be aware of it in making their decisions. Unfortunately for us, the *Annales Maximi* that contained these "prodigies" is lost, but it is supposed that Livy, Pliny and Obsequens had access to these annals and drew from them.

Among other classical sources was Boece (Boetius or Boethius), ca. 475–525 AD, Roman philosopher and statesman. There are several editions of Boece's work. An honored figure in the public life of Rome, where he was consul in 510 AD, he became the able minister of the Emperor Theodoric. Late in Theodoric's reign false charges of treason were brought against Boethius; after imprisonment in Pavia, he was sentenced without trial and put to death.

Fig. 61: Boetius

While in prison he wrote his greatest work, *De Consolatione Philosophiae* (*The Consolation of Philosophy*). His treatise on ancient music, *De musica,* was for a thousand years the unquestioned authority on music in the West. One of the last ancient Neoplatonists, Boethius translated some of the writings of Aristotle and made commentaries on them. His works served to transmit Greek philosophy to the early centuries of the middle Ages. Translations vary widely because of the unusual vocabulary used in the text. There are both prose and metrical versions and they differ in some details.

Another important source, John the Lydian (John Lydus), 490–ca. 565 AD was a bureaucrat in the praefecture in Constantinople and an antiquarian scholar. He wrote three treatises that preserve much information from earlier sources while responding to contemporary controversies. *On Offices* ('De magistratibus') is translated as Ioannes Lydus: *On Powers or The Magistracies of the Roman State* (Anastasius C. Bandy: Philadelphia, 1983). *On Months* and *On Portents* have not yet been translated into English.

Another author often quoted in our Chronology is Matthew of Paris (or "Matthew Paris"). He was an English Benedictine monk whose extensive and detailed chronicles of events in the 13th century form one of the most significant primary sources in medieval studies.

Although Paris wrote voluminously, very little of his works shed any light on his own life. We do know he was a monk at St. Albans and that he occasionally visited the royal courts. He spent most of his life at St. Albans, but he put his acquaintance with persons of import and his few trips outside the monastery to good use in acquiring news to include in his chronicles.

Fig. 62: Matthew Paris

In 1248 he went to Norway to reform the Benedictine Monastery of St. Benet Holm; on his journey he was entrusted with letters for King Haakon IV, with whom he formed a friendship. Paris was also personally acquainted with King Henry III of England and Richard, Earl of Cornwall.

Another influential chronicler is Grégoire de Tours, a sixth-century historian. Born in Clermont-Ferrand in 595, he went to Tours seeking a cure for an illness at the tomb of Saint Martin, and stayed in that city where he became a bishop. He left many treatises on history and astronomy, including a hagiography of Saint Julian and Saint Martin; a book about ecclesiastical cycles, and a tumultuous *History of the Franks* that earned him the title of first French historian.

Fig. 63: Grégoire de Tours and Salvius facing King Chilperic.

Notable among later authors we have consulted is John Howie (1735-1793) “chronicler and biographer, who lived on his ancestral farm of Lochgoin, in Renfrewshire, a noted place of refuge in Covenanting times. He early developed an interest in the Covenanters and Reformers, and went on to amass a wealth of material from manuscript and published sources he used as a basis for series of biographical sketches which he published in 1775 under the title of *Biographia Scoticana* or Scot Worthies,” (according to *Scottish Church Hist. & Theol.*, 414).

Screening

The Internet has introduced a revolution in our access to ancient texts. In particular, the network’s ability to link together groups of researchers interested in the same topics and willing to share their data has allowed us to take a giant step beyond the parsimonious and often erroneous databases available prior to this work.

Fig. 64: Three suns seen in 1492

Having assembled such a body of information, in itself a never-ending process, the challenge becomes one of validation and selection. In order to avoid creating the kind of misleading framework found in the literature, we cannot presuppose anything about the nature of the data we present. At the same time, we have to be faithful to the beliefs and statements of the participants: if they thought they were witnessing a divine manifestation or a contact with a creature from another world, we cannot censor that information, and indeed it is relevant to the way they color their testimony.

The primary phase of the selection process has to do with the elimination of what we now recognize as natural phenomena. Reliable knowledge about meteors and comets is of quite recent introduction: as late as 1803 the French Academy of Sciences didn't believe that stones could fall from the sky, and the movement of comets still baffles the average citizen today.

Reports of seemingly miraculous events, such as pillars of light in the sky or triple moons, are explained today as atmospheric effects but were understandably baffling to ancient writers. We should be grateful to them for preserving these items, even as they presented them in a supernatural context. Their contribution has augmented our ability to compute the orbit of comets by going back to sightings over the centuries. Similarly, the frequency of meteors, hence the structure and origin of our solar system, is better known because of such ancient records.

In his book entitled *L'Atmosphère*, Camille Flammarion gives numerous examples of stories based on meteorological observations misinterpreted as supernatural phenomena, and later correlated with political events. Flammarion cites such a compilation by a friend of his, Dr. Grellois, concerning "mystical meteorology."

In compiling the data for this book we have tracked down, read and ultimately rejected far more cases than we have kept. As we saw in Part II, many events listed in the contemporary literature of unusual aerial phenomena turn out to be meteors, comets, auroras or tornadoes reframed as "disks" or "craft". When medieval witnesses observed something burning in the sky they could only assume it was made of wood, hence the "flaming beam" over a German hillside in one classic illustration. Modern witnesses make similar assumptions when they jump to the conclusion that unidentified flying objects are necessarily spaceships from another world. Every century, every culture (including our own Western scientific culture) has its own myopia and peculiar obsessions.

For our own purposes, whenever we could not find compelling evidence to indicate the object was NOT a meteor, a comet or an atmospheric effect, we have generally excluded the case from our Chronology.

Rules for inclusion

Once such natural misidentifications are removed, one is left with a mix of stories that range from the factual description of puzzling phenomena (perhaps because we are missing a crucial piece of information) to extraordinary claims that are the stuff of legend, and are often embedded into religious belief systems. The problem then becomes one of setting consistent criteria. In the present book we have applied the following set of rules:

Rule 1: Credibility.

Cases that we found, to the best of our estimation, to be fictional or fraudulent were excluded from the main chronology, but some of them were kept as background reference, historical milestones or educational material in our section on "Myths, Legends, and the Chariots of the Gods" (Part II of the book).

Rule 2: Space and time.

Cases must have a specific place and time associated with them in order to be retained in the Chronology. Statements like "There were numerous reports of lights in the sky in tenth-century Asia" or "Hopi traditions allude to contact with space beings" are not helpful. They offer no historical correlation and are almost impossible to research in the context of our project. Legendary events cannot be assigned a date in a real chronology. No years can be given for the dynasties of probably fictitious kings. We expect to have at least a specific region or location and an approximate date.

We relaxed this rule somewhat for ancient cases and gradually tightened it as one got closer to the present century. We take pride in starting and ending the chronology at real, reliable dates.

Rule 3: Description of the phenomenon.

Cases must describe a specific phenomenon in sufficient detail so that common explanations (such as meteors, comets or illusions) can be recognized and excluded. The phenomenon should be linked to an aerial phenomenon or items closely related to contact with an aerial object or a non-human entity. Here again, we have relaxed these standards somewhat as we looked further back in time.

Rule 4: Witness identification.

Cases where witnesses are cited (or, even better, identified by name and function) are given greater weight than general statements about an event, especially when they make it possible to verify the existence and credibility of that particular person.

Hoaxes

We have attempted to detect and eliminate hoaxes from our chronology, but we do see such stories as important social indicators rather than spurious narratives: in order for a hoax to be credible to those who hear it, it must fit into the general belief system of the society that surrounds the author of the hoax. If we assume that actual stories of unusual observations are repressed in a given era (by Church authorities intent on fighting witchcraft, or by a "rationalist" régime determined to stamp out potentially subversive ideas) then it makes sense that we would only hear of the phenomena through the indirect channel of legends, fairy tales, and hoaxes.

The problem of false testimony becomes more complex when the authors of the hoax belong to a power system, such as a religious group or a political structure. Hoaxes then become tools for disinformation and for the shaping of society, using the credulity of common citizens to propagate a certain faith or to maintain existing structures.

Throughout history this device has been used for the convenience of emperors, kings, and Popes, and it is still being used today in disinformation and psychological warfare. For this reason we have made a special effort to track down the sources of the stories we have related, to the extent that the background could be researched.

"Explanations"

One of the secret pleasures and rewards of this work has been the study of the various "explanations" given by scholars of every era to dismiss the observations brought to them by common people.

The following figure is a case in point: on 7 March 1715, starting in the evening and lasting until 3 A.M., a strange mist arose over an English village. Inside this mist or cloud, the witnesses thought they saw the figure of a frightening giant holding a sword. This is related fully in a pamphlet entitled *A Full and True Relation of the Strange and Wonderful Apparitions, etc.*, which is kept in the British Museum.

Fortunately for rationalists everywhere, a certain expert named Doctor Flamstead was able to "explain" this phenomenon (and several following it) in terms of "the darkness of people's conscience," which "seeks to destroy Church and State."

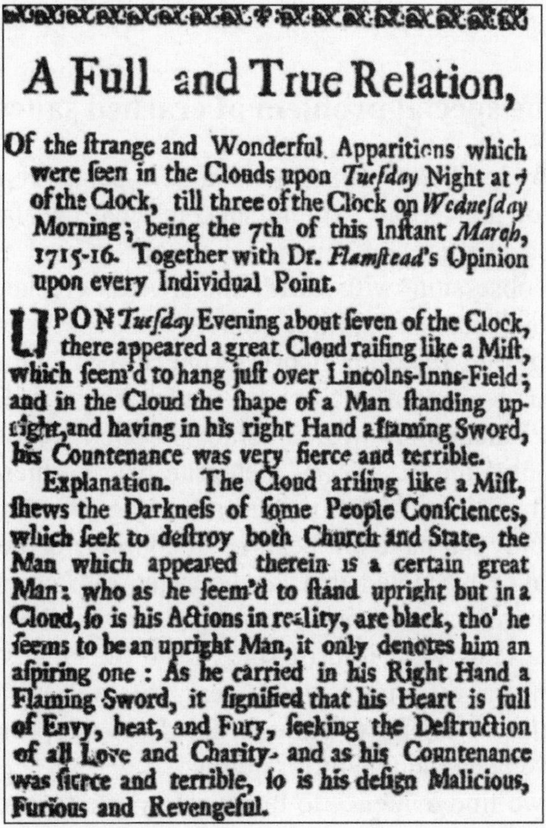

A Full and True Relation,

Of the ſtrange and Wonderful Apparitions which were ſeen in the Clouds upon *Tueſday* Night at 7 of the Clock, till three of the Clock on *Wedneſday* Morning; being the 7th of this Inſtant *March*, 1715-16. Together with Dr. *Flamſtead*'s Opinion upon every Individual Point.

UPON *Tueſday* Evening about ſeven of the Clock, there appeared a great Cloud raiſing like a Miſt, which ſeem'd to hang juſt over Lincolns-Inns-Field; and in the Cloud the ſhape of a Man ſtanding upright, and having in his right Hand a flaming Sword, his Countenance was very fierce and terrible.

Explanation. The Cloud ariſing like a Miſt, ſhews the Darkneſs of ſome People Conſciences, which ſeek to deſtroy both Church and State, the Man which appeared therein is a certain great Man: who as he ſeem'd to ſtand upright but in a Cloud, ſo is his Actions in reality, are black, tho' he ſeems to be an upright Man, it only denotes him an aſpiring one: As he carried in his Right Hand a Flaming Sword, it ſignified that his Heart is full of Envy, heat, and Fury, ſeeking the Deſtruction of all Love and Charity- and as his Countenance was fierce and terrible, ſo is his deſign Malicious, Furious and Revengeful.

Fig. 65: "Full and true relation…"

Thus the fantastic celestial apparition, instead of shaking up the existing state of knowledge, became interpreted – on the contrary – as a solemn reminder that the masses must stay in line, and always support the ruling class.

As for the man in the cloud, "his heart is full of envy, heart and fury, seeking the destruction of all love and charity."

In all periods, we are able to observe the stupidity and the arrogance of such self-styled "rationalist" scholars who seize upon the sense of wonder, terror or hope of their contemporaries to advance their own preconceived theories, and to reinforce the existing order.

The special problem of crashed saucers

Since 1947, when North American newspapers reported on dozens of mysterious "flying saucers" that had fallen into parks, backyards, and streams, there has existed an almost morbid obsession with dead aliens and wrecked space-craft. This, too, is a very old story.

During our research, Chris Aubeck has come across numerous legends of artificially made objects falling from the sky, including swords, shields, books, jewels, and statues, plus the occasional meteorite bearing hieroglyphic inscriptions. Stories of this kind are being catalogued for a future study but have not been retained in the present work.

We have also noted that a whole genre of stories about aerial travelers in trouble emerged in medieval times. Amusing tales were told of ships from the clouds that ran into technical difficulties over Great Britain, leaving behind such items as anchors. Though dated only approximately, they have been included for reference.

Until we find evidence to the contrary, we must conclude that tales involving actual UFO crashes (as we understand the term today) materialized as "factual reports" in mid to late 19th century newspapers, but the earliest crash report was described in French science fiction as early as 1775.

The special problem of "dragons"

The accounts most closely resembling UFO crashes within the scope of our chronology come from Chinese lore and describe the fall of "dragons." For example, we have mentioned the episode of 1169 AD, when dragons were seen battling in the sky during a thunderstorm and pearls like carriage wheels fell down on the ground, where they were found by herds' boys. These pearls would constitute physical proof that a phenomenon had occurred but unfortunately nothing more is said about them.

A similar situation occurred one night in the late Fourth Century AD when Lu Kwang, King of Liang, saw a black dragon in the sky: "Its glittering eyes illuminated the whole vicinity, so that the huge monster was visible till it was enveloped by clouds which gathered from all sides. The next morning traces of its scales were to be seen over a distance of five miles, but soon were wiped out by the heavy rains."

One of Kwang's attendants told him that the omen foretold "a man's rise to the position of a ruler," adding that he would no doubt attain such a rank. Lu Kwang rejoiced when he heard this, and did actually become a ruler some time afterwards. More than a century later, in 1295, two dragons fell into a lake at I Hing. This was followed by a strong wind which raised the level of the water "more than a chang," that is, some 10 feet. The fourteenth century chronicler of this incident, Cheu Mih, adds that he had personally seen the results of another 'dragonfall' himself. Seeing the scorched paddy fields of the *Peachgarden of the Ts'ing*, he interviewed one of the villagers about it. "Yesterday noon there was a big dragon that fell from the sky," he was told. "Immediately he was burned by terrestrial fire and flew away. For what the dragons fear is fire."

This raises the question of exactly what the Chinese of that era understood by the words we now translate as "dragons," obviously a term that covered a wide variety of aerial phenomena, rather than our simple contemporary image of a flying, fire-belching serpent with wings.

In cases when the circumstances surrounding the dragon are clearly stated (storms, destruction, lightning strikes, objects lifted into the sky) it seems that the terrified witnesses were observing tornadoes, with funnel clouds in the shape of giant serpents whipping around in the sky and causing widespread disaster.

Entities

Anomalies involving interaction with entities similar to those often associated with aerial phenomena, pose a special challenge. A shining being stepping outside a ball of light and addressing the witness is a valid entry in the chronology, but what about the shining being by itself, entering a room or meeting the witness, without any other aerial phenomenon reported? We excluded most of these cases from our list, keeping only instances where the interaction had a special relevance to the overall phenomenon.

This decision may be challenged by our readers. In defense, we were concerned that, the moment we added superhuman beings by themselves and suggested they communicated with humans (either to give warnings or advice or tools or instructions) every other case in our chronology would get tinged with a sense of deliberate purpose. There are anomalies and patterns here, but we should not lead the reader into believing that certain types of entities are necessarily behind the phenomena.

There is an extraordinary abundance of entity sightings (angels, demons, gods, and ghosts) in ancient chronicles. To distinguish between fictional and factual accounts now is impossible, and to use any and all would mean lumping aerial phenomena with crypto-zoological creatures willy-nilly. If we take folklore, mysticism, phantoms, fantasy, dreams, and omens as our source, entity-only sightings would easily outnumber sightings of aerial phenomena by a hundred to one, so an exhaustive catalogue containing both is not helpful.

No known criteria helps us sort "ufonauts" from other kinds of creatures (such as a mermaid, or a sea serpent) when no aerial phenomenon is present. We prefer to inform the reader that accounts involving supernatural entities were contemporaneous with aerial phenomena reports throughout history, pointing out that such stories do corroborate some aspects of the enigma as testified by modern witnesses, *but* that may imply a relationship that is beyond the scope of our compilation. Our purpose in this book is to explore an unknown phenomenon, manifesting throughout history, possibly misinterpreted by every culture in terms of its own history or religion. We suspect that the data we have compiled in our Chronology indicates the presence of a previously unknown physical element.

Biblical accounts

Religious texts such as the Bible contain many references to flying objects that are assumed to represent divine manifestations. For example, Zachariah relates that he saw such an object: "I turned, and lifted up mine eyes, and looked, and behold a flying roll. And he said unto me, 'What seest thou?' And I answered, 'I see a flying roll; the length thereof is twenty cubits, and the breadth thereof ten cubits'" (approximately 40 feet).

Descriptions of celestial chariots, visions of the Throne of God (Merkhaba), or the Shekinah generally cannot be related to specifically dated phenomena, and belong in a general analysis of religious, symbolic or mythical imagery.

Most Biblical references to UFO-like phenomena place them within a complex narrative in which divine entities intervene to assist a particular group of people for what can only be described as political and religious reasons. The difference between, say, Jane Lead's mystical experiences in the 17th century and those of the Bible is that Lead was shown spectacular things that she later interpreted in bursts of guided inspiration, whereas "celestial intervention" in the Bible had a dramatic, strategic effect. Biblical accounts show divine entities intimately working for and alongside whole communities, whereas Lead's experiences are personal and private, like those of contemporary abductees.

This means that while the physical phenomena described in the Bible resembled aerial phenomena from other historical periods, their function had a far greater impact, biased towards an ultimate goal affecting a larger number of people. This makes them stand apart from other accounts we read, whether we believe in the scriptures or not.

Given this background, the placement of the few biblical stories we quoted raised some important issues. The two authors have had many discussions and occasionally heated debates on this point. A case could be made to leave Ezekiel in the main chronology but to exclude other Biblical events. Accounts of pillars of fire and light, on the other hand, are suitable for the chronology because they don't imply any effort on the part of the phenomena themselves to become intimate with the witnesses, any more than the North Star to a traveler.

We dislike the idea of portraying aerial phenomena as having selectively aided one religious order or community above others. This has led the authors to debate what message the sightings conveyed to our readers: Is it wise, we asked ourselves, to transmit this message with its religious context when we wanted the book to be useful to a world of researchers working in different cultures? Yet the fact that certain communities, such as the Hebrews, the ancient Chinese or the followers of Clovis have interpreted unidentified aerial phenomena as divinely-ordained craft designed to help them cannot be ignored.

The contemporary belief among many ufologists that America is secretly aided by crashed saucer technology from Roswell represents a similar pattern in our own century. We can only note these beliefs and move on.

Fig. 66: The vision of Zacharias

The correlation between many unexplained sightings and religious or historical events brings up three important observations about potential biases in our data:

(1) Events that were received within a religious context were better preserved simply because witnesses, priests and monks generally could read and write. They had a tradition and techniques of preserving records. Furthermore, they thought the observation was important. (Similarly, UFOs seen over nuclear plants or missile silos are more likely to be watched and documented today.)

(2) If people attach spiritual significance to what they see, it affects their behavior and is invested with more lasting reality than witnessing a passing light in the sky.

(3) The fact that witnesses perceived transcendent images in the phenomena may be part of the mechanism of the phenomena.

Hence our argument that cases coinciding with religious dates or historical events are not necessarily the imaginative or fanciful product of obsolete belief systems. Unusual events are more likely to be recorded for posterity when they occur in important places, or on important dates, or to important people. If some kind of UFO reality is accepted (however simplistic) in such circumstances, a purely folkloric interpretation is not necessarily the best theory.

Aerial phenomena in classical art

In ancient times, up till a couple of centuries ago, religious art was the most common form of artistic expression. For centuries, painters created tapestries and pictures representing the Virgin, Christ, the Nativity, and scenes from the Old Testament. During those centuries comets and meteorites, triple suns and moons were also very commonly chronicled and taken seriously.

An excellent example of this problem arises in connection with the *Annunciation* of Carlo Crivelli displayed at the National Gallery of London because it seems to show a hovering disk-like object sending a precisely collimated beam of golden light to Mary, as she receives the message that she has been chosen to conceive the Son of God.

A modern critic named Cuoghi sees nothing unusual in this painting because "there is a vast amount of Annunciations in which a ray descends from the sky reach-

Fig. 67: Annunciation, detail

ing the Madonna. Furthermore, as far as the Crivelli painting is concerned, (…) the object in the sky is formed by a circle of clouds inside which there are two circles of small angels. It is a very common way of representing the divinity, visible in so many works of sacred art. The same particular in the Annunciation of Carlo Crivelli…."

One could point out that this argument actually brings water to the ufologist's mill: If the origin of the message to Mary is represented as a bizarre hovering disk full of celestial beings, doesn't that suggest that knowledgeable artists placed this event into the category of specific interaction between humans and intelligent forces influencing us from the sky?

This case opens an interesting discussion about the representation of unusual phenomena in art when the painting is not contemporary with the events depicted. In this case the "disk" corresponds to nothing in the biblical narrative, any more than other objects in the building such as the expensive drapes or the birds. We can only say that the story of God's selection of Mary as the mother of Christ evoked a connection in the artist's mind to a complex artifact hovering in the sky, which served as the source of a golden beam. While this connection is interesting, it tells us nothing new about Mary's actual experience.

Paintings do not offer valid evidence about the periods they represent. An image of the Virgin Mary with a disk-

Fig. 68: "Annunciation," by Carlo Crivelli

shaped object flying in the background, if painted centuries after the event, tells us nothing of the period in which Mary lived. However it does tell us how the event is being interpreted by the society surrounding the artist, which is valuable in itself and should be noted.

Fig. 69: Dialogue about Flight

This 1723 work by Pier Jacopo Martello, entitled *Del Volo Dialogo: Mattina Prima* (Bologna: Lelio dalla Volpe) is the first scientific poem, along with Antonio Conti's *Globo di Venere*, written in the eighteenth century.

We decided our best solution was to recount the history of ancient art in ufology, point out its pros and cons, and give a few examples either way. Omens have been seen in the sky for millennia, and interpreted as divine warnings, so it is not surprising to see them reflected in ancient art. This does not mean that any example of it represents an actual sighting. Sometimes the resemblance to phenomena reported today is stunning, but in the case of UFOs, as in the case of virtually any human preoccupation, art reveals more about the painter, his (or her) patron and the audience than about the subject itself.

The next engraving is a case in point. It represents an old man with a book ("Democritus ridet") at his feet. It has extraordinary importance for it foreshadows modern aeronautics with amazing insight. The old man points to two ships floating in the air. The first ship is merely a wooden boat but the second one, of more interest, represents a bird-like structure upside down with feather-wings and a small awning above. From tail to head stretches a sail; the tail acts as rudder. A figure stands inside watching another one falling through the air. On the ground lies the ruin of another ship while behind stretches an undulating landscape with bridge and tower and rows of poplars.

For further research ...

It is vital to recognize the magnitude of the progress that the opening up of archives online has made possible in this research. This fact also contributes to distorting the statistics about the data: since pre-1800 texts cannot easily be read by optical character recognition scanning software, the amount of information available to us after 1800 has become mountainous in comparison to older records. Those researchers who believe that the UFO phenomenon started manifesting on Earth around that time are simply misled by the fact that our information sources are far more abundant and more readily available after the eighteenth century.

Pre-1800 often means important but damaged records, unreadable or unpredictable script, dirty or flaking paper, fewer surviving sources, and no keyword search possibilities. It also means, in consequence, that such material is of little interest commercially, so what is available online is much less than what exists in the real world. Oddly enough, Egyptian records three or four millennia old, which were carved in the stone of stelae, preserve a more complete story than some of our yellowing nineteenth century American newspapers. We know what Akhenaton saw and heard in 1378 BC, but we have serious uncertainties about the whereabouts of the Post Office building in Jay, Ohio after 1858.

Anyone complaining that we should go out and search archives and libraries page by page has no idea how much time it takes, how hard on the eyes it is, and how hit-and-miss it can be. Some brave early researchers like Dr. Bullard have spent decades looking for old cases, compared to what the Internet-based Magoniax Project collected in just a few years. The digitizing of text has given us this gift. We can breathe new life into aerial phenomena research, just as the same sources can improve knowledge in other academic fields.

As we complete the first edition of this book, we are painfully aware that many sources remain beyond our grasp or even beyond our knowledge, because they are still buried in faraway libraries, written in languages with which we are not familiar, or even undeciphered in the dusty backrooms of museums. We are especially lacking in reliable data from Japan, China, and India, all ancient civilizations where careful records were kept down through the ages. We hope that scholars in those countries will be inspired to teach us about the knowledge preserved in their libraries. Our fervent hope is that the present book may stimulate scholars to dig out such material and bring it to the light of modern review, to inform our search for meaning among phenomena that still puzzle our best scientists today.

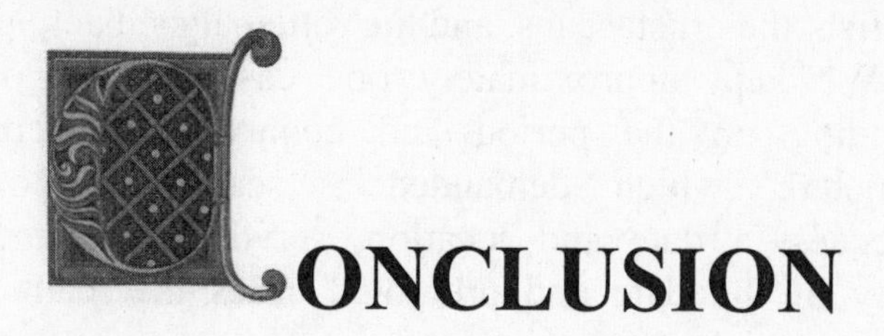

CONCLUSION

First, a word of disclosure: as authors of this compilation, we have worked with the awareness that we could not escape projecting some of our own beliefs, and those of the Western society of the 21st century to which we belong, through the case selections we have made.

The cement holding the reports together is based on two components: on the one hand, some selection and research criteria we have tried to disclose with clarity; and on the other, our faith that the majority of the sightings belong in the same group and are not fictional. This already carries a perceptible message: we believe that most of the witnesses we quote did in fact observe phenomena that have remained unexplained to this day.

Twelve important questions

At this point in our exploration of the mysteries of the past, the reader is entitled to ask: what have we learned from all this work, how significant are the findings, do they teach us anything new about the modern phenomena generally called "UFOs" and is there more yet to be discovered? We will examine these topics systematically, and present our analysis as a series of twelve specific questions.

1. How homogeneous is this Chronology of 500 cases?

The Chronology is only homogeneous by virtue of the selection criteria the authors have applied after casting a very broad net over the literature, and throwing back the little fish, the crustaceans, and the rotten algae back into the sea. We kept approximately one case in five or ten, depending on the period and context. Our screening parameters, which demanded a search for original references, a date and location, served to enhance the quality of the data and promoted cases that came from reliable records over popular rumors.

Having done this, we screened out the items where we could find no compelling reason to think the phenomena described were *other than* meteors, aurorae borealis, ball lightning, tornadoes or other unusual atmospheric effects. In spite of this effort, the Chronology remains biased across time. We have more information on 19th century incidents than medieval observations, as the following graph shows.

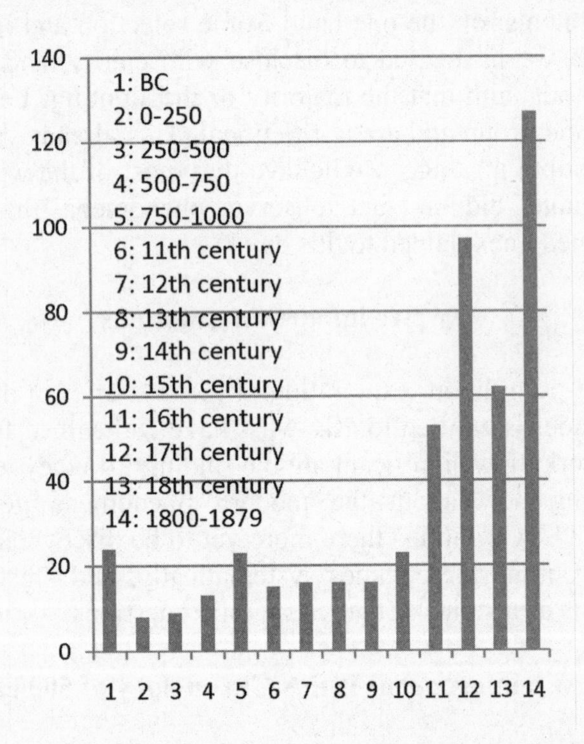

Fig. 70: Case distribution by period

The historical statistics break down as follows:

BC: 24 cases	13th century: 16 cases
0-250: 8 cases	14th century: 16 cases
250-500: 9 cases	15th century: 23 cases
500-750: 13 cases	16th century: 53 cases
750-1000: 23 cases	17th century: 97 cases
11th century: 15 cases	18th century: 62 cases
12th century: 16 cases	1800-1879: 127 cases

2. Isn't the Chronology biased by your own cultural backgrounds?

Undoubtedly it is. As disclosed above, we know much more about France or England than about Japan or China. The two authors share a background in Western humanities and the tradition of scientific enquiry that represents a selection bias against unfamiliar Asian, African, Polynesian, or Native American sources, especially those embedded in the imagery and texts of pre-1900 societies.

Most contemporary students of unidentified flying objects posit that the phenomenon is of recent origin and centers on the Anglo-Saxon world simply because it became popular in the American press after World War Two. While we have avoided this pitfall, we cannot claim expertise in using sources in cultures other than our own. Thus, given two fragmentary rumors about similar events in medieval France and in a remote part of Asia, we are more likely to invest time and effort in tracking down the more accessible French reference, because of its more familiar linguistic and historical context, than the Asian one. We are also more likely to view the Asian story through a skeptical filter because of possible mistakes in translation and an absence of cultural references. This bias could only be corrected by researchers from other cultures joining in this effort.

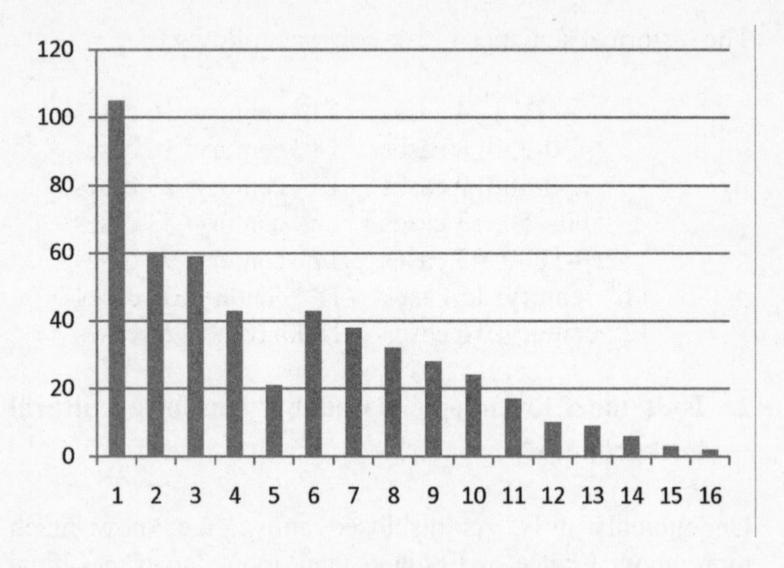

Fig. 71: Case distribution by country

The geographical statistics break down as follows:

1	U.K.	105
2	France	60
3	Italy	59
4	Germany	43
5	Spain	21
6	OtherW.Europe	43
7	N.America	38
8	Japan	32
9	China	28
10	E.Europe	24
11	Mid-East	17
12	Ocean	10
13	Unknown	9
14	Other Asia	6
15	S.America	3
16	Australia	2

3. Do these cases represent a global phenomenon?

As one reads through the chronology from end to end, two things become clear: (a) no convenient natural phenomenon or combination of natural phenomena accounts for the collection of events we have selected, although some alternative explanations may be proposed for individual cases and (b) the same unexplained features occur again and again, often in the same words in the language of witnesses.

The overwhelming fact is that we are dealing with a narrow range of anomalous objects in the atmosphere or in space, typically described as spheres, balls, or disks, capable of extraordinary trajectories, generally of significant duration, often in formation, described by multiple observers and viewed as important enough to be reported to authorities (religious, administrative, or scientific). These features do characterize the events in all countries and all periods.

4. Is this all there is?

Certainly not! Even in regions such as Western Europe, which produced most of our current data, there must be much published (but unscanned) material we have not been able to reach, and there must be large amounts of reports buried in manuscript form in local archives, unpublished personal diaries and private correspondence.

5. Who are the witnesses?

They represent a cross-section of the population, with a preponderance of scientists or enlightened amateurs of "natural philosophy" after 1750 or so. Not surprisingly, given the nature of medieval society, most of the cases involved multiple witnesses, often an entire village or two.

Other cases involved the crews of ships, even groups of soldiers at war, and in a couple of sightings, a king with his retinue. Single-witness cases did happen, of course, and religious interpretations were common, but this does not detract from the major facts of the observations.

6. Could all this be simply delusionary?

No, although delusions are a factor in the interpretation of the phenomena, as we have abundantly documented in Part II of this book. Psychological or anthropological explanations fail to account for most of the cases selected in the Chronology. In fact, the so-called "rational" explanations proposed by academic experts are often as delusionary as the most fanciful reports, and they fail to account for the observed facts in the same way.

Given the preponderance of multiple-witness cases, and the many events attested by figures in authority such as astronomers, State or Church representatives, as well as the

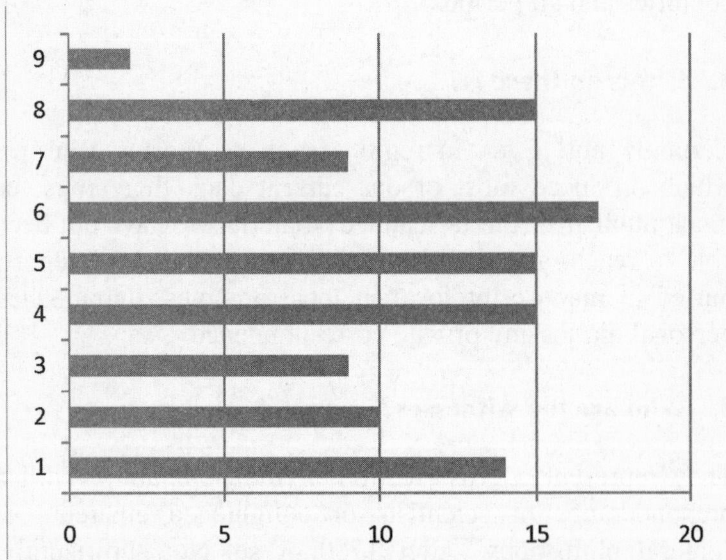

Fig. 72: Frequency of cases as a function of duration

extensive investigations some of the events have triggered, it is not reasonable to claim that no physical phenomenon of an unusual nature was present.

Furthermore, a review of the duration of the sightings (known to us in 106 cases) shows that most of the objects or phenomena were in sight for a considerable time, certainly long enough for the witnesses to have called other people, and to have ascertained the circumstances of the observation.

The duration statistics (above chart) are as follows:

1. 5 minutes or less: 14 cases
2. 6 to 10 minutes: 10 cases
3. 11 to 15 minutes: 9 cases
4. 16 to 30 minutes: 15 cases
5. 31 to 60 minutes: 15 cases
6. 1-2 hours: 17 cases
7. 2-3 hours: 9 cases
8. Over 3 hrs: 15 cases
9. Multiple days: 2 cases

Such a distribution is not typical of delusionary events or hallucinations.

7. Are there general patterns behind the sightings?

We have only begun to study this body of data for possible patterns. In particular, the time of day is known or at least estimated with some precision (plus or minus one hour) for no less than 205 cases, or 40% of the total.

The resulting distribution is consistent with the results of similar studies conducted on the basis of catalogues of contemporary UFO sightings (notably in the book *The Edge of Reality* by Dr. J. Allen Hynek and Dr. Jacques Vallee, Chicago: Contemporary Books, 1975, page 20): the frequency of reported cases rises before dawn, with a first peak about 6 A.M., and goes through a maximum between 8 P.M. and 10 P.M., returning to a low level during the night.

During most of the day, sighting frequency remains around the noise level. While this can be interpreted as an indication of the visibility of the phenomenon (most clear in contrast with its surroundings when the sun is not present) and a consequence of most people going inside after dark, it does indicate that a real phenomenon was present. Hallucinations or hoaxes would have no reason to follow the same pattern.

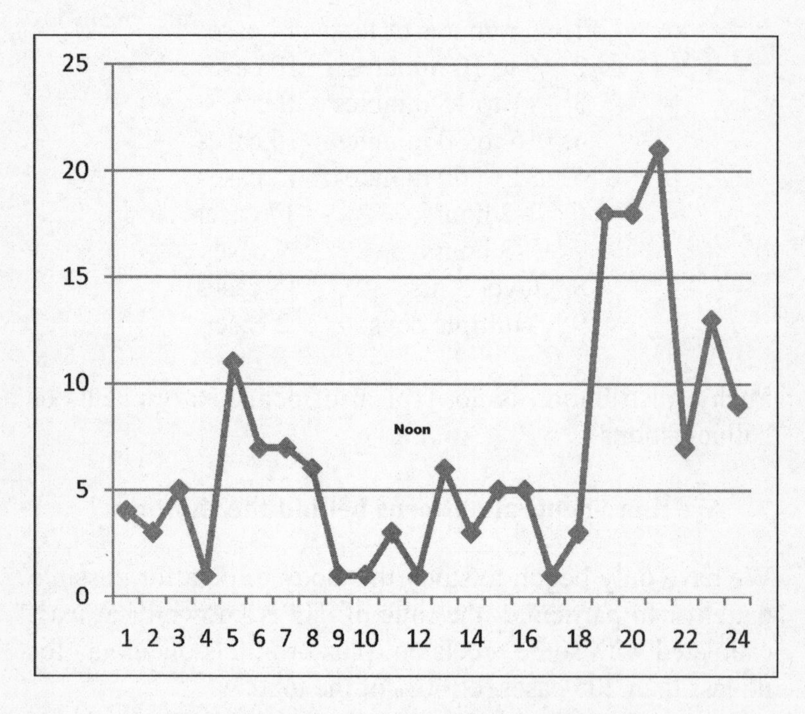

Fig. 73: Frequency as a function of time of day
(in hours, from 1 A.M. to midnight)

This is a striking result, because it only leaves us with two possible conclusions: either the same phenomenon known today under the label "UFO" has existed throughout history, or there is a massive, unrecognized mechanism that generates such imagery and propagates it through human society in all periods and in all cultures. Either conclusion represents new knowledge and has important consequences.

8. How "physical" is the phenomenon?

The witnesses are primarily describing luminous phenomena that range from "fiery globes" and "glowing forms" to vertical pillars and "towers" that occasionally emit flashes and beams or expel other objects. However, many of the cases also mention disk-shaped or globular objects that cast no light and are capable of rapid evolution in the atmosphere, reversing course, dashing and darting, or falling in zigzag patterns. In some well-documented cases the phenomenon gave off intense heat, destroyed vegetation or dropped metallic residue.

In an era when there was no radio, no radar, no awareness of radioactivity and little ability to analyze chemical substances, we know nothing about other potential effects, but it is worthy to note that experienced astronomical observers have frequently reported tracking dark objects across the disk of the sun or the moon.

All these indications converge to support the concept of an unrecognized, physical phenomenon that is relatively rare and unpredictable, but consistent in its general appearance and effects.

9. Is this relevant to the modern UFO phenomenon?

As we have already pointed out, there is little difference between the general behavior characterizing the cases in the Chronology and modern sightings of unidentified flying objects, down to sharp details such as witness paralysis, contact with forms of consciousness described as alien, and even the feeling in witnesses that a new form of communication has occurred between them and the phenomenon.

Before jumping to the simple conclusion that some extraterrestrial technology has been at work in both ancient and modern reports, we must pause and consider the contradictions this would raise: Why would this technology remain so constant? What would be its purpose? It cannot be discovery, can it, if some "Alien" race has had access to the Earth and to human civilization for centuries? And why would there be so many observations? We know that we have only detected one sighting report in ten or a hundred, others being lost to fires, revolutions, censorship or illiteracy, not to mention the vast areas of the globe where there was little or no communication with the outside world during the period we have studied.

10. Why has science ignored this body of data?

Scientific dogma dictates that any ancient observation of unexplained aerial phenomena can always be attributed to the ignorance of the populace or to simple misinterpretation of natural effects. Acting on this unproven assumption, scientists have censored their own data and intimidated their peers to silence open debate about the phenomenon.

There is a case to be made, however, for a very cautious approach. We have shown that hallucination was not a significant factor among the cases we have selected, but extreme weather and meteorite crashes do happen, as well as aurorae, globular lightning, comets and tornadoes, all fantastic phenomena that were poorly understood before modern science documented them. In particular, we asked whether meteors could play a role in the observed distribution.

The major meteor showers are the *Quarantids* (3-4 January), the *Lyrids* (21-22 April), the *Perseids* (12-13 August), the *Leonids* (17-18 November) and the *Geminids* (13-14 December). We are in a position to test this hypothesis, since 309 of our cases have a complete date (or over 60%).

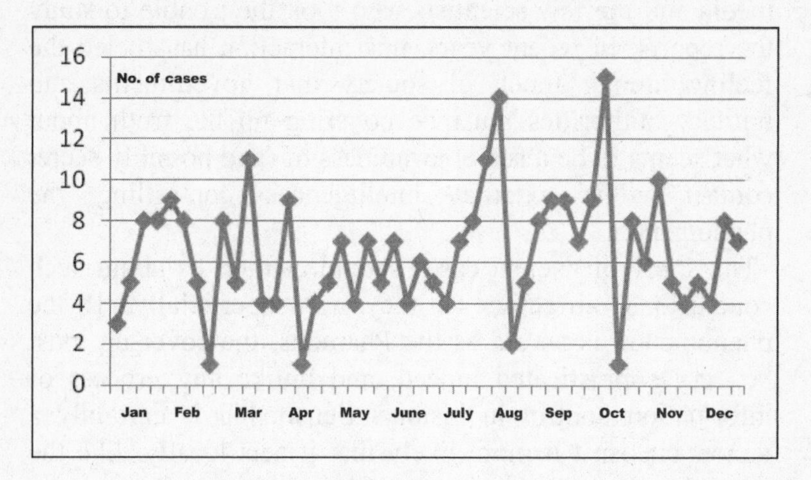

Fig. 74: Frequency during the year, with 4 points per month

We find that only one peak of the frequency curve in our Chronology shows an increase that could correlate with any cyclical meteoritic activity (the Perseids of August). The two other peaks are around March 17th and especially October 10th, as seen on the above graph, and those are not periods of meteoritic activity.

11. How can the impact on society be characterized?

Much of the attention devoted by a few courageous sociologists to unidentified phenomena has focused on the modern literature of ufology. The overwhelming majority of specialized books repeat the standard story of pilot Kenneth Arnold, who saw an apparent formation of "flying saucers" in June 1947, dating the beginning of the phenomenon from this event. Never mind that Mr. Arnold never said the objects were shaped like saucers, and that his observation came after several years of sightings of unexplained lights in the sky over Europe, Asia and America.

Such sociological research is correct, however, in characterizing the interaction between the witnesses, the media and the few scientists who took the trouble to study the reports. In recent years, this interaction has fueled the feeling among much of society that governments and military authorities must be covering up the truth about what seems to be a secret awareness of (and possibly secret contact with) external intelligences controlling the phenomenon.

The study of ancient cases should caution us about such conclusions, attractive as they are superficially. If the phenomenon is as old as the Pharaohs, the cover-up must be very sophisticated indeed, and unlike any process of information control in history. Censorship is certainly a factor, but isn't it more likely that it acts locally, like the Inquisition's efforts to impose Christianity by denying the expression of alternative beliefs?

12. What is the next step?

Better documentation is mandatory. Our limited efforts in this book have shown that much new information, and new knowledge, could be obtained by a well-organized group using modern communications technology. We did it with no money, in an environment of little interest to scientific organizations, official folklore researchers, or most publishers. We hope others will be inspired to use this model on a larger scale.

ACKNOWLEDGMENTS

This compilation, and the massive dredging of old text it implied, followed by critical study of every source, would have been impossible without the collaboration of a unique team of scholars who worked with us in this research.

The Magoniax Group, recruited and maintained by Chris Aubeck, has worked for seven years through the Internet to assemble and validate an enormous amount of raw sighting data. It has also exchanged information with interested researchers the world over on every related topic, from the fairy faith in Celtic countries to the review of archives of geophysical effects, cometography, modern astronomical records, as well as the tracking down of ancient Egyptian parchments.

Among the contributors most active in this remarkable group are: Rod Brock, Jerome Clark and Thomas Bullard from the United States, Mikhail Gershtein from Russia, Daniel Guenther from Germany, Eduardo Russo and Giuseppe Stilo from Italy, Javier Garcia Blanco and Jesus Callejo from Spain, Peter Hassall from New Zealand, and Fabio Picasso from Argentina. We are grateful for their tireless contributions.

We have also benefited from the help offered to us by many librarians, researchers, publishers and curators who were kind enough to respond to enquiries about data we were seeking to validate. Mr. Franck Marie, a tireless French researcher who has assembled a private collection of some 30,000 references on aerial phenomena and related events, many of them within our period of interest, gave us access to this valuable resource.

In our initial efforts to create an index of cases reported in the general literature we consulted the work of many private UFO researchers and writers including Desmond Leslie, Harold Wilkins, Guy Quincy, Aimé Michel, Jean Sider, Godelieve van Overmeire, Michel Bougard, Christiane Piens, Ion Hobana, Lucius Farish, Gordon Creighton, Matthew Hurley, Jean-François Boëdec, Claude Maugé, Ron Brinkley and numerous other dedicated collectors who helped us through their writing and sometimes in person, providing their own catalog data or research into specific ancient cases.

The contribution by Yannis Deliyannis, a classical scholar who tracked down and translated for us many hard-to-find references, deserves a special mention. Yannis holds a post-graduate degree in archaeology from the Sorbonne and has contributed to the development of a manuscript computer database at the *Bibliothèque Sainte-Geneviève.* In 2001 he was involved in the creation of the *Institut National d'Histoire de l'Art* in Paris.

Without such careful and critical review of the original data available on this complex subject, we would have drowned in a sea of uncorrelated and often poorly documented rumors about a variety of phenomena, many of which turned out to be explainable as natural atmospheric or physical effects.

Finally, this book owes much to the encouragements of Professor David Hufford and to the advice of Dr. Jeff Kripal, chairman of the department of Religious Studies at Rice University, who provided valuable contacts with other researchers. Michael Murphy, at the Esalen Institute, allowed us to test some of our ideas before knowledgeable and stimulating audiences, and publisher Mitch Horowitz turned the concept of this book into a reality.

Much remains to be done. We believe we express the consensus of this group of researchers when we say that the study of ancient unexplained aerial observations has only begun. We invite scholars and interested amateurs from all countries and cultures – and especially from the Middle East, Latin America, and Asia, to join this continuing effort.

An important further note from Jacques Vallee

Although my name comes first on the cover of this book, the reader should know that Chris Aubeck did the pioneering work in researching, critiquing and documenting material that had been neglected or treated with considerable inaccuracy in the literature. When we discovered we shared a passion for such ancient reports we began working together, merging our sources and catalogues and enlisting our personal networks in support of the research.

In the process Chris made me aware of many previously unknown instances of aerial phenomena, but more importantly he taught me to look at them in new ways.

Wonders in the Sky is the product of many, sometimes heated debates and compromises about the relevance of each case. In this creative interaction, which continues today, I have learned to value the high standards of authenticity and accuracy Chris has brought to the field, and I am proud to contribute in making them more widely understood for a new generation of paranormal researchers.

INDEX of places cited, by case number

LIST of ILLUSTRATIONS

Fig.23 p.218. Stralsund phenomenon. Francisci, Erasmus. Der wunderreiche Ueberzug unserer Nider-Welt/Order Erd-umgebende. Nürnberg (1680), p. 624.
Fig.24 p.220. Francisci, Erasmus. op. cit. (Front page).
Fig.25 p.222. Mittelfischach phenomenon. Abriss des Erschrecklichen wunderzeichens, so sich den 15. Novembr. 1667 beim dorff Mittelfischach …, [s.l.] (1667). [Goethe Universitätsbibliothek Frankfurt-am-Main, coll. Gustav Freytag, Einblattdr. G.Fr.11].
Fig.26 p.225. Regensburg phenomenon. Wunderzeichen, Zu Regensburg gesehen am 18.Augusti 1671, Frankfurt am Main, 1672. [Goethe Universitätsbibliothek Frankfurt-am-Main, coll. Gustav Freytag, Einblattdr. G.Fr.12].
Fig.27 p.229. Jane Lead's vision. Lead, Jane. A fountain of gardens watered by the rivers of divine pleasure…, London: J. Bradford, 1696, p. 264.
Fig.28 p.238. Mecklenburg phenomenon. Nachdencklich-dreyfaches Wunder-Zeichen..., Frankfurt am Main, 1697. [Goethe Universitätsbibliothek Frankfurt-am-Main, coll. Gustav Freytag, Einblattdr. G. Fr.13].
Fig.29 p.247. Nightwatchmen in London. The Age of Wonders: or, a farther discriptton [sic] of the fiery appartion [sic]..., London: J. Read, 1710. [British Library, 1104.a.24]. Copyright Mary Evans Picture Library, London. Reprinted by permission.
Fig.30 p.250. Bern prodigies. An Account of Terrible Apparitions…, Glasgow, 1721. (Front page).
Fig.31 p.255. London "waving rocket". Phil. Transactions, vol. 43, London, 1745, p.524.
Fig.32 p.260. Ramsberg sockens kyrkobok, E1:1 (1786-1774). Handwritten entry by Rev. Vigelius. [Landsarkivet Uppsala, Sweden].
Fig.33 p.269. Charles Messier, portrait by Ansiaume, 1771. [Bib. de l'Observatoire de Paris].
Fig.34 p.289. 1808 Moscow phenomenon. Moscow daily Komsomolskaya Pravda, 2 July 2006.
Fig.35 p.298. Arago, portrait.
Fig.36 p.300. Reverend Webb: unknown photographer.
Fig.37 p.327. Le Verrier, portrait.
Fig.38 p.361. The Dropa hoax. Das Vegetarische Universum (July, 1962), German magazine by Reinhardt Wegemann.
Fig.39 p.364. Flying Vimana at Ellora Caves, India. Ancient carving.
Fig.40 p.371. Tulli papyrus (detail). Doubt, no. 41 (1953).
Fig.41 p.373. Moses and the blue object (Plagues of Egypt, Exodus 5-9).From the Ashkenazi Haggadah, 15th century. [British Library, Add.Ms.14762].

Fig.42 p.375. Manna from Heaven (Exodus 16). Miniature from the Maciejowski Bible, 13th century. [Pierpont Morgan Library, NY, Ms. M 638].
Fig.43 p.380. Silver shields (interpretation by J. Vallee).
Fig.44 p.381. Qu Yuan: From Wikipedia.
Fig.45 p.386. Star of Bethlehem.
Fig.46 p.388. Emperor Constantine, 19th century: unknown artist.
Fig.47 p.389. Cross-shaped illusions. Flammarion, Camille. L'Atmosphère et les Grands Phénomènes de la Nature, Paris: Hachette, 1905.
Fig.48 p.395. Apparition to Mohammed. Miniature from the Jami'al-Tawarikh by Rashid al-Din. Tabriz, Persia, 1307. [Edinburgh University Library].
Fig.49 p.410. Bulletin des Antiquaires de France, Paris: Klincksieck, 1911. (Front page).
Fig.50 p.412. Miniature. Hildegard von Bingen, Liber Scivias, 12th century.
Fig.51 p.414. Celestial phenomena. Schedel, Hartman. Liber Chronicarum, Nürnberg, 1493, BSB-Ink S-195. Reproduced by permission of *Bayerische Staatsbibliothek München*, Rar.287.
Fig.52 p.419. Freiburg meteor. Mennel, Jakob. De signis, portentis, prodigiis…, 1503. [Österreichische Nationalbibliothek, Vienna, cod. 4417].
Fig.53 p.422. French *jeton*, ca. 1656 [private collection].
Fig.54 p.426. Tuscany phenomenon. Reproduced in Alata Quaderni, no. 1 (Feb 1979).
Fig.55 p.432. Casanova, portrait.
Fig.56 p.433. Goethe, portrait.
Fig.57 p.435. The meteor of August 18, 1783, as seen from Windsor Castle, painting by Paul and Thomas Sandby; 1783. [British Museum].
Fig.58 p.442. Japanese object and occupant. From the Japanese Toen-Shosetsu (1825).
Fig.59 p.445. Commodore Decatur, early 20th century reproduction: unknown artist. [Library of Congress].
Fig.60 p.454. Batman, Stephen. The Doome, warning all men…, London, 1581. (Detail).
Fig.61 p.455. Boetius: unknown artist.
Fig.62 p.456. Matthew Paris. Self portrait from the original manuscript of his Historia Anglorum, 13th century. [British Library, MS Royal 14.C.VII, folio 6r].
Fig.63 p.457. Grégoire de Tours and Salvius facing King Chilperic. Miniature from the Grandes Chroniques de France de Charles V, 14th century. [Bibliothèque Nationale de France, MS FR 2813].
Fig.64 p.458. Three suns in 1492. Flammarion, Camille. L'Atmosphère, Paris: Hachette, 1872 [1871], p. 233.

Note: The authors have made every effort to contact individuals and organizations with regards to copyright permissions prior to publication. However, many items came to us through the general literature and the Internet with limited documentation about ownership. If you feel that we have infringed on any rights or have erroneously quoted specific references, we will be grateful for information that might be useful in correcting such mistakes.

BIBLIOGRAPHY

In addition to the sources and references cited throughout this book, we have found the following works important to verify the reliability of many cases in the literature.

Burns, William E. *An Age of Wonders: Prodigies, Politics and Providence in England 1657-1727.* New York: Manchester University Press, 2002.

Christian, William A. *Apparitions in Late Medieval and Renaissance Spain*. Princeton University Press, 1989.

Corliss, William R. *Remarkable Luminous Phenomena in Nature: A Catalog of Geophysical Anomalies.* The Sourcebook Project. Glen Arm, MD, 2001.

Kronk, Gary W. *Cometography: Volume 1, Ancient-1799. A Catalog of Comets.* Cambridge University Press, 1999.

Olivyer, I. L., and J. F. Boëdec. *Les Soleils de Simon Goulart: Vague OVNI de 1500 à 1600*. Marseille: Ada, 1981.

Rasmussen, Susanne William. *Public Portents in Republican Rome*. Rome: L'Erma di Bretschneider, 2003.

Wildfang, Robin Lorsch and Isager, Jacob. *Divination and Portents in the Roman World.* University Press of Southern Denmark, 2000.

Among essential online sources are:

Newspaper Archive: www.newspaperarchive.com

Internet Medieval Sourcebook:

www.fordham.edu/halsall/sbook.html

Internet Archive: www.archive.org

For further contact:

Chris Aubeck invites comments through the Internet at:

caubeck@gmail.com

Jacques Vallee can be contacted at:

PO Box 641650
San Francisco, California 94164
USA

About the Authors

Jacques Vallee holds a master's degree in astrophysics from France and a Ph.D. in computer science from Northwestern University, where he served as an associate of Dr. J. Allen Hynek. He is the author of several books about high technology and unidentified phenomena, a subject that first attracted his attention as an astronomer in Paris. While analyzing observations from many parts of the world, he became intrigued by the similarities in patterns between moderrn sightings and historical reports of encounters with flying objects and their occupants in every culture. The result was the seminal book *Passport to Magonia*, published in 1969.

After a career as an information scientist with Stanford Research Institute and the Institute for the Future, where he served as a principal investigator for the groupware project on the Arpanet, the prototype of the Internet, Jacques Vallee cofounded a venture capital firm in Silicon Valley, where he works.

Chris Aubeck was born in London. His interest in the historical and sociological aspects of unexplained aerial phenomena began at an early age. He moved to Spain at age 19 and now lives in Madrid, where he works as an interpreter and English teacher at the Madrid Development Institute. A student of folklore and philology, he has helped compile the largest collection of pre-1947 UFO cases in the world. He has spoken on his research in many articles and on public radio. In 2008 he was awarded a prize for his contributions to the field by the Spanish organization Fundación Anomalía.

In 2003, Aubeck cofounded a remarkable collaborative network of librarians, students, and scholars of paranormal history on the Internet. This group, known as the Magoniax Project, extends from North and Central America to Russia and Germany. It has accumulated thousands of references, searched media archives in several languages, and gathered hundreds of rare documents, scientific reports, and newspaper clippings from the last four hundred years.

ABOUT THE AUTHORS

If you enjoyed this book, visit

www.tarcherbooks.com

and sign up for Tarcher's e-newsletter to receive special offers, giveaway promotions, and information on hot upcoming releases.

Great Lives Begin with Great Ideas

New at **www.tarcherbooks.com**
and **www.penguin.com/tarchertalks**:

Tarcher Talks, an online video series featuring interviews with bestselling authors on everything from creativity and prosperity to 2012 and Freemasonry

If you would like to place a bulk order of this book, call 1-800-847-5515.